Green Energy and Technology

Climate change, environmental impact and the limited natural resources urge scientific research and novel technical solutions. The monograph series Green Energy and Technology serves as a publishing platform for scientific and technological approaches to "green"—i.e. environmentally friendly and sustainable—technologies. While a focus lies on energy and power supply, it also covers "green" solutions in industrial engineering and engineering design. Green Energy and Technology addresses researchers, advanced students, technical consultants as well as decision makers in industries and politics. Hence, the level of presentation spans from instructional to highly technical.

Indexed in Scopus.

More information about this series at http://www.springer.com/series/8059

Uttam Kumar Bhui
Editor

Macromolecular Characterization of Hydrocarbons for Sustainable Future

Applications to Hydrocarbon Value Chain

Springer

Editor
Uttam Kumar Bhui
School of Petroleum Technology
Pandit Deendayal Energy University
(PDEU, Formerly Pandit Deendayal
Petroleum University-PDPU)
Gandhinagar, India

ISSN 1865-3529 ISSN 1865-3537 (electronic)
Green Energy and Technology
ISBN 978-981-33-6132-4 ISBN 978-981-33-6133-1 (eBook)
https://doi.org/10.1007/978-981-33-6133-1

This Springer imprint is published by the registered company Springer Nature Singapore Pte Ltd.
The registered company address is: 152 Beach Road, #21-01/04 Gateway East, Singapore 189721, Singapore

This book is dedicated to all those scientists and technologists of the earlier generations, who prepared the platform of scientific excellence with wealth of information, and the future generations who is really concerned about a cleaner and greener habitable Earth for the survival of human civilization.

—Uttam Kumar Bhui

Preface

Sustainability of the Earth with respect to energy security and green environment in the context of ever-increasing population is not an easy task but required a multidisciplinary team effort from many disciplines. This has been clearly evidenced by continuous decrease in fossil fuel resources and shifting towards renewable resources for production of alternative energies in greener way. Unfortunately, the 'renewable' options do not appear as attractive close up, as they do on first mention. Overall, the present picture we face is one of a gradually degrading environments and of a diminishing resource base of fossil fuels. Against this background of uncertain supply, a major long-term share for coal and heavy hydrocarbon components for power generation and as a source of chemical feedstock appears inevitable. The paradigm shift from fossil fuel to bio-based feedstock provides a promising alternative to be converted to fuels, chemicals and carbon-based materials which are strongly linked with greener and environmentally friendly processes. Coal, petroleum and biomass are the main raw materials from which synthetic fuels and feedstock for chemicals and carbon materials are obtained. The book considered the energy and environment of the earth into a multi- and interdisciplinary science looking towards the most important challenges in the twenty-first century, while individual chapters of the book take care the in-depth scientific knowledge with innovative design methodologies to maximize current available resources without compromising the future of the coming generation.

Today, hydrocarbon components from all the sources (fossil fuels and biomass), with the help of many new and evolving techniques of characterization at molecular level, are seen in various integrated approach of utilization for societal needs and benefits. With innovative ways of looking into the existing systems and materials of all the resources, active researchers and working professionals contributed their experience and output knowledge in the form of chapters of this book which makes this volume more relevant as timely knowledge for future applicability.

This book volume represents section-wise contributions on the existing systems and materials from crude oil, coal and shale, biomass- and carbon-related materials with able leadership of academicians, research scientists and practicing professionals from their area of expertise and research. This book intended to provide a strong knowledge platform to all research scientists, working professionals and managers who are aiming to provide efficient integrated solutions for energy and environment of the earth with feasible industrial applications for sustainable cleaner and greener future.

Many people helped me in this endeavour for successful completion of this volume. Professor Samir Kumar Pal, one of my young age friends cum collaborator, introduced me into the world of spectroscopy which propel me to work in the domain of crude oil and coal. Deeper insight of the crude oil and coal system through macromolecular characterization using spectroscopic methods is providing avenues to understand the problems and their optimal solutions for the operational practices of the industry. Thanks are due to those colleagues who helped by providing their remarks and comments on the original manuscript which are now the chapters in the final version of this book. An especial debt of gratitude to Rincy Anto, my doctoral student, for her skillful helping hand to complete my editorial responsibility within due timeframe. Thanks, of course, to my wife 'Rita' who supported me in the critical time of the COVID-19 pandemic to take care my two kids 'Anuran and Uddhriti' at home when I was spending long hours with related work of this volume.

Ahmedabad, India Uttam Kumar Bhui

Contents

Biomass System, Carbon Materials and Carbon Capture and Storage (CCS)

About the Editor

Dr. Uttam Kumar Bhui is currently Associate Professor at the School of Petroleum Technology, Pandit Deendayal Energy University (PDEU, Formerly Pandit Deendayal Petroleum University-PDPU), Gandhinagar, Gujarat. He obtained M.Sc. (Applied Geology—Gold Medalist) and Ph.D. (Sc.) degree from Jadavpur University, Kolkata. Dr. Bhui has both industrial (ONGC) and academic (Jadavpur University, PDPU) experience. He has working experience as Reservoir Engineer with ONGC 1999–2002. He served as a faculty member in the Department of Geological Sciences, Jadavpur University, from 2002 to 2008. Dr. Bhui, in his present academic position at PDPU, involved in teaching related to Petroleum Geoscience and Reservoir Engineering courses for Petroleum Engineering students. His present research activities are related to characterization of hydrocarbon components in molecular level from crude oil, coal and carbonaceous shale and their application in exploration, exploitation and utilization to meet the futuristic need of energy and chemical for sustainable development. His specializations and expertise involved:

Crude oil: Crude oil characterization in molecular level for application in enhanced oil recovery (EOR), flow assurance and demulsification.

Shale oil/Shale gas: Maturity assessment of hydrocarbon resources (carbonaceous shale) through molecular structure approach.

Coal: Molecular characterization of coal for utilization as cleaner and greener future fuel and chemical feedstock.

Hydrocarbon System and Characterization

Hydrocarbon Cycle for Sustainable Future: Clean Energy and Green Environment of the Earth

Uttam K. Bhui

1 Introduction

With an increasingly technology-driven civilized society, human need not only food, water, shelter within the environment-friendly habitable earth but also growing amount of energy to link with sustained development and growth. With ever-increasing thirst of energy for continual growth and development, humankind had started using wood in the initial stage followed by use of fossil fuels such as coal, petroleum and natural gas, which are the resource from the mother Earth. Fossil fuels, i.e. petroleum crudes, natural gas and coal are mainly composed of hydrocarbons with varying proportion of carbon and hydrogen, which are very much essential for the quality of modern life. The Industrial Revolution started with coal which is fuelled by oil and natural gas in twentieth century. In order to maintain the societal need and sustained development and growth, the fossil fuel is being consumed rapidly in massive scale by humankind. With the ever-increasing world population, the demand for energy is steadily increasing where fossil fuels particularly oil, natural gas and coal may take up the major share to supply the demand but at a higher socio-economical and environmental cost. Due to considerable environmental impact vis-a-vis, USA boom of shale gas use and decreasing cost of electricity from renewable sources, usage of coal has been diminishing rapidly during the last decade. It is needless to mention that majority of the present energy need as well as feedstocks of most of the chemicals are supplied by the fossil fuels. Though crude oil, natural gas and coal are non-renewable resources, the reserves of all the resources have been steadily growing over the years due to discovery of new fields and adoption of novel and optimized technologies allowing more volume of extraction in cost-effective manner. We still have significant

U. K. Bhui (✉)
School of Petroleum Technology, Pandit Deendayal Energy University (PDEU, Formerly Pandit Deendayal Petroleum University-PDPU), Raisan, Gandhinagar, Gujarat 382007, India
e-mail: Uttam.bhui@spt.pdpu.ac.in

U. K. Bhui (ed.), *Macromolecular Characterization of Hydrocarbons for Sustainable Future*, Green Energy and Technology,
https://doi.org/10.1007/978-981-33-6133-1_1

reserves of heavy oil, oil shale, tar sand and large deposit of coal, which may last for so long that there is no real resource shortage in foreseeable future of human life scale. The rapidly growing world population vis-a-viz energy consumption, compared with ever-decreasing finite fossil fuel resources, are clearly pointing towards an urgent requirement of new energy solutions for twenty-first century and beyond. Renewable resources for the production of alternative energies are believed to be the key to a future scenario in the absence of finite amount of fossil fuels. Moreover, the design of innovative methodologies to maximize current available resources without compromising the future of coming generation is one of the most important challenges in twenty-first century. In this regard, green chemical and low environmental impact technologies combined with renewable resources through innovation will be able to offer alternatives to potentially useful processes for a more sustainable society.

2 Clean Coal Technology

2.1 Conventional Use of Coal

Human civilization started using coal since ancient times and still continued to use extensively for various needs. The use of coal decreased with the discovery of oil which became the major fuel for transportation purposes as well as home heating. The widespread and extensive use of coal during Industrial Revolution also brought ecological imbalances, environmental pollution impacting human life. In the present sensible and environment conscious world, coal is partly replaced by either petroleum-derived fuels, natural gas or renewable energy sources, but consumption of coal is unlikely to reduce quickly in the near future which has been considered as backbone for many established industrial uses.

Coal was used primarily for iron and steel production, locomotive engines for transportation and household uses for heat generation. In addition, medicines, dyes, flavourings, ammonia, explosive trinitrotoluene were produced from coal. The technologies used for generating power, heat, coke and chemicals are coal combustion, carbonization, gasification and liquefaction, referred as the four 'grand processes' of coal utilization [1].

Coal Combustion Coal combustion, the most straightforward way of burning coal, is used for warmth, cooking and industrial processes. But in today's world, the major use of coal is for burning in boilers to generate electricity, while use in residential heating and industrial processes is very small.

Coal Carbonization In carbonization process, coal is heated, and both gaseous and liquid products are driven off leaving the char or coke as solid residue. Carbonization process is classified as high (temperature > 1650 F), low (T < 1350°F) and medium (T between 1350 and 1650 F) temperature based on their uses [2]. High-temperature carbonization process is generally meant for production of metallurgical coke for use in blast furnaces to smelt iron ore for the production of pig iron and in foundries to produce cast iron. Low-temperature carbonization was mainly used to provide gas for residential use. Smokeless fuels used for domestic and industrial heating while tars used as feedstock for production of chemicals.

Coal Gasification The thermal breakdown process of any carbonaceous feedstock, to the smallest stable chemical units, like hydrogen, carbon monoxide and methane, is called gasification, and the gaseous product is called synthesis gas or syngas. Depending on the quality of the carbonaceous feedstock and the condition of gasification process used, the composition of the syngas varies. The conversion process can be generalized in three steps from feedstock to product:

1. *Feedstock*: Coal is the main feedstock in gasification and expected to be its highest growth in future for clean energy. Ash contents, moisture content, ash melting temperature and available grain size [3] of coals are the most decisive properties for suitability of different gasification technologies to be used as feedstock. With advancement in the present available technology, large spectrum of coal can be used as feedstocks for gasification, while the main limiting factors are high ash content and small particle size. High-quality steam coals and caking coals have other high value utilization (combustion, coking), so cheap and low-quality coal with high moisture content, high ash, high sulphur or residue from coal beneficiation process could be used in the gasification process.
2. *Gasification process*: Gasification process is classified on the basis of the method used to bring the coal into contact with air or oxygen, the gasifying medium. There are three main types gasifier system used for large-scale industrial applications (i) fixed bed (ii) fluidized bed and (iii) entrained flow system.
3. *Product*: Most of the primary products during initial thermal breakdown are reactive where the intensity and extent of subsequent secondary reaction are very much dependent on the size of the initial feedstock as well as design of the reactor. Commercial products from coal gasification [4] are ammonia, gasoline, methanol, polypropylene, SNG (LNG competitor), power, SNG (shale gas competitor).

Coal Liquefaction Liquefaction is the conversion of coal into liquid products mainly through pyrolysis, indirect and direct liquefaction process. In pyrolysis process, the liquids are a by-product during coke production, while in liquefaction (direct or indirect), liquid is the primary product during conversion of the coal. In indirect liquefaction, the coal is gasified into syngas, a mixture of carbon monoxide and hydrogen, which is then processed into liquid products using Fischer–Tropsch synthesis. In direct liquefaction, coal is mixed with a hydrogen donor solvent and reacted at elevated pressure–temperature condition to produce a liquid fuel. These methods referred to as coal hydrogenation.

2.2 Clean Coal Technology for Future

Coal is the solid fossil fuel and is the largest source of fuel for power generation. Along with increased coal usage, the environmental concern has come into the forefront against its usage as 'dirty' fuel. System efficiencies must be improved for every usage of coal by ramped-up implementation of 'clean coal technology' considering environmentally friendly acceptable usage.

Coal as a Source of Cleaner Fuels

Liquid and gaseous fuels are easier to handle than solid coal, whether it is used for heating, cooking, transportation or power generation. Coal is also being considered as a way to produce gaseous or liquid fuels via direct and indirect liquefaction technology. Understanding the molecular architecture, binding energy viz-a-viz their conversion process to gas and liquid of solid combustible coal is not very obvious to us till date. The following advantages could be visualized after conversion of coal either gas or liquid fuels:

- Impurities in coal can be readily removed through gasification.
- Synthetic fuels (gas or liquid) burn more cleanly than coal with less evolution of toxic gases (sulphur and nitrogen oxide).
- The pollutant CO_2 can be more readily removed from the gas stream.

It is quite well known that when coal is heated progressively in an inert atmosphere, coal starts to decompose and evolves a mixture of hydrogen and/or oxygen-rich products, and aromatics units until coal comes to resemble a microcrystalline graphitic solid at a high enough temperature. This process of thermal decomposition, on one hand, depends upon the nature and prior history of the coal; on the other hand, is influenced by the particular conditions under which the process proceeds. Pyrolysis is more usually applied to a process that involves extensive thermal decomposition of coal. Above a threshold temperature, the complicated structure of coal is dissociated into aromatic ring units by rupture of bridge carbons with formation of 'free radicals'. Recombination of the smaller radicals results in the formation of small aliphatic gas molecules, and water and tar components of medium molecular weight. The volatile compounds diffuse from the interior of the coal/char particle into the bulk gas phase and react with the surrounding atmosphere. The diffusion can also happen from the surrounding atmosphere to the interior of coal. In principle, the thermal cleavage of the covalent bond in coal may generate two radical fragments where one of this is volatile and another one is non-volatile or both may be volatile [5–8]. In general, the couplings of these radical fragments form a stable product [9–12]. The radical concentration in chars is always greater than tar which suggests that volatile radical fragments underwent coupling reaction before the volatiles being considered as products [13]. As radical concentration of chars is much higher than that of commercial fuels, crude oils and direct coal liquefaction liquid, it may lead to good coupling event for formation of liquid or gaseous hydrocarbons. Recent work has emphasized rapid heating of dilute coal samples, technically termed as 'flash pyrolysis'. Behind this shift in emphasis, there has been the desire to elucidate the role of pyrolysis in coal combustion and gasification and hence the possibility of conversion of coal to liquid fuels or other desired chemicals. Recognition of immediate and long-range potential of coal has resulted in substantial efforts towards the development of new or modified technologies in coal gasification, liquefaction and direct combustion.

The goal of most coal-to-liquid process is to convert coal to a liquid that have weak dispersive Van der Waals forces just like typical petroleum liquids. The production of liquid fuels from coal is a major application of coal utilization in much cleaner form. Petroleum hydrogenation is typically done at 4–20 MPa with a large stoichiometric excess of hydrogen.

Catalytic Process: The catalysts for coal hydrogenation are typically sulfidic bi-minerallic Co/Mo, Ni/Mo or Ni/W in a silica, alumina or silica-alumina base. Coal slurry in a hot heavy aromatic recycle liquid is mixed with catalysts. The dissolved coal molecules can adsorb in the catalytic pores and then diffuse out of the pores.

Donor solvent process: An aromatic compound with partial hydrogenated saturating one of the aromatic rings is used in the process. This hydrogenated molecule then reacts with the solid coal surfaces transferring the hydrogen to the coal surfaces.

Coal as a source of carbon materials

Carbon materials are described as materials that are composed of a high percentage of carbon. The science and technology of carbon materials are in the process of continuous development and are constantly finding ways to improve the properties of the already existing materials even at the nanometric scale. Coal, petroleum, organic chemicals and biomass are the main raw materials from which precursors of carbon materials are obtained. Since all coals are carbon-rich solids, they can be used as precursors for new carbon-based materials. Advanced knowledge of aromatic compounds presents in the coal helped for the production of high-value synthetic dyes, pharmaceuticals, explosive trinitrotoluene (TNT) from coal tar. Depending on the rank and composition, raw coals or its by-products may be used directly as precursors to obtain some platform chemicals as well as solid carbon materials.

3 Processing of Heavy Hydrocarbons

Due to the decline of light crude oil supply, increasing production of heavy crude oil becomes trend to the supply chain. Crude oils of heavy and extra-heavy nature and bitumen are characterized first by viscosity and then by density. To differentiate between crude oil and bitumen, 10,000 cP viscosity is considered as border where crude oil has viscosities less than or equal to 10,000 cP while bitumen has >10,000 cP. Among crude oils, extra-heavy crude oils have °API less than 10, while heavy crude oils have API 10°–20°. Besides the higher viscosity and lower API gravity, heavy crudes are characterized by high asphaltene contents, high level of heteroatomic (N, S, O) compounds and high Conradson carbon (CCR) and microcarbon residue (MCR). Presence of larger molecules with high boiling points poses the challenges and opportunities for the processing of heavy and extra-heavy crude oils, which are considered to be so-called opportunity crudes (OC's) due to their lower prices in the crude market. But there are physical, chemical, logistic and economic constraints associated with processing of this type of crudes.

The oil industry is a molecular trader whose business is based on the price of such molecules and their properties [14]. The market value of crude and its molecular composition together with processing technology determine the profit margin of the historically known low profitability 'refining' industry. As crude price is the most significant element contributing to this profit margin of refinery, refiners turn towards processing of the 'opportunity crudes' (OC), due to their low prices but faced enormous technical challenges to be refined into high-quality products. Consequently, the processing of heavy crudes is inevitable to the refiner to increase their flexibility in the refining processes. Therefore, understanding the molecular management along with performance of a refining process is the key for processing and refining of a crude oil to its final products. In general, OCs are known to cause problems all along the refinery process. For optimizing refining and processing operation of unconventional opportunity crude oil, modifications, improvements have been adopted in commercially available technologies, but new or improved technologies are still needed to solve the existing issues.

3.1 Downhole in Situ Upgrading

Thermal recovery: The advancement in technology and method development for thermal recovery of heavy crude oil and oil sand opens up opportunities for upgrading in downhole environment, which could not be possible in earlier days. Due to the presence of crude oil, water and oxygen (of air) in subsurface high temperature–pressure environment, a certain level of upgrading occurred through breaking of bonds during thermal recovery processes through pyrolysis and oxidation reaction [15]. This downhole in situ upgrading reduces the cost involved in production from subsurface reservoirs and follow-up surface upgrading.

Bioprocessing: Downhole in situ bioprocessing by microorganism of heavy crude oils and bitumen may find more opportunities with considerable R&D efforts focusing in reservoir condition following the advancement achieved so far. Oxidation of high molecular weight aliphatics, aromatics, napthenic acids, N-S-O compounds as well as hydrogenation or dehydrogenation of lower molecular compounds promoted by microorganism might upgrade the crude oil in the downhole reservoir condition.

3.2 Radiation-Thermal Upgrading

Though the process of radiation cracking of hydrocarbon has attracted research scientist since 1960, its application for the upgrading of the heavy oil has taken a new dimension in recent times [16]. Ionizing irradiation of different forms (neutron, electron, x-ray, γ-ray, etc.) is used in the processes which induce chain cracking reactions of both C–C and C–H bonds in hydrocarbons with considerable changes in the oil fractional contents and its chemical composition. But all forms of irradiation have not succeeded in adding value to the upgrading process of crude oils. Progress in radiation technologies for oil processing is associated with more detailed theoretical elaboration of cracking reaction with serious difficulties in applying them to the interpretation of experimental data and results. A correct description of different relationship like (i) dependence of the cracking start temperature on the irradiation dose rate, (ii) dependence of the chain length on temperature and dose rate, (iii) dependence of cracking rate on temperature and dose rate and (iv) the nature of the reaction and molecular mechanisms of bond breaking still remain unclear. The presently available knowledge and technologies could be ready for industrial applications with inputs from further research which may provide solutions to overcoming many problems of the oil industry. These areas seem to be a very unpopulated field with huge opportunities for further research and development with advancement in technological innovation which is really needed for the future.

3.3 Most Recent Upgrading Concepts

Sonochemistry: It was demonstrated that the effect of ultrasound for breaking C–C and C–H bonds is localized in space and time while surface-active agents and solvents were used to make it more permanent. Efforts were made with this process for conversion and more specifically oxidative desulphurization. Since heavy oil and bitumen contain sulphur in percentage range, and as ultrasound did not sufficiently enhance the action of peroxide to partially oxidize the sulphur compounds, the required (OH) radicals yield increases [17] which required an additional step to extract the formed oxi-sulphide compounds.
Mechanochemical: In this process, a specific region would undergo chemical transformation at any instant of time with sequential accumulation of such sites at nearby region with progression of time by the mechanical energy transfer to complete the process in the whole system. A potential exists for a mechanochemical process for upgrading opportunity crudes (OC's) with MCP-treated or manufactured catalysts.

3.4 Concluding Remarks

The requirements and regulations, in the context of present and future days, continue to become increasingly stringent imposing tougher quality specification on the product side with environment-friendly processing. Extraordinary advancement has been achieved in almost all areas of concern, but conventional thinking does not seem to have any immediate future scope to find out the solution with unconventional opportunity crudes (UOC). Some of the most recent upgrading concepts were out-of-box methods using energy sources like electron beam, γ-rays, microwaves, plasma and ultrasound but failed in demonstrating any potential benefit with commercial applicability. New out-of-box ideas with multidisciplinary synergic teams are needed for reducing cost in the processing operation optimizing energy balances with minimal environmental impact.

4 Conversion of Biomass

Biomass is one of the most promising and widely available renewable feedstocks that have significant potential to offer a number of alternatives to be converted to materials, fuels and chemicals. Biomass consists of many different polymeric molecules, which are hard to depolymerize, separate, transform selectively by employing various chemical, thermal or catalytic pathways. The high structural and chemical complexity of most biomass feedstocks, from the simplest polysaccharides to more complex lignocellulosic and lignin, is a key factor to consider the design and development of novel processes aimed to facilitate their conversion to simple fraction. A series of (bio) chemical as well as thermochemical processing strategies includes gasification, pyrolysis, fermentation, anaerobic digestion, catalytic transformation are the focus for the process development and application. A recent emerging area in biomass valorization to high added value compounds includes bio-chemicals, biomaterials and biofuels entailing the use of customized nanomaterials for biomass conversion. A lot of technological innovation and process improvement is necessary towards more efficient utilization of biomass, either for production of fuels, source feedstock of chemicals or for use in other applications for energy efficient environmentally friendly habitable earth.

4.1 Biofuels from Biomass

Due to decline of fossil fuel resources, alternatives are being intensively investigated globally, out of which generation of biofuel production from biomass showed an attractive proposal.

Biogas: Rapid expansion of established processes and technology for biogas production open up the avenues of utilization for production of electricity and thermal energy in modern power plants, decreasing dependency on fossil fuels [18]. Biogas acts as a source of methane and CO_2, since agricultural waste as well as municipal waste or purpose-grown energy crops are being dedicatedly used for transformation into this energy-rich gas mixture [19]. Biogas can be used directly after appropriate purification, as an individual source of biomethane or carbon dioxide [20].
Biodiesel: Biodiesel is a liquid biofuel that is defined as fatty acid methyl ester matching with certain standard like (EN14214 (European Commission) and ASTM6751 (ASTM). It is obtained by transesterification or alcoholysis of natural triacylglycerol or triglycerides found in vegetable oil, animal fats, waste cooking oils, etc. Methanol is the most common alcohol used in the transesterification followed by ethanol, reported in few cases [21–23]. Biodiesel is now produced from edible oil with methanol and alkaline catalysts, like NaOH, KOH, etc. [24, 25].

4.2 Chemicals from Biomass

Lignocellulosic biomass consists of three main compounds: cellulose, hemicellulose and lignin. Cellulose is a linear polymer of glucose monomer representing 40–70% of a plant; hemicellulose is a similar biopolymer accounting 10–40% of a plant, while lignin is a biopolymer based on aromatic building blocks. Since cellulose consists of the main fraction of inedible plant material, it is considered the most promising feedstock for various platform chemicals like hydroxy-methyl furfural (HMF) and furfural (FF). The development of processes that utilize these platform chemicals to produce biomass-derived substitutes and analogs for chemical intermediates is of the utmost importance in the present research. These products can in turn be applied as polymer intermediates [26], fine chemicals, pharmaceutical precursors, solvents and fuel additives [27].

Catalytic oxidative and reductive transformation of these platform chemicals can result in a wide variety of value-added products with necessary steps using diverse catalysts, which can be implemented in future biorefinery schemes. Selective photocatalytic biomass transformation holds significant promise for environmentally friendly and economically cheap synthesis processes to produce important chemicals [28].

5 CO_2 Capture and Sequestration (CCS)

Carbon dioxide (CO_2) together with methane (CH_4), nitrous oxide (N_2O), ozone (O_3) and water vapour are greenhouse gases which are vital for life on the Earth since they regulate the temperature of the Earth's atmosphere favourable for life.

Atmospheric CO_2 concentration is steadily increasing from the start of the industrial era (1740) due to combustion of fossil fuels, deforestation and changes in land use. According to the World Metrological Organization (WMO), CO_2 is considered as the single most important man-made greenhouse gas in the atmosphere which contributes about 26% to the total greenhouse effect for global warming, depletion of ice cover, changes in snow cover and a decrease in upper oceanic pH. In response to global warming, the United Nations Framework Convention on Climate Change (UNFCCC) proposed a policy aiming at mitigating CO_2-related problems. Intergovernmental Panel on Climate Change (IPCC) suggests the importance for an instant reduction in CO_2 emissions 50–70% to stabilize global CO_2 concentrations of 1990 levels by 2100 (IPCC Report, 1995). In the past two and half decade, carbon capture and storage (CSS) has emerged as a feasible technology to reduce CO_2 emission to the atmosphere which involved capturing CO_2 at its source, transporting it through pipeline and storing in geological formation or oceanic sites.

5.1 *CO_2 Capture*

There are two main approaches to CO_2 capture:

1. As pure or near pure CO_2 stream, from an existing and/or re-engineered industrial processes or concentration of discharge from an existing industrial process,
2. Direct air capture, into a pure CO_2 stream or into a chemically stable end product.

Fossil fuel-based power plant accounts for the majority of stable emitting sources of CO_2 while other industries like cement and steel production industry, oil refining, natural gas processing plants also account for carbon capture technologies, as the concentration level of every cases is very high (>30%). At the other end, the possibility of capture from ambient air with $\sim$0.04% has also been proposed using various chemical absorption approaches [29]. Capture from industrial processes will reduce future emissions and limit the resulting increase in atmospheric CO_2. Direct air capture has the potential to accelerate the natural decline in atmospheric CO_2 below a future peak level.

5.2 *Geological Carbon Storage (GCS)*

Geological sequestration of CO_2, known as geological carbon storage (GCS), is a proposed technology to store CO_2 into deep, porous and permeable rock formation for permanent storage [30]. The captured CO_2 can be compressed and sequestrated underground in geological formations, deep saline water aquifer, depleted oil or gas reservoirs or even at the bottom of the sea [31]. Deep ocean storages constitute the largest potential storage capacities, but because sequestrated CO_2 in this way would eventually increase ocean acidification, it is not considered to be feasible. Mineral CO_2 sequestration is another storage option that is believed to be effective. Carbonates produced by CO_2 reaction with minerals containing magnesium and calcium are thermodynamically stable. This phenomenon occurs naturally during carbonate weathering of such rocks, such as peridotite and serpentine over geological time periods. The end product of such reaction produces stable minerals for millions of years. The essence of this process is that these rocks act as a natural sink to CO_2 and the re-release of the same into the atmosphere is not an issue. However, natural carbonation is very slow under ambient P-T conditions. Accelerating this process for commercial and environmental use is underway, but little has been done so far. Currently, the worldwide focus is placed on geological CO_2 sequestration (GCS), known as CO_2-geosequestration, involves injection of CO_2, normally in a supercritical state into a carefully selected underground formation site which has to satisfy some basic criteria, such as, it must be (1) highly permeable (2) surrounded by impermeable cap rock layer (3) environmentally safe and (4) economically feasible. The geological storage of CO_2 poses the advantage of short-to-medium term option of CO_2 storage for significantly enhancing CO_2 sinks, thus reducing net carbon emissions into the atmosphere. Geological sequestration of CO_2 is the process of long-term storage of CO_2 in depleted or depleting oil/gas reservoirs, deep unminable coal seams/gas shale formations, deep saline formations and basalts:

1. **CO_2-EOR in oil fields**: Injecting CO_2 into declining oil field has been utilized to enhance oil recovery (EOR). Though the volumetric sweep efficiency in CO_2-EOR is low, sufficient CO_2 injection rates were achieved with retention of significant portion of CO_2 in the subsurface reservoirs [32]. However, much of the injecting CO_2 is only temporarily stored, as significant part of it released due to the blow-down of the reservoir pressure to increase oil recovery. So, more effort could be applied after CO_2 injection to confirm that reservoir fluids are remaining within the reservoir for considerable timescale, concerned to GCS.
2. **Unminable Coal seam/gas shale formation**: The organic phases in coal and shale consist of nano to microporous disordered organic framework which allow adsorption (desorption as well) of gases including CO_2, CH_4, etc. Coal seams and organic-carbonaceous matter rich shales are potential sources of CH_4 and sink for CO_2 [33, 34]. Abandoned coal seam can be utilized to store CO_2, as that CO_2 molecule physically diffuses the pore spaces and adsorb onto the coal carbon. This adsorption takes place at approximately twice the rate of adsorption of methane, thus allowing for potential enhanced coal bed methane (CBM) recovery. Sequestration of CO_2 in coal beds and gas shale formations holds an additional advantage of enhanced methane recovery/production (e.g. ECBM) which can help offset the costs of storage making it an economically viable alternative.
3. **Deep saline formations**: Deep saline formations might have the largest CO_2 geosequestration potential due to their large storage volumes and common

availability. Following the thermodynamics of CO_2, if it is injected deeper than 800 m, the CO_2 is in liquid or supercritical state where CO_2 basically undergoes four trapping mechanisms. During the first couple of years after injection, CO_2 be trapped underneath an impermeable cap rock layer, known as structural and stratigraphic trapping. Residual trapping simultaneously occurs in this period where a significant amount of CO_2 molecules resides in the formation pores. At the later stage, the CO_2 starts to dissolve in the brine water, and brine water dissolves in CO_2. In the long term, chemical reaction between CO_2-water mix and the formation solid phase will take place.

4. **Crystalline basement/Basalt formation**: Recent research proved these rock types to be potential geological formations, but more data needs to be acquired to predict the injection of CO_2 on crystalline basement or basalt formations.

The injection of supercritical CO_2 into the subsurface system perturbs the existing stress system along with disturbances in chemical, thermal and biological equilibrium conditions. Perturbation of stress and geological response plays a crucial role in the performance of the CGS either in short-term or long-term basis. Comprehensive assessment of CCS requires accounting of different processes occurring in different spatial scale (nanometre to kilometre) as well as temporal scale (nanoseconds to thousands of years). Establishing quantitative relationship of different processes operative in different direction and dimension into the heterogeneous reservoir with thermal, chemical and stress gradient for CCS remains a fundamental research challenge till date [35].

5.3 Bioenergy with CCS (BECCS)

BECCS is defined as carbon capture and storage in which feedstock is biomass [36]. BECCS enables negative emissions because CO_2 absorbs during the growth of the biomass are not released after the biomass is combusted to produce biofuel. The CO_2 captured and stored underground outweighs the emission related to biomass production, land use change and transformation into the final product [37]. BECCS is the most mature negative emission technology option over other alternative technology, which has now been recognized for many years, yet progress with deployment has remained stagnant.

6 Methanol—The 'Chemical Link' Between Present and Future

The ever-increasing energy need of humankind in the foreseeable future will be fulfilled using any available sources including alternate sources. Recycling of CO_2, evolving from human activities, industrial and natural resources to useful products like fuels and synthetic hydrocarbon, opens up a new vista to mitigate global warming problem and at the same time provides an inexhaustible renewable carbon source instead of diminishing fossil fuel resources. As we still cannot store energy efficiently on a large scale, production of methanol from CO_2 offers two ways of saving the humankind by storing energy in form of methanol and saving the environment of the earth for sustainable future [38].

Methanol or dimethyl ether (DME) can be produced from still existing fossil fuel resources or from CO_2 produced from burning of fossil fuels or from other industrial and natural resources, through chemical, catalytic and electrochemical methods alternative to photosynthesis [39–41]. Methanol or DME, an excellent transportation fuel, converted into ethylene and/or propylene, the essential building block in the petrochemical industry for preparation of hydrocarbons and wide variety of derived chemical products. Methanol is suitable for fuel cell, being capable of producing electric energy by reacting with atmospheric oxygen.

Methanol has already proven its worth as one of the most important industrial base materials for replacement of oil-based products, for energy storage and synthetic proteins and carbohydrates for animal feed. The production of synthetic regenerative methanol from CO_2 and H_2O is a proven concept that will gain more logical ground for technological development with industrial applications as environmental concerns and issues related to use of fossil fuels in different industrial applications are mounting. Methanol could be looked as a 'chemical link' between a fossil fuel dominated present and a sustainable future earth in terms of energy and environment.

7 Concluding Remarks

All living being on the earth are based on carbon, and thus, carbon has to be managed efficiently for future generations where hydrocarbons share an important place and role in the whole cycle of food, energy and environmental security for sustainable future. Sun is the source of majority of our energy. The chlorophyll of the plant, through photosynthesis, recycles CO_2 and water into glucose, the fundamental structural unit of all hydrocarbons, with initiation of new plant life. Conversion of old plant biomass to fossil fuel through natural cooking and cracking processes within the mother Earth is very slow, while biofuel production from contemporary biomass is also possible through innovative green technology with much faster rate.

Despite their non-renewable nature with diminishing resource base on human timescale, adverse effect on environment, fossil fuels will maintain their leading role in the foreseeable future as they are readily available and convenient to use with available infrastructure. Continuous efforts are being made to make use of 'clean coal' technology to produce gas (coal to gas) and liquid (coal to liquid) as well as in situ upgrading of heavy crude oils for cleaner energy. The increase of atmospheric CO_2 content due to excessive use of fossil fuel and other industrial

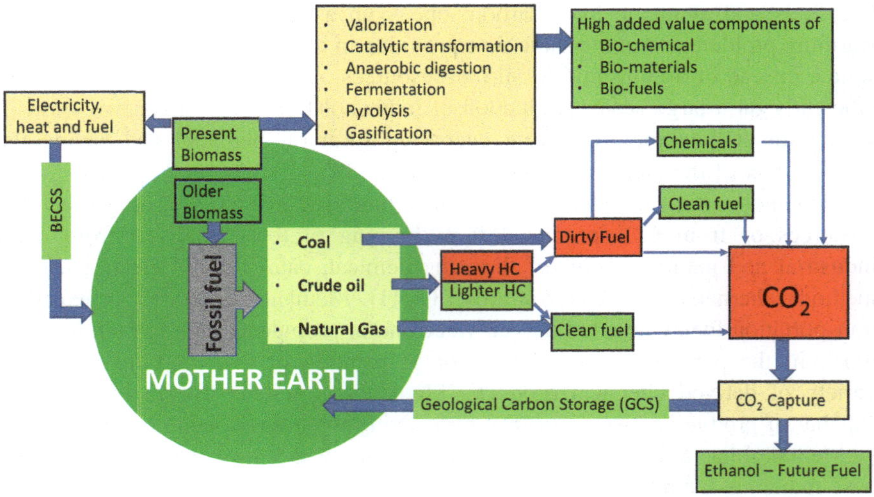

Fig. 1 Hydrocarbon cycle for clean energy and green environment of the Earth

activities causes global warming superimposed on nature's own long- and short-term cyclical changes. In the backdrop of fossil fuel-based energy system, CO_2 is a one carbon problem, which needs to be solved by relying on all possible solutions available to satisfy its energy needs. Selective absorption, adsorption and other separation methods are being developed to capture liberated industrial CO_2 or atmospheric CO_2, and either proposed to permanently store into subsurface geological formation or chemically recycled into methanol, which could be served as a bridge fuel for a seamless transition towards a sustainable future. With an optimistic view, we need to find technologically advanced solutions to continue our lives at a comparable or even higher standard of living following the hydrocarbon cycle (Fig. 1) for the cleaner and greener Earth.

Acknowledgements The author acknowledges the support of PDPU to carry out carbon-based energy and allied research particularly crude oil and coal system and consider him as a member of 'CO_2 research group' of PDPU. Discussions of 'CO_2 research group' enhance the thinking horizon of the author in the domain of carbon-based research.

References

1. Van Krevelen DW, (1993) Coal: Typology-physics-chemistry-constitution. Elsevier Science
2. Berkowitz N (1979) An introduction to coal technology. Academic Press
3. Hillr H, Reimert R, Marschner F, Renner HJ, Boll W, Supp E, Brejc M, Liebner W, Schaub G, Hochgesand G, Higman C, Kalteier P, Muller WD, Kriebel M, Schlichting H, Tanz H, Stonner HM, Klein H, Hilin W, Gronemann V, Zwiefelhofer U, Albrecht J, Cowper CJ, Driesen HE (2000) Gas production, in Ullmann's Encyclopedia of industrial chemistry, Wiley-VCH Verlag GmbH

4. Morehead H (2013) Siemens gasification: progress and innovation. In Proceedings of the gasification technologies conference, 13–16 Oct 2013, Colorado Springs, CO. www.gasification.org
5. Petrakis L, Grandy DW (1981) Formation and behaviour of coal free radicals in pyrolysis and liquefaction conditions. Nature 289:476–477
6. Seehra MS, Ghosh B, Mullins SE (1986) Evidence for different temperature stages in coal pyrolysis from in situ esr spectroscopy. Fuel 65:1315–1316
7. Bolton C, Snape CE, O'Brien RJ, Kandiyotit R (1987) Influence of carrier gas flow and heating rates in fixed bed hydropyrolysis of coal. Fuel 66:1413–1417
8. Fowler TG, Bartle KD, Kandiyoti R (1989) Limitations of electron spins resonance spectroscopy in assessing the role of free radicals in the thermal reactions of coal. Energy Fuels 3:515–522
9. McMillen DF, Malhotra R, Nigenda SE (1989) The case for induced bond scission 330 during coal pyrolysis. Fuel 68:380–386
10. Shi L, Liu Q, Guo X, He W, Liu Z (2014) Pyrolysis of coal in TGA: extent of volatile 332 condensation in crucible. Fuel Process Technol 121:91–95
11. Shi L, Liu Q, Guo X, Wu W, Liu Z (2013) Pyrolysis behavior and bonding information of coal-A TGA study. Fuel Process Technol 108:125–132
12. Solomon PR, Fletcher TH, Pugmire RJ (1993) Progress in coal pyrolysis. Fuel 72:587–597
13. He W, Liu Z, Liu Q, Ci D, Lievens C, Guo X (2014) Behaviours of radical fragments in tar generated from pyrolysis of 4 coals. Fuel 134:375–380
14. Ramirez-Corredores MM (2017) The science and technology of unconventional oils: finding refining opportunities. Academic Press, p 761
15. Kapadia PR, Kallos MS, Gates ID (2015) A review of pyrolysis, aqua thermolysis, and oxidation of Athabasca bitumen. Fuel Process Technol 131:270–289
16. Zaikin Y, Zaikina R (2014) Petroleum radiation processing. CRC Press, p 364
17. Gogate PR, Shirgaonkar IZ, Sivakumar M, Senthilkumar P, Vichare NP, Pandit AB (2001) Cavitation reactors: efficiency assessment using a model reaction. AIChE J 47(11):2526–2538
18. Thrän D, Seiffert M, Müller-Langer F, Plättner A, Vogel A (2007) Möglichkeiten einer Europäischen Einspeisungsstrategie. Leipzig, Germany. Deublein D, Steinhauser A (2008) Biogas from waste and Renewable Resources. Wiley, Weinheim
19. Santosh Y, Sreekrishnan TR, Kohli S, Rana V (2004) Enhancement of biogas production from solid substrates using different techniques—a review. Bioresource Technol 95:1–10
20. Cavenati S, Grande CA, Rodrigues AE (2006) Separation of $CH_4/CO_2/N_2$ mixtures by layered pressure swing adsorption for upgrade of natural gas. Chem Eng Sci 61:3893–4106
21. Perego C, Ricci M (2012) Diesel fuel from biomass. Catal Sci Technol 1:1776–1786
22. Okoronkwo MU, Galadima A, Leke L (2012) Advances in biodiesel synthesis: from past to present. Elixir Appl Chem 43:6924–6945
23. Veljkovic VB, Avramovic JM, Olivera SS (2012) Biodiesel production by ultrasound-assisted transesterification: state of the art and the perspectives. Ren Sust Energy Rev 16:1193–1209
24. Arzamendi G, Arguinarena E, Campo I, Zabala S, Gandia LM (2008) Alkaline and alkaline-earth metals compounds as catalysts for the methanolysis of sunflower oil. Catal Today 133:305–331
25. Santori G, Di Nicola G, Moglie M, Polonara F (2012) A review analyzing the industrial biodiesel production practice starting from vegetable oil refining. Appl Energy 92:109–132
26. Davis SE, Houk LR, Tamargo EC, Datye AK, Davis RJ (2011) Oxidation of 5-hydroxymethylfurfural over supported Pt, Pd and Au catalysts. Catal Today 160:55
27. Grabowski G, Lewkowski J, Skowronski R (1991) The electrochemical oxidation of hydroxymethylfurfural with the nickel oxide/hydroxide electrode. Electrochim Acta 36:1995
28. Quintero JCC (2014) Tunable biomass transformations by means of photocatalytic nanomaterials. In Luque R, Balu AM (eds) Producing fuels and fine chemicals from biomass using nanomaterials, Chap 11, pp 283–313
29. Rackley SA (2010) Carbon capture and storage. Elsevier

30. Newell P, Ilgen AG (2019) Overview of geological carbon storage. In Newell P, Ilgen AG (eds) Science of carbon storage in deep saline formations. Elsevier, pp 1–10
31. Socolow RH (2005) Can we bury global warming? Aci Am 293(1):49
32. Lake LW, Lotfollahi M, Bryant SL (2019) CO_2 enhanced oil recovery experience and its messages for CO_2 storage. In Newell P, Ilgen AG (eds) Science of carbon storage in deep saline formations. Elsevier, pp 11–23
33. Busch A, Alles S, Gensterblum Y, Prinz D, Dewhurst DN, Raven MD et al (2008) Carbon dioxide storage potential of shales. Int J Greenh Gas Control 2:126–140
34. Kang SM, Fathi E, Ambrose RJ, Akkutlu IY, Sigal RF (2011) Carbon dioxide storage capacity of organic rich shales. SPE J 16:842–855
35. Gao J, Xing H, Tian Z, Pearce JK, Sedek M, Golding SD et al (2017) Reactive transport in porous media for CO2 sequestration: pore scale modeling using the lattice Boltzmann method. Comput Geosci 98:9–20
36. IEA (2016) 20 years of carbon capture and storage: accelerating future deployment. OECD/IEA, Paris
37. IEA (2015) World Energy outlook special report: energy and climate change. OECD/IEA, Paris
38. Olah GA, Goeppert A, Prakash GKS (2018) Beyond oil and gas: the methanol economy, Third updated and enlarged version, Wiley-VCH, p 472
39. Olah GA, Goeppert A, Prakash GKS (2009) Chemical recycling of carbo dioxide to methanol and dimethyl ether: from greenhouse gas to renewable, environmentally carbon neutral fuels and synthetic hydrocarbons. J Org Chem 74:487
40. Goeppert A, Czaun M, Jones JP et al (2014) Recycling of carbon dioxide to methanol and derived products—closing the loop. Chem Soc Rev 43:7995–8048
41. Olah GA, Prakash GKS, Goeppert A (2011) Anthropogenic chemical carbon cycle for a sustainable future. J Am Chem Soc 133:12881–12898

Exploration of Molecules in Hydrocarbons with an Interdisciplinary Approach: Current Status and Future Implications

Uttam K. Bhui and Samir Kumar Pal

1 Introduction

Human civilization is presently facing huge challenges for food, water and energy security which may lead to an uncertain future in terms of sustainability. Facing all these challenges in this Earth with ever-increasing population is not an easy task but required an interdisciplinary team effort from many disciplines to come up with suitable alternative for the future. In case of energy, this has been clearly evidenced by continuous decrease in fossil fuel resources and shifting towards renewable resources for production of alternative energies, which are believed to be the key for future generation. During the last decade, the oil industry faced a scenario with records of high oil prices, price volatility and increasing supplies of emerging unconventional crude oil with support from renewable resources. No doubt, large proportion of the readily extractable oil has already been used up. The present rush to gas can indeed delay the end of plentiful hydrocarbon supplies. Unfortunately, the 'renewables' options do not appear as attractive, as they appear on first mention. Many of the alterations suffered by the planet Earth are of an irreversible nature. The paradigm shift from fossil fuel to bio-based feedstock provides a promising alternative to be converted to fuels, chemicals and carbon-based materials which are strongly linked with greener and environmentally friendly processes. Overall, the

U. K. Bhui (✉)
School of Petroleum Technology, Pandit Deendayal Energy University (PDEU, Formerly Pandit Deendayal Petroleum University-PDPU), Raisan, Gandhinagar, Gujarat 382007, India
e-mail: Uttam.bhui@spt.pdpu.ac.in

S. K. Pal
Department of Chemical, Biological and Macromolecular Sciences and Technical Research Centre (TRC), S. N. Bose Bose National Centre for Basic Science, JD Block, Sector III, Salt Lake, Kolkata (Calcutta) 700106, India
e-mail: skpal@bose.res.in

U. K. Bhui (ed.), *Macromolecular Characterization of Hydrocarbons for Sustainable Future*, Green Energy and Technology,
https://doi.org/10.1007/978-981-33-6133-1_2

present picture we face is one of gradually degrading environments, diminishing fossil fuel resource base with uncertain supply from renewables.

With the help of many new and evolving techniques of characterization in molecular level, today coal and petroleum components are seen in various ways of utilization other than conventional methods. Against this background of uncertain supply, a major long-term share for coal and heavy hydrocarbon components for power generation and as a source of chemical feedstock appears inevitable. Synthetic fuels are generally understood to include liquid and gaseous fuels as well as clean solid fuels produced by the conversion of coals. Since all coals and heavy crude oil components are carbon-rich, they may be used directly as precursors to obtain solid carbon materials of various uses. Coal, petroleum, organic chemicals and biomass are the main raw materials from which precursors of carbon materials are obtained, and presently, the field of carbon materials has turned into a multi and interdisciplinary science. The science and technology of carbon materials are in the process of continuous development and are constantly finding ways to improve the properties of the already existing materials even at the nanometric scale.

2 Molecules in Hydrocarbon System

2.1 Fossil Fuel System

Kerogen is the source of all hydrocarbons in the fossil fuel system. Fossil organic matters present in different form, e.g. coal, kerogen, bitumen and crude oils within the mother Earth after being modified due to the effect of temperature and pressure through geologic time. Organic matter (OM) is primarily defined as organic compounds that are directly or indirectly derived from cells or tissues of living organisms and from plant materials. Sedimentary organic matter (SOM) is progressively transformed, during the burial of sediments, into smaller molecules by thermal cracking with the formation of crude oil/natural gas as by-product and retention of residual organic matter in source sedimentary rocks as dispersed organic matter for 'source shale' or as 'coal' when OM is greater than 50%. Increasing thermal cracking of complex organic molecules leads to relative condensation and aromatization in the residual OM, whereas by-product crude oil/natural gas shows small aromatic structures diluted within saturated hydrocarbon fractions. Tissot and Welte [1] categorized kerogen as type-I, II and III based on H/C and O/C ratio values. Type-I is typically algal and often lacustrine in composition and includes boghead coal. Type-II is marine in origin, includes channel coal. Type-III is dominated by humic materials from terrigenous sources. Most coals are formed from terrigenous higher plant matter with $H/C < 1$ of type-III kerogen, while most petroleum source rocks are algal-planktonic organic matter with $H/C > 1$ of type-I kerogen.

Crude oil: Examination of pyrolysis products of algal-dominated residue deposits ranging in age from Permian to modern day [2] indicates that functional groups are lost during maturation and the poly-methylenic structure is retained along with small amount of aromatic and/or olefinic compounds. It has shown that algal biomass has a high potential to produce crude petroleum oil [3, 4].
Coal: Coal is a complex assemblage of transformed plant remains. It is clearly evident from various studies that there is a direct chemical structural link between biologically recalcitrant biopolymer in plant and microscopically recognizable remains in coal [5]. Most of the component organic matter in coal is complex, macromolecular and insoluble in most organic solvent. Thus, chemical characterization was done on a very small fraction which could not provide detailed structural arrangement in molecular level keeping the information intact. Hence, the structure and reactivity of coal can be characterized as the sum of the chemistry of all contributors [6].

2.2 Biomass System

Biomass is one of the most promising, easy and widely available renewable resources to be converted to fuels, chemicals and materials without compromising the future of coming generations. A major focus of hydrocarbon from fossil fuels and biomass is to relate their fundamental molecular structure with their utilization for energy source as well as chemical feedstocks. Lignocellulosic biomass materials are made up of mainly of cellulose ($\sim$40%), hemi-cellulose (35%) and lignin (25%) with relatively small amounts of extractable organic matter and inorganic mineral matter [7]. Compared to coal, most biomass and waste have lower mass and energy densities as because they generally contain less carbon (45–55% range), higher hydrogen (5.8–6.2%) and higher oxygen (35–45%) [8], nitrogen less than 0.5%, while sulphur content is in the order of 0.1%. The higher oxygen content of lignocellulosic biomass makes these material more reactive, while thermal breakdown takes place at about 100–120 °C below the characteristic temperatures for most coal.

3 Characterization of Hydrocarbon Components: Conventional Techniques for Structural and Molecular Profiling

Fuel science is very much related to the understanding of organic geochemistry, while development of new instrumentation helped the scientist to deal into the chemical structure of the source materials from which they derive. The first ideas about the molecular structure of sedimentary organic matter (SOM) were from coal study based on elemental composition, pyrolysis products and solubility fractions with different solvents. The structural complexity of fossil organic matter as well as biomass of the present day has provided a formidable challenge to the fuel scientist.

Characterization of components present in the fossil fuels as well as biomass feedstock is very much important for both process development and proper utilization as well as developing efficient operation practices. Macromolecular characterization of hydrocarbon components is very much important for fruitful and efficient use of coal, crude oil, shale and biomass as source of energy and chemicals in a much cleaner and greener way. The two most important attributes of any chemical component/compound are its elemental constituents, molecular weight and exact molecular structure. For a mixture of relatively smaller molecular mass, the analytical methods are simple and easy to handle while characterizing structural architectures of larger molecular mass materials have very much complex process and instrumentation. Analytical techniques for examining the materials recovered during conversion of coal, petroleum and biomass through pyrolysis, gasification, liquefaction are really complex and required thorough understanding of process and method development.

The chemical complexity of heavy crude oil, coal, biomass and other fuel-derived materials required detailed characterization of their constituents by conducting special separation methods followed by characterization via spectroscopic methods. Depending on the molecular mass present in the mixture along with their polarity, a variety of chromatographic methods is generally used to separate components with gas, supercritical fluid or liquid mobile phases in tandem with different spectroscopic identification and characterization method, either coupled or offline. Proper separation and characterization of fuel-derived samples and their components are very much dependent on their physical, chemical and electrical properties particularly molecular size, molecular mass, polarity of electrical charges which should be matched to the capabilities and ranges of available analytical techniques. There are larger molecules in coal, petroleum liquids and liquid produced from biomass materials by pyrolysis or simple solvent extraction method. Methods are generally applied initially for sample fractionation which involves several different combinations of sample introduction methods, into mass spectroscopic system. Combination of GC with mass spectroscopic analysis is the final steps for identification and characterization of hundreds of compounds from the complex mixture of hydrocarbon components. Capillary column flame ionization chromatography may be used for identification with the help of retention indices, calibrated from GC-MS system of standardization. Standard methods and practices are followed for characterization which has a great deal of overlap among various scientific disciplines.

Optical Microscopy: It is the most common method to characterize sedimentary organic matter. With the help of optical microscopy, morphology at a resolution down to about 1 μm could be observed for differentiation of 'maceral' which relate to the kind of cells that occurred in the deposited organic matter. The important optical properties namely the reflectance of light in polished section, the colour of transmitted light in thin sections and fluorescence in either transmitted or reflected light are commonly observed in coal [9] and petroleum source rock for source identification and maturity indication [10].

Elemental Analysis: Typical elements used to characterize organic matter are C, H, N, O, S. For characterizing kerogen and coal, H/C versus O/C ratio is used for getting the information on both source and maturity by plotting in van Krevelen diagram. During thermal cracking of maturation, oxygen tends to be eliminated first from the structure as water and CO_2 followed by hydrogen as oil and gas.
Rock-Eval pyrolysis: Rock-Eval pyrolysis is conducted on whole rock and is used for petroleum source rock analysis. The CO_2 yield (OI) during Rock-Eval pyrolysis from kerogen correlates with its O/C ratio, and the hydrogen index (HI) correlates with its H/C ratio. A plot of HI versus OI mimics the traditional van Krevelen diagram and provides information regarding source information along with maturity level for hydrocarbon generation. It has an advantage over elemental analysis as conducted with whole rock rather than on kerogen isolated from rock through mineral dissolution by acid treatment.
Spectroscopic methods: Branch of science the deals with atoms and molecules. Spectroscopy is actually the study of interaction between molecules and their radiation. The measure of the intensity and the nature of radiation as a function of its wavelength is a characteristic feature of all molecules. Spectroscopic methods are classified on the basis of the region of the electromagnetic spectrum involved in measurement. The electromagnetic spectrum can be broadly divided into six regions, namely the gamma-ray, ultraviolet (UV), visible, infrared (IR), microwave and radio frequency (RF). The nature of interaction between the molecules and their radiation as a function of their wavelength can be used to divide the spectroscopic techniques into three broad categories, namely the absorption spectroscopy, emission spectroscopy and scattering spectrometry. Applications of spectroscopic approach for hydrocarbon industry are discussed in brief below while optical to IR spectroscopy have been dealt in Sect. 4.

X-Ray Raman Spectroscopy (XRRS) To gain information regarding local structure of carbonaceous compound particularly for polycyclic aromatic hydrocarbon and asphaltenes, Raman spectroscopy is used. X-ray Raman spectroscopy (XRRS) is the energy loss version of X-ray absorption spectroscopy (XAS) [11]. XAS is commonly divided into near edge X-ray absorption fine structure (NEXAFS) or equivalently X-ray absorption near edge structure (XANES) and extended X-ray absorption fine structure (EXAFS) regions. All these methods could not provide solution to probe the bulk properties of heterogeneous concentrated compounds like large PAH, asphaltenes and coals. In such cases, XRRS directly probe the local structure with the help of synchrotron X-ray sources and innovation in spectrograph design. In such specific cases, XRRS with more penetrating hard X-ray (between 6-10 keV) by synchrotron radiation (SR) provide the unique structural information and sensitivity of low Z NEXAFS.

NMR Spectroscopy NMR (^{1}H &^{13}C) spectroscopy is a well-known technique for establishing structural formulas with electronic structures of organic compounds and has been used extensively for various aspect of hydrocarbon industry [12–17]. NMR method can help to determine aromatic and aliphatic hydrogen and carbon atoms in hydrocarbon-based samples. ^{1}H and $^{13}CNMR$ help us to provide general information regarding functional group with presence of olefins but not been used for analysis of high molecular weight complex hydrocarbon components. ^{13}C NMR technique provides spectroscopic information allowing determination of principal carbon types. The quantitative 13C NMR spectra is the only direct method for measuring aromaticity (C_{ar})—relative percentage of aromatic carbon atoms in hydrocarbon [18, 19]. ^{13}C NMR provides a quantitative representation of the structure with simple spectra in case of small molecules. But for complex mixture of macromolecules such as in coal, only average structural representation could be discerned. In cross-polarization magic angle spinning (CPMAS) ^{13}C NMR, peak intensity in the spectra is approximately proportional to the fraction of carbon with its specific structural fingerprint assigned in the abscissa (chemical shift). High resolution NMR spectroscopy is used to derive information about polynuclear hydrocarbon compounds based on superconducting magnet technologies.

Mass-Spectroscopy Pyrolysis method is linked to some sort of characterization method provides much more information about the SOM. Py-GC is used for effective characterization of coal and kerogen aromaticity, maturity, alkane chain length, sulphur speciation and biomarker content [20, 21]. In py-GC-MS, the assumption is that the pyrolysis yield volatile molecules are representatives of the reconstructed original macromolecular network but the pyrolysis yields are dependent on numerous factors of the pyrolysis system. Flash pyrolysis combined with gas-chromatography, mass-spectrometry (py-GC-MS) provides molecular-level structural details and framework.

4 Optical to IR Spectroscopic Approach

Due to the introduction of Internet of things (IoT) with inputs of artificial intelligence (AI) and machine learning (ML) in the petroleum industry, online remote characterization of crude oil with real-time understanding of the situation is the most important challenges faced by the industry throughout its value chain. A rapid, cost-effective, easy to handle in situ and reliable methods for remote chemical characterization of crude oil from reservoir to refinery is of great importance for real-time adjustment of critical operational parameters which results in economic and environmental benefits. A variety of techniques including the spectroscopic techniques have been used over the last few decades for classification, characterization for in-depth understanding and generalization for better operational practices in the day-to-day activities of the petroleum industry. Of all the optical spectroscopic techniques used, UV–visible to infrared (IR) optical spectroscopic study has shown the highest potential due to their rapid accurate response, low cost easy to handle, small instrumentation, in situ analysis without losing any vital information about the system.

4.1 UV–Visible Absorption Spectroscopy

UV–visible spectroscopy is a powerful method to investigate molecular structure of complex aromatic materials, not amenable to be characterized by chromatography, mass-spectroscopy for their larger molecular weights and/or aromaticity and polarity. Optical absorption spectroscopy is increasingly being used for crude oil because they are low cost, offer high speed in analysis, non-contact and non-destructive in nature. The absorption behaviour of crude oils in the UV–visible wavelength is due to the presence of wide range of polycyclic aromatic hydrocarbons (PAH), while the intensity and spectral behaviour of this absorption being directly related to chemical composition [22]. Evdokimov et al. [23] with the help of near-UV/visible absorption spectra clearly demonstrated the asphaltene aggregation phenomenon in toluene solution of crude oil. The absorption coefficients and the shape of the UV–visible spectra from different carbon materials namely aromatic oil, naphthalene pitch, carbon soot, carbon nanotubes, carbon black and standard PAH have been compared by Apicella et al. [24] and applied the technique on the aromatization process during soot inception and growth along a flame.

4.2 UV–Visible Fluorescence Spectroscopy

Fluorescence spectroscopy provides a wealth of information for a variety of purposes related to the organic matter [1], while fluorescence studies of crude oil provide valuable insight into their fundamental and dynamic properties for a variety of applications [25]. Optical measurements on crude oils with the help of commercially available optical instruments help to understand the crude oil environment from reservoir to refinery for developing better operational practices. The rapid miniaturization with increasing sensitivity and capability of fluorescence instrumentation is further expanding the range of onsite and in situ sensor-driven measurement. Fluorescence spectroscopy can provide two-dimensional and time-resolved signals with more detail information than one-dimensional absorption and steady-state emission spectrum.

Emission (fluorescence) spectroscopy: UV–visible emission (fluorescence) spectroscopy has been used extensively for identifying complex aromatic mixtures [26–28]. For understanding the smallest chromophores, we need to see the fluorescence emission spectra because only small chromophores emit short wavelength light. All fluorescence spectra of asphaltenes lack much emission from aromatics with one or two rings at 290 and 320 nm, respectively [22]. Asphaltenes lack fluorescence emission because most part they lack one to three fused aromaticrings [29]. UV–visible fluorescence cannot identify precise size of PAH and structural features within complex mixtures.

Synchronous fluorescence spectroscopy: The synchronous fluorescence spectrum is very much distinctive for differentiating individual components from complex mixture as because a single response is expected from each type of aromatic system, corresponding to excitation by the shortest wavelength and emission by the longest wavelength for a chromophore molecule. Total synchronous fluorescence scan (TSFS) has been extensively used for the analysis of crude oil classification [30] aromatic hydrocarbons [31], petroleum photodegradation [32] and petroleum product [33, 34]. Ryder [30] with crude oil TSFS plots show a general diagonal contour trend from short excitation wavelength/large wavelength interval (large $\Delta\lambda$) to long excitation wavelength/short wavelength interval (small $\Delta\lambda$).

Time-resolved fluorescence depolarization (TRFD): Time-resolved fluorescence depolarization method allows interrogation of different polycyclic aromatic hydrocarbon and classes to be tested regarding molecular weight and molecular structure [35]. TRFD studies have already been proven to be a powerful tool to unravel asphaltene complexities. With this technique, a polarized laser of specific (selected) wavelength excites a subset of PAH in the sample solution. The TRFD results are very much clear where independent interpretation is very much correct to convey the meaningful information [35, 36] TRFD data of crude oil indicates that there is one chromophore per asphaltene molecule and they are not crossed-linked with mean molecular weight ~750 gm/mole with a range of 500–1000 gm/mole [35]. TRFD studies confirm that coal asphaltenes are found to be smaller ring system with small alkyl substitution compared to petroleum asphaltene [35, 36]. 'The TRFD studies pave the way towards new simplifying causality in asphaltene science and build expectation for integration of fundamental chemical principles with petroleum science, thereby realizing the vision of Petroleomics' [35]. The TRFD results together with optical absorption data predict 4–10 fused aromatic ring, on an average, in individual asphaltene molecule. Ryder and his group [37–39] measured the lifetime of crude oil of different types by using fixed emitting wavelength [37, 38] and fixed excitation wavelength [39, 40] that the mean lifetime was linearly correlated with °API of the crude oil and its aromatic hydrocarbon concentration. Based on the study on crude oil, Ryder et al. [38] further concluded that the most significant factor influencing the fluorescence lifetime is the concentration of quenching species like sulphur, polar and asphaltene fractions. Time-resolved fluorescence signals are represented in terms of contour diagram of four different crude oils and successfully distinguish all four crude oils, even between more closely similar crudes [41].

4.3 Fourier Transform Infrared Spectroscopy

IR spectroscopy measures the types of chemical bonds present in a material by absorption of either transmitted or reflected light. It is an important method since long for determining the chemical composition and structures of hydrocarbon components. The FTIR spectroscopy analysis is very rapid in nature, low cost, easy sample preparation, broad applicability with easy to handle methods provided a preference to use by the scientist. The infrared spectra cover between 4000 and 400 cm^{-1} wavelength regions where it is used quantitatively to measure the composition of the hydrocarbon mixture particularly to determine the number of CH_3 and CH_2 groups in hydrocarbons. Most of the functional group such as methylene, olefinic or aromatic C–H give rise to various C–H stretching vibrations that are mainly independent of the rest of the molecules. The NIR region is attractive for hydrocarbon components for crude oil analysis because many of the absorption bands observed in this region arise from overtones or combinations of carbon–hydrogen stretching vibration. The first study of FTIR analysis of hydrocarbon-related materials from pitches, coal and petroleum distillation residue was done in 1953 [42]. Quantitative FTIR analysis of aromatic and aliphatic hydrogen in the carbon materials namely asphaltenes, carbonaceous particulate matter, naphthene pitch and carbon black has been developed involving the spectral deconvolution of peak-to-peak of C–H stretching (3100–2800 cm^{-1}) and bending (900–700 cm^{-1}) [43]. The curve resolved spectra of acquired FTIR spectra facilitated the interpretation of bands present and provided significant information about the substitution linked to functional group present in the asphaltenes [44].

5 Combinations of Optical Study for Exploration of Hydrocarbon Molecules in Crude Oil and Coal

Bhui et al. [45] used TG, GCMS, FTIR and fluorescence spectroscopic studies for determining the entrapped crude oil components and their state of aggregation inside the formed micelles during surfactant flooding for enhanced oil recovery. FTIR and steady-state emission spectroscopy study was used by the authors to pinpoint the PAH present in the crude oil components with functional group, while emission spectra with time-resolved fluorescence data of crude oil-aqueous-surfactant solution clearly demonstrate the trapping mechanism of PAH for microemulsion formation for selecting suitable surfactant for EOR. Chemical characterization of different types of crude oil (from heavy oil to light oil) and their interaction study with saline surfactant (anionic, cationic and non-ionic) solution was studied by Bhui with his research group [46, 47] with UV–visible and fluorescence spectroscopic approach. These studies clearly demonstrated that the anionic surfactant entraps smaller fused aromatic ring (FAR) polycyclic aromatic hydrocarbon (PAH) while the non-ionic surfactant entraps medium to large FAR PAH, which has deep-rooted implications for designing surfactant solutions for EOR from crude oil reservoirs.

Sanyal et al. [48–50] demonstrated the interaction among clay (kaolinite, montmorillonite)-crude oil-saline water with different cation and concentration, for designing injection water for enhanced oil recovery where FTIR, UV–Visible and fluorescence spectroscopic studies have been extensively used for characterization. FTIR and fluorescence (steady-state and synchronous) spectroscopic studies were used by the authors to characterize the individual PAH groups present with their functional group, while UV–visible absorption, steady-state and synchronous fluorescence spectra of filtered out saline water, after interaction with oil treated clay (kaolinite, montmorillonite) demonstrated types of dislodging PAH by the saline water having different cation and concentration.

Anto et al. [51] used FTIR, UV–Vis and fluorescence spectroscopy to show the compositional difference between two crude oils in terms of presence of PAH and their functional group and tried to correlate their rheological behaviour. The authors used FTIR and synchronous fluorescence study to show the interaction behaviour of crude oil and nanoparticles (silica and alumina nanoparticles) and their applicability as flow improver depending on the crude oil composition system.

With the help of UV–visible and fluorescence spectroscopic study, Bhui et al. [52] characterized different PAH structures present in solvent extracted shale samples and their demineralized counterpart which could provide a potential scale of thermal maturation for source rock evaluation. UV–visible and fluorescence spectroscopic study of solvent extracted organic matter with FTIR spectroscopic data of whole rock powder sample of shale is also used by Bhui et al. [53] to understand the maturity of source shale in comparison with already established maturity data like vitrinite reflectance (VRo), Rock-Eval pyrolysis data and thermal alteration index.

UV–visible and fluorescence (steady-state and synchronous) spectroscopic study of solvent extracts of coal of different rank (lignite, subbituminous and bituminous) by Sarkar et al. [54] clearly demonstrated rank-wise population distribution of smaller and larger fused aromatic ring (FAR), which could lead to understand the innate macromolecular structure of coal. Based on UV–visible and fluorescence spectroscopic study, Bhui et al. [55] compared the structural components present in solvent extracted and residual fraction of coal and used the information for understanding the behaviour of coal under hydrogen plasma for future use as cleaner fuel. Sarkar et al. [56] clearly recognize the molecular signatures of different components in bituminous and subbituminous coal extracts using optical spectroscopic study and a relationship between bond strength and thermal maturity was established and correlated with the existing established maturity parameters.

6 Concluding Remarks for Future Implications

In summary, we reviewed the importance of electronic and vibrational spectroscopy which is interdisciplinary by nature for exploration of hydrocarbon molecules in crude oil and coal system. The UV–visible absorption and fluorescence spectroscopy of crude oils can be used to identify the nature of the aromatic hydrocarbons present in them, which can be directly correlated with their chemical composition. The time-resolved fluorescence spectroscopy provides an arsenal of information about the type of the molecules in the crude oil and coal extracts, through excited state lifetime study. The use of vibrational spectroscopic methods like the Fourier transform infrared spectroscopy gives us information about the interacting bonds linked to the functional groups leading to huge aromatic structures in both crude oils and coal. The usage of all these optical techniques for understanding the molecular-level interaction mechanisms has clearly demonstrated for customized selection of injection fluid particularly during surfactant flooding and low-saline water flooding EOR methods. Rank-wise characterization, through all these optical characterization methods, could help to understand the macromolecular network with their bonding characteristics of coal for future cleaner use. This combination is beneficial and even crucial to study the bulk properties of hydrocarbons of economically important systems like coal, heavy hydrocarbons and biomass to use them in an environmentally friendly manner for sustainable future. The present review would definitely improve our knowledge and stimulate further discussion in future to provide a multidisciplinary approach for characterizing hydrocarbon components and their applications for progress and sustainable development of human civilization.

Acknowledgements UKB acknowledge the support of SNBNCBS for developing his interest into the field of spectroscopic study where the role of my co-author, SKP is noteworthy. Financial support of PDPU for establishing Reservoir Characterization Laboratory (RCL), equipped with UV–visible and Fluorescence spectrophotometer, had a far-reaching future implication in the domain of carbon-based research for PDPU. SKP thanks the Indian National Academy of Engineering (INAE) for the Abdul Kalam Technology Innovation National Fellowship, INAE/121/AKF.

References

1. Tissot BP, Welte DH (1984) Petroleum formation and occurrence. Springer-Verlag, New York
2. Derenne S, Largeau C, Behar F (1994) Low polarity pyrolysis products of Permian to Recent *Botryococcus*-rich sediments: first evidence for the contribution of an isoprenoid algaenan to kerogen formation. Geochim Cosmochim Acta 58:3703–3711
3. Collinson ME, van Bergen PF, Scott AC, de Leeuw JW (1994) The oil generating potential of plants from coal and coal bearing strata through time: a review with new evidence from Carbonaceous plants. In Scott AC, Fleet AJ (eds) Coal and coal bearing strata as oil-prone source rocks? vol 77. Geological Society Special Publication, pp 31–70
4. Tegelaar EW, de Leeuw JW, Derenne S, Largeau C (1989) A reappraisal of kerogen formation. Geochim Cosmochim Acta 53:3103–3106
5. Mukhopadhyay PK, Hatcher PK (1993) Composition of coal. In Law BE, Rice DD (eds) Hydrocarbons from coal American Association of Petroleum Geologists. Studies in Geology Series, vol 38, pp 79–118
6. Hatcher PG, Clifford DJ (1997) The organic geochemistry of coal: from plant materials to coal. Org. Geochem 27(5/6):251–274
7. Vassilev SV, Baxter D, Andersen LK, Vassilev CG, Morgan TJ (2012) Fuel 94:1
8. Vassilev SV, Baxter D, Andersen LK, Vassilev CG (2010) Fuel 89:913
9. Hoffmann E, Jenkner A (1932) Die inkohlung und ihreerkennungimmicrobild. Glückauf 68:81–88
10. Dow WG (1977) Kerogen studies and geological interpretation. J Geochem Explor 7:77–79

11. Bergmann U, Mullins OC (2007) Carbon X-ray raman spectroscopy of PAHs and Asphaltenes. In Mullins OC, Sheu EY, hammami A, Marshall AG (eds) Asphaltenes, heavy oils, and petroleomics. Springer, pp 139–155
12. Salim OK, Mal TK (2019) NMR spectroscopy analysis of asphaltenes. Energy Fuels 33:10391–10414
13. Parlov Vukovic J, Novak P, Plavec J, Friedrich M, Marinic Pajc L, Hrenar T (2015) NMR and chemometric characterization of vacuum residues and vacuum gas oils from crude oils of different origin. Croat Chem Acta 88:89–95
14. Poveda-Jaramillo J-C, Molina-Velasco D-R, Bohorques Toledo N-A, Torres M-H, Ariza-Leon E (2016) Chemical characterization of the Asphaltenes from Colombian Colorado light crude oil. CT&F Cienc Tecnol Futuro 6:105–122
15. Rakhmatullin IZ, Efimov SV, Margulis BY, Klochkov VV (2017) Qualitative and quantitative analysis of oil samples extracted from some Bashkortostan and Tatarstan oilfields based on NMR spectroscopy data. J Pet Sci Eng 156:12–18
16. Gao Y, Zou Y-R, Liang T, Peng P (2017) Jump in the structure of type I kerogen revealed from pyrolysis and 13CDP MAS NMR. Org Geochem 112:105–118. https://doi.org/10.1016/j.orggeochem.2017.07.004
17. Mondal S, Yadav A, Kumar R, Bansal V, Das SK, Christopher J, Kapur J (2017) Molecular-level structural insight into clarified oil by nuclear magnetic resonance (NMR) spectroscopy: estimation of hydrocarbon types and average structural parameters. Energy Fuels 31:7682–7692. https://doi.org/10.1021/acs.energyfuels.7b00994
18. McBeath AV, Smernik RJ, Schneider MPW, Schmidt MWI, Plant EL (2011) Determination of the aromaticity and the degree of aromatic condensation of a thermosequence of wood charcoal using NMR. Org Geochem 42:1194–1202. http://dx.doi.org/10.1016/j.orggeochem.2011.08.008
19. Fergoug T, Bouhadda Y (2014) Determination of Hassi Messaound asphaltene aromatic structure from 1H & 13C NMR analysis. Fuel 115:521–526. https://doi.org/10.1016/j.fuel.2013.07.055
20. Horsfield B (1989) Practical criteria for classifying kerogens: some observations from pyrolysis-gas chromatography. Geochim Cosmochim Acta 53:891–901
21. Horsfield B (1997) The bulk composition of first-formed petroleum in source rocks. In Welte DH, Horsfield B, Baker DR (ed) Petroleum and basin evolution: insights from petroleum geochemistry, geology, and basin modeling. Springer, pp 335–402
22. Mullin OC (1998) Optical interrogation of aromatic moieties in crude oils and asphaltenes. In Mullin OC, Sheu EY (Eds) Structure and dynamics of asphaltenes, Chap 2. Plenum Press, New York, pp 21–77
23. Evdokimov IN, Elisev NY, Akhmetov BR (2003) Assembly of asphaltene molecular aggregates as studied by near-UV/visible spectroscopy I. structure of the absorbance spectrum. J Petrol Sci Eng 37:135–143
24. Apicella B, Alfe M, Barbella R, Tregrossi A, Ciajolo A (2004) Aromatic structures of carbonaceous materials and soot inferred by spectroscopic analysis. Carbon 42:1583–1589
25. Downare TD, Mullins OC (1995) Visible and near-infrared fluorescence of crude oils. Appl Spectrosc 49(6):754–764
26. Lakowicz JR (1986) Principles of fluorescence spectroscopy. Plenum Press, New York
27. Apicella B, Ciajolo A, Tregrossi A (2004) Anal Chem 76:2138–2143
28. Herod AA, Bartle KD, Morgan TJ, Kandiyoti R (2012) Chem Rev 112:3892–3923
29. Ralston CY, Mitra-Kirtley S, Mullin OC (1996) Small population of one to three fused-ring aromatic molecules in asphaltenes. Energy Fuels 10:623
30. Ryder AG (2004) J Fluoresc 14:99–104
31. Taylor TA, Patterson HH (1987) Anal Chem 59:2180–2187
32. Guedes CLB, Di Mauro E, De Campos A, Mazzochin LF, Bragagnolo GM, De Melo FA, Piccinato MT (2006) Int J Photoenergy 1–6
33. Patra D, Mishra AK (2002) Anal Chim Acta 454:209–215
34. John P, Souter I (1976) Anal Chem 48(3):520–524

35. Groenzin H, Mullins OC (2007) Asphaltene molecular size and weight by time-resolved fluorescence depolarization. In Mullins OC, Sheu EY, hammami A, Marshall AG (eds) Asphaltenes, heavy oils, and petroleomics. Springer, pp 17–62
36. Badre S, Goncalves CC, Norinaga K, Gustavson G, Mullins OC (2006) Molecular size and weight of asphaltene and asphaltene solubility fractions from coals, crudeoils and bitumen. Fuel 85:1–11
37. Ryder AG (2004) Time-resolved fluorescence spectroscopic study of crude Petroleum oils: influence of chemical composition. Appl. Spectrosc 58(5):613–623
38. Ryder AG, Glynn TG, Feely M, Barwise AJG (2002) Characterization of crude oils using fluorescence lifetime. Spectrochim Acta Part A 58:1025–1037
39. Ryder AG (2002) Appl Spectrosc 56(1):107–116
40. Ryder AG, Glynn TJ, Feely M (2002) Influence of chemical composition on the fluorescence lifetimes of crude petroleum oils. In Proceedings of SPIE, vol 4876. https://doi.org/10.1117/12.463915
41. Hegazi E, Hamdan A, Mastromarino J (2001) New approach for spectral characterization of crude oil using time-resolved fluorescence spectra. Appl Spectrosc 55(2):202–207
42. Yen TF, Wu WH, Chilingar GV (1984) A study of the structure of petroleum asphaltenes and related substances by proton nuclear magnetic resonance. Energy Sources 7:275–304
43. Russo C, Stanzione F, Tregrossi A, Ciajolo A (2014) Infrared spectroscopy of some carbon-based materials relevant in combustion: qualitative and quantitative analysis of hydrogen. Carbon 24:127–138
44. Asemani M, Rabbani AR (2019) Detailed FTIR spectroscopy characterization of crude oil extracted asphaltenes:curve resolve of overlapping bands. J Petrol Sci Eng. https://doi.org/10.1016/j.petrol.2019.106618
45. Bhui UK, Sanyal S, Saha R, Rakshit S, Pal SK (2018) Steady-state and time-resolved fluorescence spectroscopic study of petroleum crudes in aqueous-surfactant solutions: its implications for enhanced oil recovery (EOR) during surfactant flooding. Fuel 234:1081–1088
46. Bhui UK, Anto R, Ghosh R, Pal SK (2020) Interaction mechanism between crude oils and surfactant solutions: implications for designing injection fluids during enhanced oil recovery from crude oil reservoirs. AVHKolleg 2020 on 'FLOW', Abstract Volume, pp 34–35
47. Anto A, Sanyal S, Bhattacharjee S, Bhui UK (2020) Optical characteristics of petroleum crudes-surfactant-brine solutions: understanding molecular level interaction for designing injection fluids for enhanced oil recovery (Chapter 5 of this volume)
48. Sanyal S, Bhui UK, Saurabh Kumar S, Balaga D (2017) Designing injection water for enhancing oil recovery from kaolinite laden hydrocarbon reservoirs: a spectroscopic approach for understanding molecular level interaction during saline water flooding. Energy Fuels 31:11627–11639
49. Sanyal S, Bhui UK, Balaga D, Saurabh Kumar S (2019) Interaction study of montmorillonite-crude oil-brine: molecular level implications on enhanced oil recovery during low saline water flooding from hydrocarbon reservoirs. Fuel 254:115725. https://doi.org/10.1016/j.fuel.2019.115725
50. Sanyal S, Anjirwala H, Bhatia M, Lalnuntluanga, Shah D, Vyas D, Bhui UK (2020). Interaction of clay-crude-injection brine: an experimental approach for understanding the effectiveness of low saline water (LSW) during enhanced oil recovery (EOR) (Chapter 4 of this volume)
51. Anto R, Deshmukh S, Sanyal S, Bhui UK (2020) Nanoparticles as flow improver of petroleum crudes: study on temperature-dependent steady-state and dynamic rheological behavior of crude oils. Fuel 275:117873. https://doi.org/10.1016/j.fuel.2020.117873
52. Bhui UK, Basu A, Sanyal S (2018) Maturity assessment of Shale from Kachchh Basin of India for hydrocarbon prospect: an optical spectroscopic approach. In Tu P4 07, 80th EAGE Conference and Exhibition 2018

53. Bhui UK, Ariketi A, Sarkar A, Sanyal S, Anto A (2020) Maturity assessment of cambay shale formation (CSF) for hydrocarbon prospect: a molecular structure approach with optical spectroscopy study (Chapter 14 of this volume)
54. Sarkar A, Anto R, Bhui UK (2019) UV–Visible and fluorescence spectroscopic study of solvent-extracts of coal: macromolecular investigation based on polycyclic aromatic hydrocarbons. In 71st annual ICCP meeting, symposium on organic petrology with special focus on oil generation from coals and carbonaceous shales, p 21
55. Bhui UK, Sarkar A, Sanyal S, Pal SK, Ghosh J (2019) Characterization of coal in molecular level for future use: study based on Gondwana and Tertiary coal of India. In EMAAT 2019, Abstract volume, OL-31, pp 49–50
56. Sarkar A, Patel U, Dutta S, Singh BD, Pal SK, Bhui UK (2020) Molecular insight of coal: a spectroscopic approach for evaluating maturity parameters (Chapter 12 of this volume)

Crude Oil

Challenges in Mature Field Redevelopment

Omkar Nath Gyani and Sujit Mitra

1 Introduction

Mature fields contribute around two-thirds of global daily average oil production. And as the number of new major conventional discoveries is declining steadily for the last few decades, the contribution from mature fields is increasing over time. In addition, the size of discoveries over the last decade has become much smaller, and this trend is not going to change in the future. All the recent major discoveries are in the deep water. But this discoveries need to overcome significant techno-economic hurdle to be monetised, whereas significant volume in world oil production has been added in last decade from the unconventional and shale reservoirs. But that too is techno-economically challenging to produce under volatile oil price scenario. So mature fields are significant to maintain global energy demand in the near future.

Mature field is one of the most commonly used words in the oil industry. But if you look at the literature, there is no objective definition of mature field exists. It is more of a perception, either based on the producing age or the recovery achieved of any field or the combination of both. As it is well understood that recovery of a field depends predominantly on geology, petro-physical parameters, geographical location, the company/companies operate the field, the fiscal regime, etc., and not only on the producing age. So, what is common in all the mature fields? The only thing that is common that they have all past their peak production and on the decline, irrespective of their producing age or recovery.

O. N. Gyani · S. Mitra (✉)
Institute of Reservoir Studies, Oil and Natural Gas Corporation Ltd., Chandkheda Campus, Ahmedabad 380005, India
e-mail: Mitra_Sujit@ongc.co.in

O. N. Gyani
e-mail: gyani_omkarnath@ongc.co.in

U. K. Bhui (ed.), *Macromolecular Characterization of Hydrocarbons for Sustainable Future*, Green Energy and Technology,
https://doi.org/10.1007/978-981-33-6133-1_3

Every fields in their life cycle go through the following stages:

1. Discovery
2. Appraisal and delineation
3. Development: commercial production.

Conventionally, development phase has three stages,

- Primary: Pressure depletion, including natural water or gas drive
- Secondary: Mostly, water or gas injection, including pressure maintenance
- Tertiary: Mainly Enhanced Oil Recovery (EOR).

But with the evolution in the industry, the concept of life cycle development is becoming popular, and in the process, sequential development concept has lost its original meaning, as many operators include EOR in very early stage of development to address multiple techno-economic challenges of mature fields.

But one thing has not changed, most of the fields in the world reached mature stage during secondary phase of development; i.e., it reached peak production and started to decline.

2 Mature Field Management

Major advantage of a mature field is that significant amount of oil still remains to be produced and the availability of existing production and evacuation facilities. But these advantages come with fair amount of challenges. The major challenges of mature field redevelopment are as follows:

- Managing production decline over short term and increasing recovery over the long term.
- Integrate new development within current facilities and/or to add new facilities to the existing one.
- Management of increasing water production.

2.1 Managing Production Decline and Increasing Recovery

In the past 25 years, significant progresses have been made in advanced logging tools, 4D seismic evaluations, 3D geo-modelling and fine-scale numerical simulation all that help locate bypassed oil. Another advantage is the industry's ability to drill more complex wells and precisely reach targets containing untapped oil. All these infill drilling efforts led to acreage reduction and along with optimization of ongoing water flood improve sweep efficiency. These efforts are termed as Improved Oil Recovery (IOR) methods. All these advance techniques come at significant cost. With gradual decline in reservoir productivity under primary and secondary recovery techniques, these IOR methods also become economically and technically challenging with time.

The IOR methods are efficient in arresting the declining production in the short/medium term but not very effective in increasing well productivity and recovery in the long term. So to address these issues, the option is enhanced oil recovery (EOR) methods.

EOR projects are generally complex, technology-heavy, capital and resource intensive. EOR has its inherent risk, challenges and benefits. Most importantly, any successful EOR applications need to be tailored to each specific reservoir. The implementation of EOR is intimately tied to the price of oil and overall economics. In spite of all the risk and challenges, the biggest benefit of a successful EOR process has a cascading effect in increasing the economic life of the field. It re-energizes the reservoir which leads to the increase in well productivity and improves the economics of the infill drilling. Hence, a combination of push (EOR) and pull (infill/IOR) [1] is the key to the revitalization of mature fields in terms of production enhancement and increasing the recovery.

It is important to mention that the realization of IOR/EOR potential can only be achieved through long-term commitments, both in capital and human resources, a vision to strive towards ultimate oil recovery, research and development, and a willingness to take risks and excellence in operational practices.

2.2 Management and Integration of Process Facilities

Availability of processing and evacuation facilities is one of the advantages of mature field redevelopment. But with time most of the mature fields need to produce more liquid to lift the same amount of oil as water cut kept on increasing. Especially, in natural aquifer-driven or mature water-flooded reservoir, this problem becomes acute as the existing facilities decays due to ageing. Even, for any liquid-based EOR processes like any type of chemical EOR, low-salinity water flood or steam-based EOR, creation of water treatment for injection as well as disposal requires significant facility creation. In India, in onshore due to rapid urbanisation availability of land for additional facilities is a critical problem. In offshore, logistics is always a challenge due to limited space. Innovative ways, like vertical integration of facilities rather than lateral growth is to be adopted to address these problems.

2.3 Produced Water Management

Implementation of IOR/EOR schemes necessitates drilling of infill wells and increasing the liquid uptake from the reservoirs. As in most of the mature fields, water cut kept on increasing with time, so to maintain production and enhance recovery, more amount of water needs to be produced and managed with time. In mature field redevelopment, significant cost is incurred for produced water management. It is ironical but true; oil companies produce more water than oil. In 2013, Boysen et al. [2] mentioned that oil industry generates $\sim$70 billion bbl of contaminated water from crude oil and natural gas extraction. Traditionally, this water is being treated and disposed in deep aquifer or reinject for water injection or liquid-based EOR processes. But with time, environmental norms of water disposal have become stringent, and so the technology of treatment has also improved. Many companies now treat this water for societal benefit like creation of water bodies or even for farm use.

3 Indian Scenario

In Indian scenario, the role of IOR/EOR has assumed prime importance as around two-third of ONGC's production since last two decades are from 15 major fields including India's largest field of Mumbai High in offshore and major fields like Gandhar, North Kadi, Kalol, from western onshore and the fields of Lakwa, Geleki, from Assam. All these fields have long production history and are past their peak production. As an important step towards arresting the production decline and simultaneous increase in production from mature fields, around 19 IOR/EOR schemes were initially formulated in the year 2000–01 as structured programme. This has led to arrest the decline in these fields to 2% as compared to global average of 8%. Moreover, these schemes have also led to augmentation and sustenance in oil production. Success of these schemes has led to multiple IOR schemes. All the major fields, especially in offshore, are developed through rolling development schemes, i.e., IOR schemes implemented in successive phases so that the learning and improved understanding from one phase can be incorporated in the next phase. Along with the infill drilling and water flood optimization, the following efforts are also made as IOR on regular basis to improve performance of the mature fields:

- Installation of new/optimization of existing surface facilities and artificial lift systems
- Liquidation of candidate sick wells thro rigless or workover services
- Flow assurance in case of adverse quality of oil: for waxy and/or low pour point oil
- Hitech drilling including horizontal wells
- Appraisal wells to assess field growth
- Installation of compressor to reduce back pressure (Gas fields)
- De-liquification of wells (Gas fields)

But contribution from acreage reduction through in-fill drilling is likely to boost the recovery factor by 2–3%, and it will take the average overall recovery factor to around 28%. Further increase of recovery factor to a level of 35% and beyond can be facilitated with EOR technology. EOR processes bring twofold benefits to a mature field. It not only increases recovery factor by itself but also enables the operator to drill more infill wells (IOR) by increasing the individual well productivity. IOR and EOR have symbiotic relationship; one without the other is not economically attractive.

Over the last 40 years with the setting up of Institute of Reservoir Studies at Ahmedabad, ONGC, had been vigorously pursuing R&D in EOR irrespective of global oil price scenario. The result of this R&D has seen the country and ONGC in particular being placed in the world map of EOR with the commercialization of in situ combustion process in the heavy oil belt of Mehsana, polymer flooding in Sanand field and miscible gas injection in Gandhar GS-12 sand of Gujarat.

In 1985, the first EOR pilot on polymer flood was initiated in medium gravity reservoir of Sanand [3]. The encouraging results of the pilot led to its expansion on semicommercial phase in January 1993 and subsequent commercialization in October 1995 with an aim to enhance recovery to about 30%. In March 1990, in situ combustion pilot was initiated in southern part of Balol [4] heavy oil field in challenging conditions. Teething problems encountered at this stage were mitigated by in-house efforts, and the pilot was expanded to semicommercial phase in January 1992. Subsequently in 1997 and 2000, the process was not only extended to entire Balol but also in adjoining Santhal field in two phases with an envisaged enhanced recovery from about 10–15% to about 40%. Currently, the implementation of the process has resulted in more than twofold increase in oil production with substantial decrease in average field water cut.

Miscible hydrocarbon gas injection was conceived for GS-12 sand of Gandhar field through laboratory investigation and simulation studies in 1996. Its field implementation in 1996 has resulted in significant increase in oil production and successful in enhancing recovery beyond 50%.

The success of in situ combustion in Balol and Santhal [5, 6], miscible hydrocarbon in Gandhar and polymer flood in Sanand has placed India and ONGC in particular in the world EOR map. Over the years, there has always been a thrust on EOR. In the recent years, it has only shifted to the fore front.

Apart from the ONGC operated fields as mentioned above, the Joint Venture fields like Ravva in offshore and Mangala, Bhagyam and Aishariya fields in Barmer basin have witnessed significant IOR/EOR efforts to increase recovery and enhance production. Ravva field in eastern offshore is an example of efficient and optimized waterflood and Mangala field in Barmer basin is one of the world's largest ongoing polymer flood projects.

4 Journey Ahead

4.1 Onshore

Since 2015, with a renewed focus on EOR, processes like polymer flood and cyclic steam stimulation (CSS) in heavy oil field, gas injection in the fields of Assam, high pressure air injection (HPAI) in light oil field, commercial ASP flood in medium gravity oil field and miscible CO_2 injection as carbon capture utilization and storage (CCUS) in Gandhar field have been studied and planned.

Pilot polymer flood in heavy oil field of Bechraji is the India's first heavy oil polymer flood project. Similarly, miscible CO2 injection project of Gandhar [7] shall be the India's first EOR project as CCUS utilizing industrial CO_2.

4.2 Offshore

Globally, EOR is most popular in onshore fields and predominantly in sandstone reservoir, but carbonate has its share as well. But in offshore, the EOR footprints are few. And it becomes rarity in case of carbonate. The major challenges in offshore are logistics and large well spacing leading to delayed response.

Among the few available cases, most popular is gas injection, both hydrocarbon and non-hydrocarbon in various modes: Miscible/Immiscible/Water Alternate Gas (WAG)/Simultaneous Water and Gas (SWAG), etc. In the recent past, polymer pilots in captain field in North Sea and Bohai Bay have been tested.

In Mumbai High, immiscible SWAG was tested but techno-economic viability could not be established. Moreover, high temperature (>90 °C), high salinity (>30,000 ppm) and the presence of bivalent ions makes chemical EOR a technically challenging prospect.

Hence, a modified variety of water flood, i.e. low-salinity water flood, has been planned. Low-salinity water injection is an EOR process which involves injecting brine, with a low salinity (5000 ppm or less). It improves microscopic displacement efficiency by modifying the wettability. This has the effect of mobilizing more of the oil behind the displacement front and increasing recovery. In addition to the increased recovery, there are additional benefits of injecting lower salinity water:

- Reduced scaling tendency
- Less sulphate injection could help reduce H_2S generation in the reservoir

These benefits significantly help the economic case for a low-salinity flood. While low-salinity EOR generally requires a lower upfront capital investment than a gas injection project, there is still a significant addition to the standard waterflood system. Chemical EOR may be the long-term planning for Mumbai High as a natural step up from low Sal Water flood in terms of quantum jump in increasing displacement efficiency and ultimate recovery.

5 Conclusion

- Mature fields are the one which has past their peak production and on the decline. Globally, around two-third of oil production comes from mature fields. As the discovery of major conventional fields have diminished over the last 25 years, the dependence on maintaining and even increasing output from mature fields have gained significance.
- The advantages of mature fields are that significant amount of oil remains to be produced and facilities for processing and evacuation of the crude oil already exist.
- Challenges are mainly to arrest the declining production in short term and increase recovery in the long term, management of ageing facilities and creation of new ones and management of produced water.
- In India, the redevelopment of mature fields is more significant as $\sim$95% of our domestic oil productions are from mature fields.
- ONGC is working since 1980, in planning and implementation of IOR/EOR schemes for mature field redevelopment and could manage to arrest decline in its $\sim$20 major fields at $\sim$2%, whereas the global average decline in mature fields is $\sim$8%.
- ONGC has already implemented in situ combustion in heavy oil, polymer flood in medium gravity oil and miscible hydrocarbon in light oil since 1985. Recently, polymer flood pilot has been implemented in heavy oil field.
- ONGC planned to implement air injection in light oil, cyclic steam injection in heavy oil and miscible CO_2 injection as CCUS in near future as its commitment towards improving recovery and production from India's mature fields.

Acknowledgements Authors express their sincere appreciation to Oil and Natural Gas Corporation Limited (ONGC) for permission to publish this work. The authors thankfully acknowledge the assistance rendered by Heavy oil Development Group of IRS.

References

1. Ogbuagu F, Okani N, Silpngarmlers L, Strebelle S (2020) Infill opportunities validation for a mature Niger Delta Reservoir: lessons learned and best practices. SPE-203688
2. Boysen B, Henthorne L, Johnson H, Turner B (2013) New water-treatment technologies tackle offshore produced-water challenges in EOR. SPE-0613-0017-OGF
3. Tiwari D, Marathe RV, Patel NK, Ramachandran KP, Maurya CR, Tewari PK (2008) Performance of polymer flood in sanand field, India—a case study. SPE-114878
4. Dayal HS, Pandey V, Mitra S, Bhusan BV, Bhandari AC, Dwivedi MM (2011) Monitoring of in situ combustion process in Southern Part of Balol field through analysis of produced fluids. SPE-150310
5. Ghosh BN, Sarkar SD, Lohia JP, Das TK (2004) Improved oil recovery by in-fill drilling in a mature field, a success story. SPE-89368

6. Pratap V, Sood RK, Ram B (2006) IOR strategy in brownfield of Mehsana Asset, North Cambay Basin, India. SPE/IADC-102298
7. Mishra G, Meena RK, Mitra S, Saha K, Dhakate VP, Om P, Singh RK (2019) Planning India's first CO2-EOR project as carbon capture utilization & storage: a step towards sustainable growth, SPE-194629

Interaction of Clay–Crude Oil–Injection Brine: An Experimental Approach for Understanding the Effectiveness of Low Saline Water (LSW) During Enhanced Oil Recovery (EOR)

Saheli Sanyal, Harsh Anjirwala, Meet Bhatia, Lalnuntluanga, Darsh Shah, Divyanshu Vyas, and Uttam K. Bhui

1 Introduction

Since 1990, low saline water flooding (LSWF) is used as a potential technique for enhancing oil recovery from the sandstone reservoir rock. The past researches prove that lowering the salinity of injection brine gives rise to substantial improvement in oil recovery from the hydrocarbon reservoir [1–4]. Several reasons are put forward to explain the most pertinent interaction mechanisms for the additional oil recovery such as the initial wettability, the presence of clay minerals [5, 6] and the polar crude oil components [7, 8] in the hydrocarbon reservoir rocks. The role of clay minerals is very undeniable as they present in the sandstone reservoir in a considerable amount and retain permanent negative charges. In addition to this, another crucial factor is the molecular constituents of crude oil composition [9, 10]. It is a

S. Sanyal · H. Anjirwala · M. Bhatia · Lalnuntluanga · D. Shah · D. Vyas · U. K. Bhui (✉)
School of Petroleum Technology, Pandit Deendayal Energy University (PDEU, Formerly Pandit Deendayal Petroleum University-PDPU), Raisan, Gandhinagar, Gujarat 382007, India
e-mail: Uttam.bhui@spt.pdpu.ac.in

S. Sanyal
e-mail: sahelisanyal87@gmail.com

H. Anjirwala
e-mail: harshanjirwala@gmail.com

M. Bhatia
e-mail: meet@shreevinayakgases.com

Lalnuntluanga
e-mail: nuntluanga64@gmail.com

D. Shah
e-mail: darshshah13596@gmail.com

D. Vyas
e-mail: dvyas13ad@gmail.com

U. K. Bhui (ed.), *Macromolecular Characterization of Hydrocarbons for Sustainable Future*, Green Energy and Technology,
https://doi.org/10.1007/978-981-33-6133-1_4

complex mixture of a large variety of molecules that contain bound aromatic and non-aromatic rings, with several alkyl side chains and various functional groups such as S=O, C=O, COOH, –NH2 and SH attached to the outermost rings [11]. Such compounds are concentrated in the heaviest polar petroleum fractions, i.e. resins and asphaltenes [12, 13]. As the clay minerals are negatively charged particles, they have a strong affinity towards adsorption of resins and asphaltenes [14] which results in the wettability alteration of reservoir rocks. The adsorption phenomenon also changes with the changing clay species [15–17]. The clay minerals that are usually present in most of the sandstone reservoir rocks are kaolinite, montmorillonite, illite, illite-smectite mixed layer and chlorite, etc. Initially, the sandstone is water-wet and over time the clay minerals adsorbed the polar oil components and alter the wettability from water-wet to oil-wet. Austad et al. [18] explain that depending on the cation exchange capacity (CEC) the amount of oil recovery during LSWF would change with the change of clay species (the CEC varies from montmorillonite > illite/mica > kaolinite in descending order). Thus, the oil recovery may vary depending upon the clay mineral/s present in the reservoir. So, for successful implementation of the LSWF, molecular level understanding of clay minerals is very necessary.

Numerous mechanisms have been described for successful oil recovery during LSWF such as clay particle (fine) migration [7], osmosis [5], increased pH and reduced interfacial tension (IFT) similar to the alkaline flooding [3], multicomponent ion exchange (MIE) [19–22], expansion of the double layer [20, 23, 24], mineral dissolution [25], the salting-in effect [26] and salting-out effect [27]. According to Ligthelm et al. [20], the wettability alteration from oil-wet to water-wet during LSWF would result in incremental oil recovery. The wettability alteration depends upon the complex nature of interaction among brine–oil–rock. Lager et al. [19] proposed multicomponent ion exchange (MIE) which is explained as the root cause of wettability change of rock surface. During MIE, the cation exchange and cation bridging mechanism [14, 19] play a vital role where either the positively charged crude oil components adsorbed by the clay surfaces replacing the inorganic cations on the polar crude oil components are bonded with divalent cations present in the clay surface. The alteration of oil-wet to water-wet clay surfaces takes place due to replacement of the divalent cations. These cations are acting like bridging component between the clay surface and the polar crude oil components and can be quickly exchanged by the monovalent cations as it expands the electric double layer of clays. Thus, the presence of the common divalent cations, i.e. calcium, magnesium in the injected saline water has a significant impact during LSWF. It has been reported [28] that 1500 ppm low saline water with divalent ions recovered additional oil. On the contrary, Lager et al. [19] show the concentration of the divalent ions should be low but not zero. Thus, there is a gap in understanding the or estimating the concentration of the divalent ions in the injection water for optimization of the oil recovery during LSWF.

In the present work, an attempt is made to understand the molecular level insight into clay–crude oil–injection saline water interactions occurring during EOR and thereby investigating the role of NaCl as well as $CaCl_2$ and $MgCl_2$ brine of different salinity. Moreover, the type of crude oil components removed during the LSWF has also been discussed. The study was done with the help of X-ray diffraction (XRD), thermogravimetric analysis (TGA), wettability measurements, Fourier transform infrared (FTIR) spectroscopy, UV−visible absorption spectroscopy, and fluorescence spectroscopy for their cost-effectiveness and preciseness.

2 Materials, Experiments and Instrumentation

2.1 Samples and Materials Used

Bentonite powder sample (purchased from vendor), labelled here as B4, was used in this study as a representative clay mineral. The sample was oven-dried and characterized using XRD, FTIR and TG.

Dead crude oil sample (P1), collected from the wellhead of the Gamij field of Oil and Natural Gas Corporation (ONGC) of the Cambay basin, was used in this study. For separation of SARA components, first, the asphaltenes were separated from the crude oil (P1) following ASTM D2007-80 (standard asphaltene separation method from crude oil).

Saline water samples were prepared by dissolving an appropriate amount of salts, i.e. sodium chloride (NaCl), calcium chloride ($CaCl_2$) and magnesium chloride ($MgCl_2$) in distilled water. Eight different salinities (1500, 3000, 5000, 10,000, 15,000 and 50,000 ppm) of each salt were prepared and interacted with treated clay samples for understanding molecular-level interaction mechanism.

2.2 Sample Preparation and Experimental Procedures

2.2.1 Oil–Brine-Treated Clay Samples

The clay powder was placed in a drying oven for 24 h in which moisture has to be removed. The oven-dried clay was mixed with the crude oil (P1) and stirred properly for 1 h. Beaker was taken and filled with the required amount of clay sample and into which 30,000 ppm saline water was added which can represent the formation water and also added crude oil (P1) to mimic the reservoir condition. However, as it is not possible to keep the clay–crude oil–brine mixture for geological period ageing, the beaker was packed and left for 2 months in the oven at 65 °C to make them oil-wet. The oil-treated bentonite sample was washed thoroughly by n-Pentane, followed by washing with distilled water for the complete removal of excess surficial oil. The washed samples were then characterized using XRD, FTIR and TG-DTA.

2.2.2 Interaction Study Between Oil-Treated Clay and Saline Water

Two gm. of crushed oil-treated sample were taken in six 100 ml graduated cylinders. NaCl solutions of 1500, 3000, 5000, 10,000, 15,000 and 50,000 ppm were added to the 30 ml mark. All the cylinders were shaken vigorously and allowed to settle for the next 48 h. The effluent brine was collected after 48 h of settlement and filtered with the help of filter paper to separate solid and liquid phases. The separated liquid phase was collected in a liquid sampler. The procedure was repeated for the other two ($CaCl_2$, $MgCl_2$) salts. These liquid samples were investigated with UV–visible and fluorescence spectroscopy analysis.

2.3 *Instrumentation and Analytical Details*

2.3.1 X-ray Diffraction (XRD)

XRD analysis of the untreated B4 was carried out for the identification of different mineral phases using a Panalytical X-ray powder diffractometer was used with Ni-filtered Cu Ka radiation employing a range of 5°–50° at a step size of 0.05°/2θ and a count time of 1.5 s per step. The mineral identification was done with the standard 2θ value from the existing literature.

2.3.2 Fourier Transform Infrared Spectroscopy (FTIR)

FTIR spectra of untreated and oil-treated B4 samples were recorded in the range 4000–400 cm^{-1} using a Perkin Elmer FTIR spectrometer (spectrum two model) version 10.4.2 using KBr pellet technique. The powdered sample was combined with spectroscopic grade KBr (1:20) in a mortar, and pellets were prepared with a hand-held KBr press machine. All the transmittance spectra were taken by considering an average of 20 scans at room temperature with a resolution of 4 cm^{-1}. With the same instrument, the infrared spectra of a neat crude oil sample are captured using the ATR mode which is equipped with a diamond crystal having a refractive index of 2.4.

2.3.3 Thermogravimetric Analysis (TGA)

The thermal analysis of B4 and P1 was carried out by a Mettler thermal analyser, TA4000, in the range of 28–790 °C under nitrogen flow (100 cm^3/min). As the temperature is increased, the sample loses weight successively due to dehydration, component loss (due to oxidation) and material decomposition. Hence, the mass loss was recorded in the data.

2.3.4 UV–Visible Spectroscopy

Optical absorption spectra of the filtered saline water samples after interaction with oil-treated B4 were recorded using Perkin Elmer Lambda 35 spectrometer employing 260–750 nm wavelength range. Optical quartz cuvettes (10 mm × 10 mm) were used to hold the sample and reference (neat saline water solutions of the same concentration) solutions.

2.3.5 Fluorescence Spectroscopy

A Perkin Elmer-LS55 fluorescence spectrophotometer was used for recording the steady-state fluorescence emission spectra of the filtrate resulted from the interaction between oil-treated bentonite over a wavelength range of 200–700 nm employing 300 nm excitation wavelength. The excitation and emission slit width was kept at 5/5 nm, and the scanning speed was 500 nm min^{-1}.

2.3.6 Powder Wettability

Wettability was measured as water–oil adsorption by a digital penetration meter (Model DS210) instrument, prepared by the Instrumentation and Control Department, Dharmsinh Desai University, with a specially designed cylinder of 80 mm height and 18.5 mm diameter, plotting changes in sample mass as a function of time in one second of the interval.

2.3.7 Contact Angle Measurement

The contact angle measurement was performed on an optical tensiometer (Theta of Attension, Biolin Scientific). For this, the specimens were prepared under hydraulic presses, having a maximum working pressure of 260 kg/cm^2, with a sample Teflon holder of 20 mm diameter.

3 Results and Discussion

3.1 Characterization of Bentonite

The raw bentonite sample was characterized by the analysis of X-ray diffraction (XRD), Fourier transform infrared spectroscopy (FTIR), TG-DTA.

The XRD pattern of the clay sample (B4) is presented in Fig. 1 with the characteristic d-space values and the symbols of the minerals against each respective peak. The powder is predominantly composed of smectite, however, quartz,

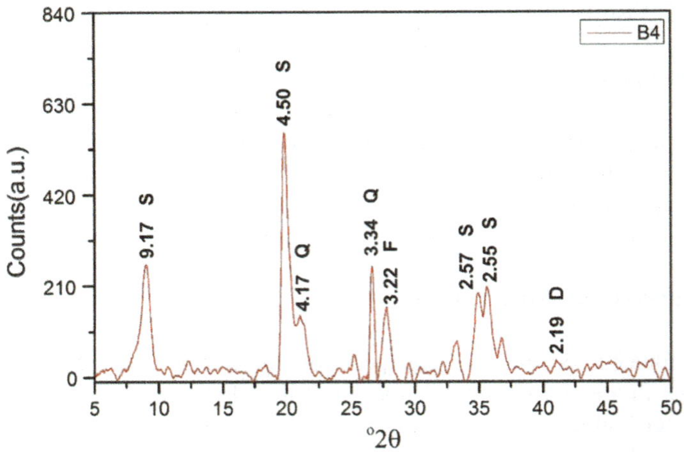

Fig. 1 XRD pattern of the untreated B4 used in the experiments. The mineral symbols and d-space values are assigned for corresponding peaks. The assignments of the minerals are as follows: S = smectite, Q = quartz, F = feldspar and D = dolomite

Table 1 Percentage composition of minerals present in bentonite sample

Sample name	Percentage composition of minerals present (%)			
	Smectite	Quartz	Feldspar	Calcite
B4	55.51	24.60	16.06	3.81

feldspar and dolomite are present as impurity [29–32]. Moreover, semiquantitative mineralogy has been tried to establish for the study B4 sample using the standard procedure of peak heights from the XRD results (Table 1).

The FTIR transmission spectrum (Fig. 2) of B4 rock reveals the presence of characteristic transmission bands [33–38] at 3690, 3620 cm^{-1} denoting the O–H stretching vibration and 1627 cm^{-1} for the H–O–H deformation. The peaks appeared at 1118 and 998 cm^{-1} are assigned for Si–O–Si stretching vibration. The peaks at 911 and 872 cm^{-1} indicate AlAlOH bending and AlFeOH bending, respectively. At 790 cm^{-1}, a weak peak has appeared showing the Si–O–Al compound vibration. The peak at 686 cm^{-1} presents the SiO in quartz, and 528 cm^{-1} is associated with the Si–O–Si bending. The peak at 467 cm^{-1} presents the Si–O–Si bending (Table 2).

The thermogravimetric analysis (TGA) of the B4 clay powder was carried out to study their thermal properties (Fig. 3). It shows three-step weight loss on the thermal treatment. The first stepmass loss took place at 30–130 °C which is attributed to the desorption of absorbed water on the clay. The second step weight loss was at the range of 135–240 °C due to the elimination of the water species coordinated to the interlayer cations. The mass loss in the range of 300–670 °C originates from the dehydroxylation of amorphous smectites [39] (Table 3).

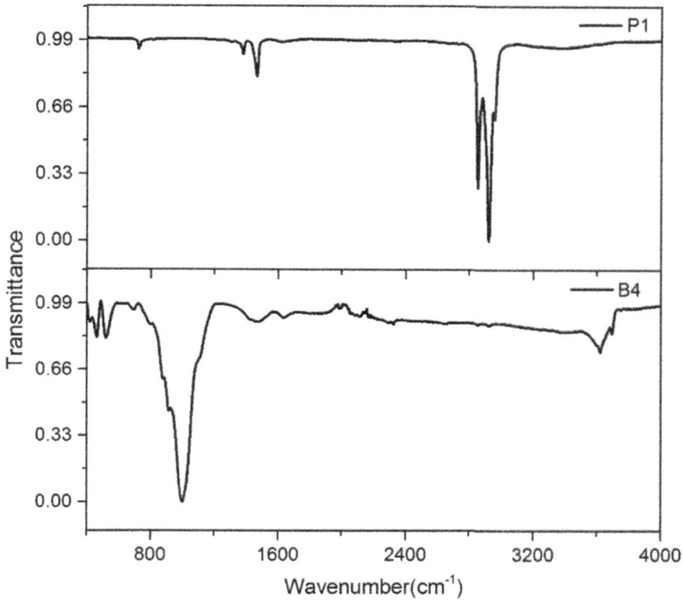

Fig. 2 FTIR spectra of untreated B4, neat crude oil P1 samples

Table 2 Infrared band positions (cm^{-1}) and their assignments in clay powder (B4) and crude oil (P1)

B4	P1	Vibration assignments
3690, 3620	–	O–H stretching vibration
–	2953	CH_3 the aliphatic hydrogen in asymmetric stretching
–	2918	Asymmetric stretching of CH_2 groups in aliphatic chains
	2849	Symmetric stretching of CH_2 groups in aliphatic chains
1627		H–O–H deformation
	1463	CH_3 antisymmetric deformation
	1376	CH3 symmetric in aliphatic compounds
1118, 998	–	Si–O–Si stretching vibration
911	–	Al–Al-OH bending
872	–	Al–Fe-OH bending
790	–	Si–O–Al compounds vibration
	719	OH out-of-plane deformation
686	–	Si–O quartz
528	–	Si–O–Si vibration
467	–	Si–O–Si bending

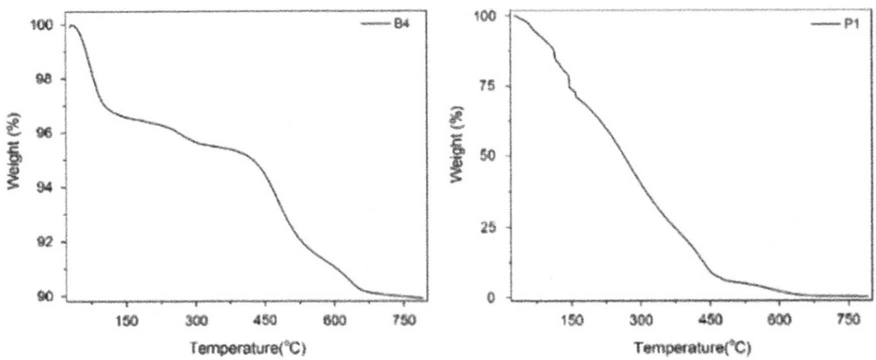

Fig. 3 TG curves of B4 and P1 samples

Table 3 SARA % of the study (P1) crude oil

	API	S	A	R	A
P1	23.4	56.85	14.57	6.35	12.64

3.2 *Crude Oil Characterization*

Figure 3 exhibits only one prominent weight loss zone between 50 and 473 °C temperature. Above 473 °C, it shows an insignificant weight loss. The first weight loss occurred at underpinning the release of lighter aliphatic and aromatic components present within them along with the decomposition of complex asphaltenes [40, 41] associated with different sizes of PAH structures [41]. The weight loss zone between 100 and 400 °C is known to be the decomposition of different sizes of PAH structures [41]. Moreover, a constant TG profile is found above 500 °C for the residual carbonaceous matter [40].

3.3 *Oil and Brine-Treated Bentonite*

XRD patterns for raw bentonite clay (B4) and oil (P1)-treated bentonite clay (B4 +P1) are shown in Fig. 4. The intensity of the smectite peak, near 9.17 Å, is decreased slightly after oil treatment indicating a decrease in the stacking order which may be due to crude oil absorption [42].

The IR spectrum of oil-treated bentonite clay samples shows strong bands around 2956, 2926 and 2854 cm^{-1} attributing to the CH stretching, which indicates the presence of organic component.

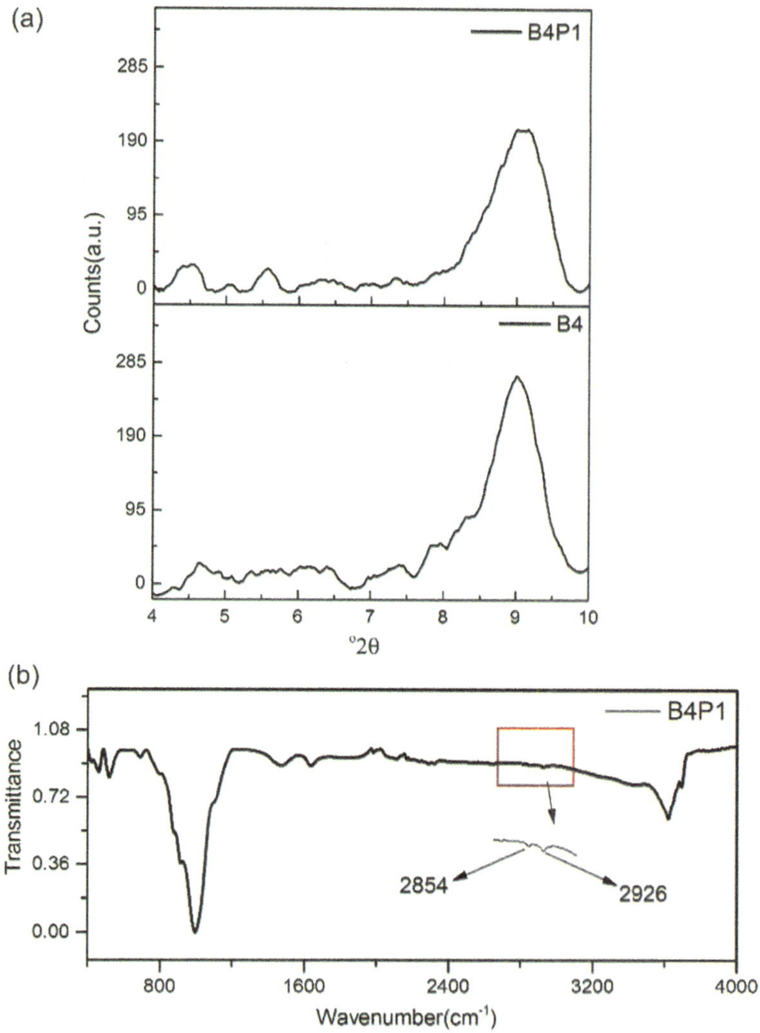

Fig. 4 **a** Enlarged view of the primary peak (9.17 Å) of smectite before and after oil treatment. **b** FTIR spectra of oil-treated B4 samples exhibiting the presence of new sets of bands near 2956, 2926 and 2854 cm^{-1}. (Inset) Enlarged view of the marked portion of the spectrum

3.4 Interaction Study of Oil-Treated Bentonite and Saline Water

Bentonite study:

As can be seen from Fig. 5, there is a considerable change in the peak position as well as an increase in the d-spacing in presence of saline water (1500 ppm)

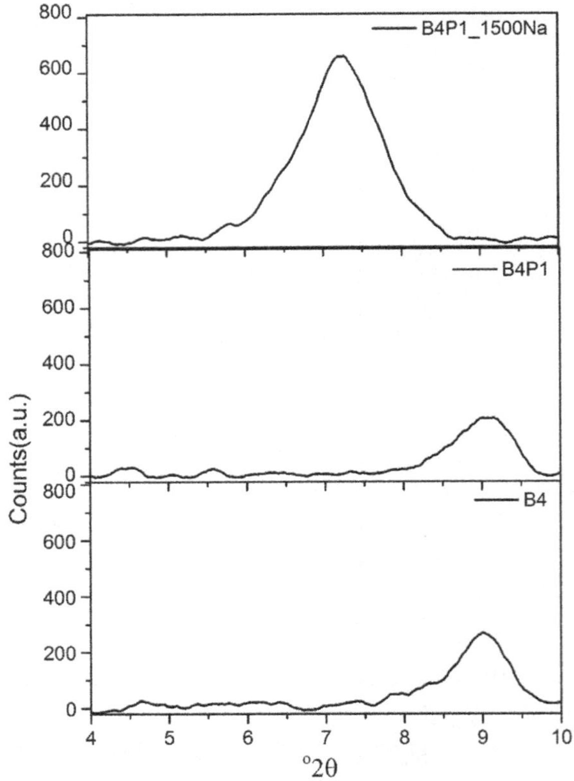

Fig. 5 XRD pattern of untreated clay sample (B4), oil-treated clay sample (B4P1) and LSW treated B4P1 sample. After interaction with LSW, the d-space value is increased concerning B4 and B4P1 samples

prepared by NaCl. This increase in d-spacing corresponds to the crystalline swelling where water enters into the interlayer of smectite leads to an abrupt jump in d-spacing at 12.35 Å which could be related to the removal of the oil components.

3.4.1 Wettability Status of the Bentonite

By Powder Wettability

Figure 6 shows liquid adsorption or liquid penetration curve which is sharply increased initially and approached saturation. The graph of mass versus time for two B4P1 samples is given after interaction with low saline and high saline water. The mass of B4P1 is increased with time after interaction with low saline water contrarily the mass of the same sample is not increased as the former after treated with high saline water. As the B4P1 sample after LSW treatment alters to water-wet by releasing the adsorbed crude oil components, it will allow more penetration of water into the sample and increases the mass. On the other hand, the wettability remains unchanged after the interaction with HSW and B4P1 as HSW has a lesser

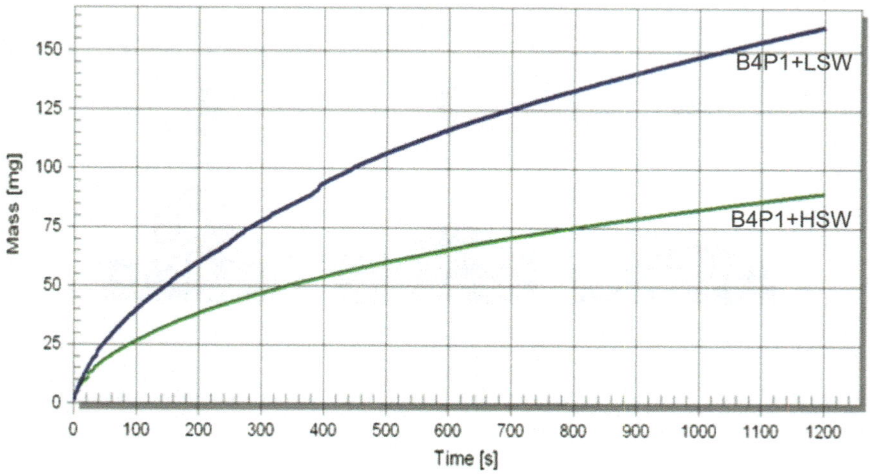

Fig. 6 Wettability measurements by water penetration method of B4P1 sample after interaction with LSW and HSW solution

ability to remove oil from the clay surface. The sample shows less mass gain because of its hydrophobicity. This also supports that the decrease in salinity of injection water results in the removal of more crude oil thus making the clay surface more water-wet.

Contact Angle Measurement

It is seen in Fig. 7 that the contact angles are decreasing as the salinity decreases. For the high saline water (50,000 ppm)-treated clay, the contact angle is 134°–137° however with the lowering of the salinity (5000 to 1500 ppm), and the contact angle decreases to 21°–27°. Thus, a clear trend of wettability alteration is noticed with a change in salinity of injected water, and as the salinity of injected water decreases, contact angle also decreases, thus water wetness increases. As the low saline water removes more crude oil components, the clay sample is more water-wet. It also proves the higher oil recovery in low saline water.

3.4.2 Study with Filtrate

UV Analysis

The concentration of samples whose absorbance was measured for the preparation of the absorption curve is 1500, 3000, 5000, 10,000, 15,000 and 50,000 ppm.

A typical UV spectrum of extracted saline water (NaCl) sample of B4 with P1 oil is presented in figure. The UV spectra for these samples in figure support that

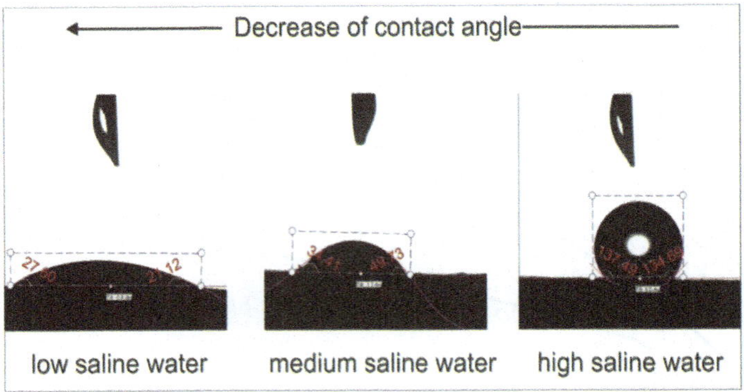

Fig. 7 Contact angle pictures oil-treated B4P1 sample after interaction with low saline, medium saline and high saline water

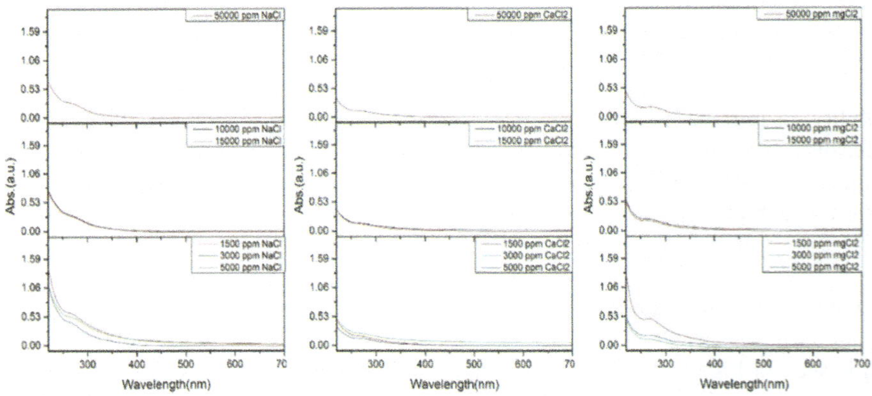

Fig. 8 Absorption spectra of filtered out saline waters (1500, 3000, 5000, 10,000, 15,000 and 50,000 ppm of NaCl, $CaCl_2$ and $MgCl_2$) after interaction with oil-treated B4P1

some compounds have been extracted from the oil-wet clay to the saline water phase regarding the absorption intensity. From the deflection, we can conclude the presence of a 2 and/or 3 rings polycyclic aromatic hydrocarbon (PAH) structure in the saline water. Figure 8 shows that the low salinity domain (1500–5000 ppm) of all the filtered saline water samples is showing the high absorbance values compared to their middle salinity and higher salinity domain. This absorbance comes as the saline water releases crude oil components from the oil-wet clay surfaces and low saline water always has a greater potential to remove these crude components. Moreover, Na-rich saline water has the highest capability to extract crude oil from the clay than the other two salts ($MgCl_2$ and $CaCl_2$). However, Mg saline water has also removed oil components but not as much as Na-rich saline water. Ca-rich saline water is not showing such significance absorbance values than the other two salts. In the low saline domain, 1500 ppm works better for oil removal as seen from

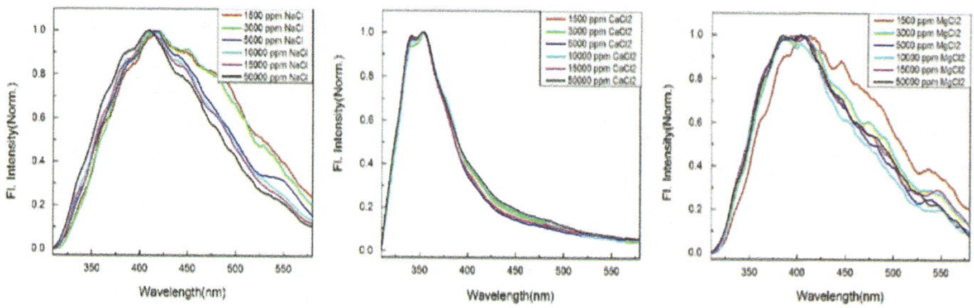

Fig. 9 Normalized steady-state emission spectra (at 300 nm excitation wavelength) of filtered saline waters (1500, 3000, 5000, 10,000, 15,000 and 50,000 ppm of NaCl, $CaCl_2$ and $MgCl_2$) after interaction with oil-treated B4P1. All the samples are showing similar broad emission pattern extending from 310 to 585 nm range

its higher absorbance values. So, the capability of oil removal varies from type of cation as well as the concentration of the cation in the injection water.

Fluorescence Spectroscopic Analysis

Fluorescence spectroscopy has been used for the complementary technique of UV–visible spectroscopy for the further analysis of structures with fused aromatic rings (FAR) present in the filtered out saline water. The inference is drawn from the UV–vis spectra can be supported by the normalized fluorescence emission spectra (Fig. 9) where it is shown that the emission range is broader (310–580 nm) in the case of NaCl and $MgCl_2$ salt solutions. This implies that the solution prepared from these two salts contains significant crude oil components than $CaCl_2$ salt solutions where the emission range is narrow (310–450 nm). The fluorescence emission range extending from 310 to 580 nm suggested the presence of different polycyclic aromatic hydrocarbons (PAH) containing various sizes of the fused aromatic ring (FAR) within the saline water samples [43]. The recorded emission spectra demonstrated that the saline water samples can remove the polar PAH components associated with 2–5 FAR [40, 44]. Thus, for hydrocarbon reservoir where the major clay phase is smectite Na^+ and Mg^{++} low saline water ranging from 1500 to 5000 ppm rich in would help to increase the oil recovery by removing the polar oil components rich in 2–5 FAR from the clay layer surface. However, Ca^{++}-rich low saline water is only capable of removing 2–4 FAR structures from the clay surface.

4 Conclusions

The optimum range of salinity for any salt is 1500–3000 ppm. However, in this range, the salinity that can produce maximum oil depends on the composition of oil and the amount of smectite (clay species) in a reservoir. The ionic solution extracts different polyaromatic structures of the oil sample and hence a detailed study has to be performed to determine the compatibility of different ions and their mixtures with the oil and the rock before any application of the low saline water flooding. From the fluorescence spectrum, it is identified that the monovalent cations like Na^+ are more compatible to exploit oil as compared to divalent cations like Ca^{++} and Mg^{++}.

Acknowledgements SS thankfully acknowledging PDPU for fellowship during doctoral research. UKB acknowledges the support of ONGC for providing the crude oil samples for this work. Petroleum Engineering Laboratory and Reservoir Characterization Laboratory (RCL) of SPT and PDPU are thankfully acknowledged for the optical characterization data. Office of Research and sponsored Projects (ORSP), PDPU, is thankfully acknowledged for partial financial support of this work.

References

1. Agbalaka CC, Dandekar AY, Patil SL, Khataniar S, Hemsath JR (2009) Coreflooding studies to evaluate the impact of salinity and wettability on oil recovery efficiency. Transp Pourous Med 76:77–94. https://doi.org/10.1007/s11242-008-9235-7
2. Hadia NJ, Hansen T, Tweheyo MT, Torsæter O (2012) Influence of crude oil components on recovery by high and low salinity water flooding. Energy Fuels 26:4328–4335
3. Mcguire PLL, Chatham JRR, Paskvan FKK, Sommer DMM, Carini FH (2005) Low salinity oil recovery: an exciting new EOR opportunity for Alaska's North Slope. SPE West Reg Meet 1–15. https://doi.org/10.2118/93903-ms
4. Webb KJ, Black CJJ, Al-Ajeel H (2003) Low salinity oil recovery—log-inject-log. In 13th middle east oil show conference, Society of Petroleum Engineers
5. Morrow N, Buckley J (2011) Improved oil recovery by low-salinity waterflooding. J Pet Technol 63:106–112. https://doi.org/10.2118/129421-JPT
6. Austad T (2013) Enhanced oil recovery field case studies. Elsevier
7. Tang GQ, Morrow NR (1999) Influence of brine composition and fines migration on crude oil/brine/rock interactions and oil recovery. J Pet Sci Eng 24:99–111. https://doi.org/10.1016/S0920-4105(99)00034-0
8. McMillan MD, Rahnema H, Romiluy J, Kitty FJ (2016) Effect of exposure time and crude oil composition on low-salinity water flooding. Fuel 185:263–272
9. Sanyal S, Bhui UK, Kumar SS, Balaga D (2017) Designing injection water for enhancing oil recovery from kaolinite laden hydrocarbon reservoirs: a spectroscopic approach for understanding molecular level interaction during saline water flooding. Energy Fuels 31 (11):11627–11639
10. Sanyal S, Bhui UK, Balaga D, Kumar SS (2019) Interaction study of montmorillonite-crude oil-brine: molecular-level implications on enhanced oil recovery during low saline water flooding from hydrocarbon reservoirs. Fuel 254:115725
11. Cadena-Nava RD, Cosultchi A, Ruiz-Garcia J (2007) Asphaltene behavior at interfaces. Energy fuels 21(4):2129–2137
12. Gonzalez G, Moreira MB (1994) The adsorption of asphaltenes and resins on various minerals. Dev Pet Sci 40:207–231. https://doi.org/10.1016/S0376-7361(09)70256-0
13. Speight JG (1998) Asphaltenes and the structure of petroleum.In Petroleum chemistry and refining, pp 103–120

14. Soraya B, Malick C, Philippe C, Bertin HJ, Hamon G (2009) Oil recovery by low-salinity brine injection: laboratory results on outcrop and reservoir cores. In SPE annual technical conference and exhibition. Society of Petroleum Engineers
15. Clementz DM (1976) Interaction of petroleum heavy ends with montmorillonite. Clays Clay Miner 24(6):312–319
16. Clementz DM (1977) Clay stabilization in sandstones through adsorption of petroleum heavy ends. J Petrol Technol 29(09):1–061
17. Clementz DM (1982) Alteration of rock properties by adsorption of petroleum heavy ends: implications for enhanced oil recovery. In SPE enhanced oil recovery symposium. Society of Petroleum Engineers
18. Austad T, Rezaei Doust A, Puntervold T (2010) Chemical mechanism of low salinity water flooding in sandstone reservoirs. In SPE improved oil recovery symposium. Society of Petroleum Engineers
19. Lager A, Webb KJ, Black CJJ, Singleton M, Sorbie KS (2008) Low salinity oil recovery-an experimental investigation. Petrophysics 49(01)
20. Ligthelm DJ, Gronsveld J, Hofman JP, Brussee NJ, Marcelis F, van der Linde H (2009) Novel waterflooding strategy by manipulation of injection brine composition. In EAGE annual conference and exhibition, SPE. SPE Europec
21. Lee SY, Webb KJ, Collins I, Lager A, Clarke S, O'Sullivan M, Routh A, Wang X (2010) Low salinity oil recovery: increasing understanding of the underlying mechanisms. In SPE improved oil recovery symposium. Society of Petroleum Engineers
22. Lee SY, Webb KJ, Collins I, Lager A, Clarke S, O'Sullivan M, Routh A, Wang X (2010) Low salinity oil recovery: increasing understanding of the underlying mechanisms. In SPE improved oil recovery symposium. Society of Petroleum Engineers
23. Xie Q, Liu Y, Wu J, Liu Q (2014) Ions tuning water flooding experiments and interpretation by thermodynamics of wettability. J Pet Sci Eng 124:350–358
24. Nasralla RA, Nasr-El-Din HA (2014) Double-layer expansion: is it a primary mechanism of improved oil recovery by low-salinity waterflooding? SPE Reserv Eval Eng 17(01):49–59
25. Pu H, Xie X, Yin P, Morrow NR (2010) Low-salinity waterflooding and mineral dissolution. In SPE annual technical conference and exhibition. Society of Petroleum Engineers
26. Rezaei Doust A, Puntervold T, Strand S, Austad T (2009) Smart water as wettability modifier in carbonate and sandstone: a discussion of similarities/differences in the chemical mechanisms. Energy Fuels 23(9):4479–4485
27. Lashkarbolooki M, Riazi M, Hajibagheri F, Ayatollahi S (2016) Low salinity injection into asphaltenic-carbonate oil reservoir, mechanistical study. J Molecul Liq 216:377–386
28. Zhang J, Yuan K, Wang YP, Gu SJ, Zhang ST (2007) Preparation and properties of polyacrylate/bentonite superabsorbent hybrid via intercalated polymerization. Mater Lett 61 (2):316–320
29. Moore DM, Hower J (1986) Ordered interstratification of dehydrated and hydrated Na-smectite. Clays Clay Miner 34(4):379–384
30. Proust D, Caillaud J, Fontaine C (2006) Clay minerals in early amphibole weathering: tri-to dioctahedral sequence as a function of crystallization sites in the amphibole. Clays Clay Miner 54(3):351–362
31. Özgen S, Yildiz A (2010) Application of box—Behnken design to modeling the effect of smectite content on swelling to hydrocyclone processing of bentonites with various geologic properties. Clays Clay Miner 58(3):431–448
32. Worasith N, Goodman BA, Neampan J, Jeyachoke N, Thiravetyan P (2011) Characterization of modified kaolin from the Ranong deposit Thailand by XRD, XRF, SEM, FTIR and EPR techniques. Clay Miner 46(4):539–559
33. Shah KJ, Mishra MK, Shukla AD, Imae T, Shah DO (2013) Controlling wettability and hydrophobicity of organoclays modified with quaternary ammonium surfactants. J Colloid Interface Sci 407:493–499

34. Alver BE, Alver Ö, Günal A, Dikmen G (2016) Effects of hydrochloric acid treatment on structure characteristics and C 2 H 4 adsorption capacities of Ünye bentonite from Turkey: a combined FT-IR, XRD, XRF, TG/DTA and MAS NMR study. Adsorption 22(3):287–296
35. Madejová J (2003) FTIR techniques in clay mineral studies. Vib Spectrosc 31(1):1–10
36. Ikhtiyarova GA, Özcan AS, Gök Ö, Özcan A (2012) Characterization of natural-and organobentonite by XRD, SEM, FT-IR and thermal analysis techniques and its adsorption behaviour in aqueous solutions. Clay Miner 47(1):31–44
37. Tyagi B, Chudasama CD, Jasra RV (2006) Determination of structural modification in acid activated montmorillonite clay by FT-IR spectroscopy. Spectrochim Acta Part A Mol Biomol Spectrosc 64:273–278. https://doi.org/10.1016/j.saa.2005.07.018
38. Saikia BJ, Parthasarathy G (2010) Fourier transform infrared spectroscopic characterization of kaolinite from Assam and Meghalaya, Northeastern India. J Mod Phys 1(4):206–210
39. Hassan MS, Abdel-Khalek NA (1998) Beneficiation and applications of an Egyptian bentonite. Appl Clay Sci 13(2):99–115
40. Merola MC, Carotenuto C, Gargiulo V, Stanzione F, Ciajolo A, Minale M (2016) Chemical–physical analysis of rheologically different samples of a heavy crude oil. Fuel Process Technol 148:236–247
41. Bhui UK, Sanyal S, Saha R, Rakshit S, Pal SK (2018) Steady-state and time-resolved fluorescence spectroscopic study of petroleum crudes in aqueous-surfactant solutions: its implications for enhanced oil recovery (EOR) during surfactant flooding. Fuel 234:1081–1088
42. Sugahara Y, Satokawa S, Kuroda K, Kato C (1990) Preparation of a kaolinite-polyacrylamide intercalation compound. Clays Clay Miner 38(2):137–143
43. Groenzin H, Mullins OC (2000) Molecular size and structure of Asphaltenes from various sources. Energy Fuels 14:677–684. https://doi.org/10.1021/ef990225z
44. Kasha M (1960) Paths of molecular excitation. Radiat Res Suppl 2:243–275

Optical Characteristics of Petroleum Crudes–Surfactant–Brine Solutions: Understanding Molecular Level Interaction for Designing Injection Fluids for Enhanced Oil Recovery

Rincy Anto, Saheli Sanyal, Shubhankar Bhattacharjee, and Uttam K. Bhui

1 Introduction

Petroleum crudes are complex mixtures of organic compounds with a wide range of chemical composition, structures and physical properties which show complex behaviour with a change of physicochemical conditions in the whole value chain of the petroleum industry. As significant amounts of crude oil remain trapped as residual oil inside the oil reservoirs, a small increment in the oil recovery efficiency would yield enormous economic benefits. With increasing energy demand viz-a-viz declining production from the existing reservoir, the need of the time is to characterize petroleum crudes of complex nature at the molecular level for designing efficient production methods for enhancing the recovery. Enhanced oil recovery (EOR) techniques are being used widely in order to recover the trapped crude oil from the subsurface reservoir. The most investigated and used enhanced oil recovery methods include injection of chemicals mainly surfactants, polymers and alkalies where surfactants-assisted chemical methods are expected to play a dominant role in enhancing the crude oil production in future days. Chemical injection enhances the microscopic and macroscopic displacement efficiency to effectively reduce the residual oil saturation [1, 2].

R. Anto · S. Sanyal · S. Bhattacharjee · U. K. Bhui (✉)
School of Petroleum Technology, Pandit Deendayal Energy University (PDEU, Formerly Pandit Deendayal Petroleum University-PDPU), Raisan, Gandhinagar, Gujarat 382007, India
e-mail: Uttam.bhui@spt.pdpu.ac.in

R. Anto
e-mail: rincyanto19@gmail.com

S. Sanyal
e-mail: sahelisanyal87@gmail.com

S. Bhattacharjee
e-mail: Shubhankar.bmtpe14@spt.pdpu.ac.in

U. K. Bhui (ed.), *Macromolecular Characterization of Hydrocarbons for Sustainable Future*, Green Energy and Technology,
https://doi.org/10.1007/978-981-33-6133-1_5

Surfactant flooding is based on injecting an aqueous solution of low surfactant concentration into the hydrocarbon reservoirs where interaction took place between reservoir brine, crude oils, injected water and mineral surfaces of the reservoir rocks. When water and crude oil mix, water in crude oil emulsion is generally created because the surface-active amphiphilic molecules present in the crude oil are mostly oil-soluble due to their low polarity and molecular weight. As the viscosity of an emulsion is always higher than the continuous phase, it is obvious that for moving viscous residual oil through capillary pore channels it is necessary to make an emulsion with water as a continuous phase. There are basically two methods by which the natural tendency to form a water in oil emulsion is overcome. One is to add a high hydrophilic–lipophilic balance (HLB) surfactant. In general, non-ionic emulsifiers are a good choice as they are economically viable, not affected by the salinity of the produced water, do not produce any undesirable organic residue that can affect the oil properties [3–5]. The other method is to 'activate' the natural surfactant which is part of the heavy crude oil composition. These surfactants can be made more hydrophilic by ionizing the acid functional group carried by the fatty acids and asphaltenes with a strong base like soda. Addition of strong bases is a well-known process in enhanced oil recovery by decreasing interfacial tension to very low values [6–12]. Surfactants are special classes of molecules generally used for the reduction of interfacial tensions forming either micelle or reverse micelles. Micellization in the aqueous system is a process where interfacial energy between the water molecules and the hydrocarbon chains of the surfactant is the primary factor that leads to the formation of aggregates of surfactants, which do not separate from the solution. The micellization in the aqueous system means a reduction of the high energy water/hydrocarbon interface by moving the hydrocarbon chain to the interior of the micelle where the growth of the micellar size is moderated by the repulsion of the polar head group [13, 14]. Micelles formed by the injected surfactants trap the oil components inside their micelles and reduce the interfacial tension which effectively enhances the crude oil recovery from reservoirs. Research studies have established the fact that both asphaltenes and resin molecules are present as nano-sized aggregates inside the micelles formed by the surfactants in an aqueous solution [15, 16].

Surfactant flooding is a potentially promising EOR technique which is carried out by injection of fluids containing surfactants or a combination of surfactants (surfactant and co-surfactants). Although the surfactants flooding technique is promising, its application has been limited due to its high cost, its adsorption into the reservoir rock surfaces [17] and vague understanding of the interaction phenomenon at the molecular level. To analyse and understand the effectiveness of a surfactant, it is very crucial and critical to have the knowledge of the interaction of crude oil with the surfactant and brine solution. Optical techniques are nowadays used widely in the application of exploration and production of upstream and downstream petroleum-related activities [18–22] for its sensitivity, rapidity, easy to handle, economical and non-destructive nature. The fluorescence behaviour of petroleum crudes derives largely from aromatic hydrocarbon fraction which is very much dependent on the chemical composition (presence of fluorophores and quencher concentration) and physical properties such as viscosity and optical density [18, 23, 24]. In this regard, fluorescence spectroscopy can be one of the useful techniques to investigate the interaction of luminescent petroleum materials with various macromolecules/assemblies including surfactants in different physicochemical conditions.

We have studied optical characteristics of three types of petroleum crudes of Cambay Basin, India, in surfactant–brine solutions, using a combination of optical study. Based on the interaction behaviour of the crude oil, saline–surfactant aqueous solutions, we have established that customized designing of injection fluid is very much important for efficient oil recovery from reservoirs with specific crude type.

2 Materials, Methods and Instrumentation

2.1 Materials

The petroleum crude oil samples, namely S1, S2 and S3 were obtained from the wellheads of three different oil fields of Cambay Basin of India, operated by Gujarat State Petroleum Corporation (GSPC). Three types of surfactants, namely cationic–cetyltrimethylammonium bromide (CTAB), anionic–sodium dodecyl sulphate (SDS) and non-ionic–Triton X 100 (TX-100) of research grade were used as received. Sodium chloride (NaCl) from Merck Chemicals was used in the brine solution.

2.2 Methods

The three crude oils were characterized both physically and chemically. Crude oil was mixed with n-heptane in a ratio of 1:40 and kept for 48 h. Asphaltenes were separated by vacuum filtration using 0.45 μm Millipore filter paper. The obtained asphaltene was dried in the hot air oven for 24 h and weighed. API gravity of the crude oil samples was determined with the help of density bottle.

Surfactant–brine solutions were prepared by mixing 100 g water, 1 g of NaCl and 3 g of three different surfactants. The crude oil samples were mixed with the surfactant–brine solution in the ratio of 1 (oil): 9 (surfactant–brine solution). The content of the mixtures was mixed vigorously by the agitator and kept for stirring overnight using a magnetic stirrer. The whole solution was then centrifuged, and the supernatant clear solution was used for the optical measurements.

2.3 Instrumentation and Analytical Procedures

2.3.1 Crude Characterization

Physical properties and chemical compositions of the study crude oils were analysed in the laboratory and presented in Table 1. Steady-state emission spectra of diluted bulk crude oil samples (diluted into toluene in 1:1000 dilution factors) were recorded on a PerkinElmer-LS55 fluorescence spectrophotometer at a constant excitation wavelength of 300, 350, 400 and 450 nm. Synchronous fluorescence emission spectra of diluted samples (bulk crude and asphaltenes) were captured by applying a simultaneous scanning of excitation and emission wavelengths keeping $\Delta\lambda$ constant at 30 nm.

2.3.2 Interaction Study of Crude Oil in Surfactant-Brine Solutions

The characteristics optical properties of the crude oil-surfactant-brine solutions were studied using DLS, UV–Vis and fluorescence spectroscopic approach.

Particle size analyser: Malvern Zetasizer was used to determine the particle size of clear solutions. Its measurement is based on the principle of dynamic light scattering (DLS). It is equipped with a laser of 633 nm and has detection at a scattering angle of 90°. The minimum sample volume required for the study is 12 μL and its measurement range is 0.3 nm–10.0 μ (diameter). The particle size was analysed of the three surfactant solutions separately and also in combination of the surfactant with brine and the surfactant with brine and crude oil.

UV–visible spectroscopy: UV–Vis spectroscopic analyses of the samples were investigated with PerkinElmer Lambda 35 UV–Vis spectrophotometer with double beam recording and 1 nm resolution scale. All the measurements were performed at room temperature and ambient pressure using 10 mm path length quartz optical cuvette. The UV–vis data was collected and stored automatically in a computer system with support software.

Fluorescence spectroscopy: During the recording of the fluorescence emission spectrum, the excitation wavelength is kept fixed and the emission monochromator is run and for different excitation wavelength, the emission spectrum was recorded. Steady-state emission spectra of crude oil–surfactant–brine solutions were measured over a wavelength range of 200–800 nm, on a PerkinElmer-LS 55 fluorescence spectrophotometer employing 350 nm excitation wavelengths. The excitation

Table 1 Physical characteristics of crude oil samples

No.	Crude	API	Asphaltene (%)
1	S1	38.10	4.43
2	S2	36.15	3.4
3	S3	35.6	0.42

and emission slit width were kept 5/5 nm, and scanning speed was 5 nm/s^{-1} for recording the spectra.

3 Results and Discussions

3.1 Crude Oil Characterization

3.1.1 Physical Properties and Chemical Composition

The API and asphaltene percentage of all the three crude oils are given in Table 1. The crude oils have wide variation in the asphaltene percentage.

3.1.2 Spectroscopic Properties

Fluorescence spectroscopy can be used as a tool to identify the polycyclic aromatic hydrocarbon (PAH) structures present in crude oil since aromatic structures present in crude oil are known to fluoresce when excited at different wavelengths. The steady-state fluorescence emission spectra of asphaltene recorded within the window of 200–800 nm at excitation wavelengths of 300, 350, 400 and 450 nm, respectively, are shown in Fig. 1. The emission spectra of S2 and S3 are almost similar at all the excitation wavelengths while the emission spectrum of S1 shifts towards the red wavelength demonstrating the presence of larger aromatic moieties. This demonstrates the presence of different size of polyaromatic hydrocarbons in the crude oil samples. Synchronous fluorescence spectroscopy is an important tool to identify the size of fused aromatic rings (FAR). The synchronous spectra of crude oil asphaltene at $\Delta\lambda$ of 30 nm are shown in Fig. 2. The spectra indicate the presence of 3–7 FAR in S2 and S3 and the presence of 4–7 FAR in the asphaltene of S1 crude. Though all the crude oils are of medium API gravity, it has variation in the FAR present in the asphaltene.

3.2 Characterization of Crude Oil in Surfactant–Brine Solutions

Surfactants are known to self-assemble into micelles at a concentration above the critical micelle concentration. The micelle formation is a multistep and gradual process. In the aqueous solution of the crude oil–brine–surfactant solution, the micelle structures entrap the crude oil components in it depending on the polarity of the surfactant. The size of micelle varies with the concentration of NaCl and surfactant [25]. Small-angle X-ray and neutron scattering (SANS) studies have revealed detailed structural information of a variety of micelles [26–29].

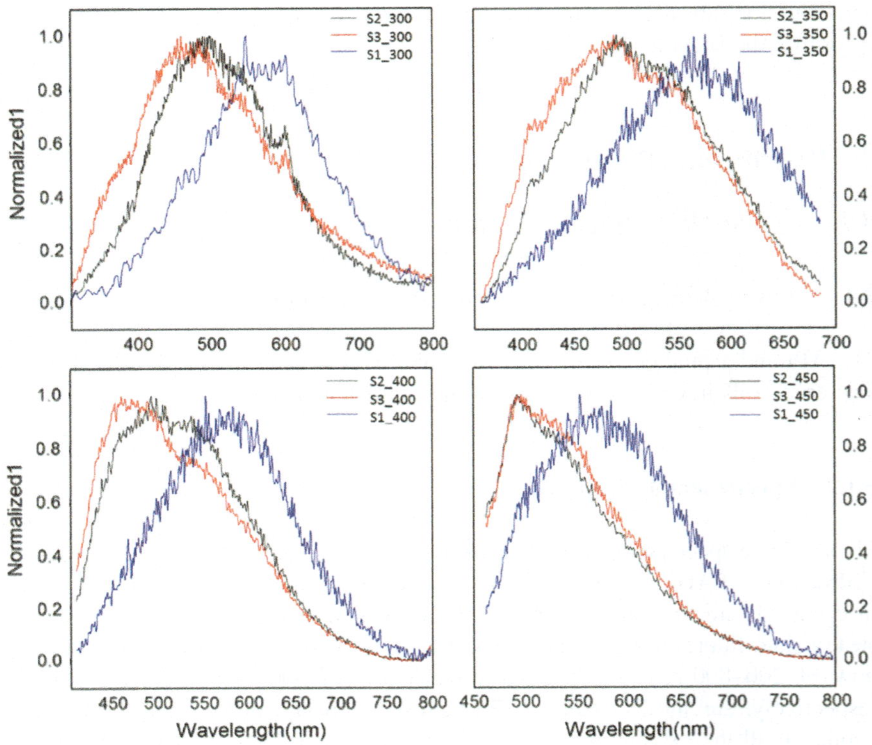

Fig. 1 Fluorescence emission spectra of asphaltenes from crude oil S1, S2 and S3 at an excitation of 300, 350, 400 and 450 nm

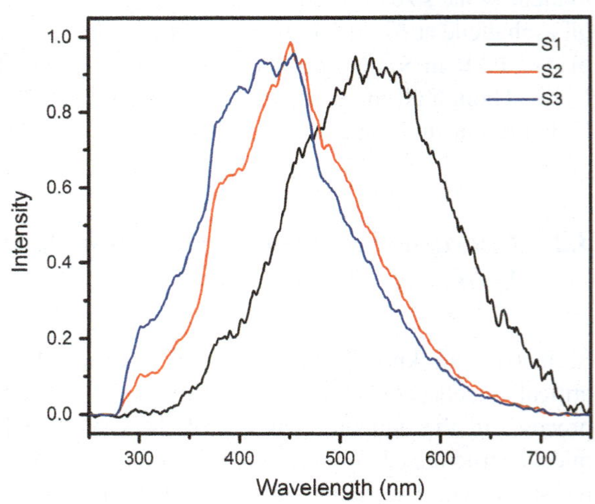

Fig. 2 Synchronous fluorescence spectra of asphaltenes from crude oil S1, S2 and S3 at constant Δλ of 30 nm

3.2.1 Particle Size Analysis

From the size distribution in Fig. 3 and Table 2, it can be seen that the micelle size of anionic surfactant decreases on the addition of NaCl showing a bimodal distribution. Also it can be seen that on mixing the crude oil with the surfactant–brine solution leads to an increase in the size of the micelles. The increment in the particle size may possibly be due to the entrapment of crude oil particles into the micelle of surfactant [16, 30, 31]. A similar trend is observed for the cationic surfactant (Fig. 4). The micellar size of pure cationic surfactants (CTAB) shows a bimodal distribution and on the addition of NaCl salt, the average particle size decreases similar to the anionic surfactant. The particle size tends to increase for all type of surfactant with S2 crude as compared to the other two crude oil samples demonstrating that the micelle formed with same surfactant have dissimilar size with

Anionic surfactant:

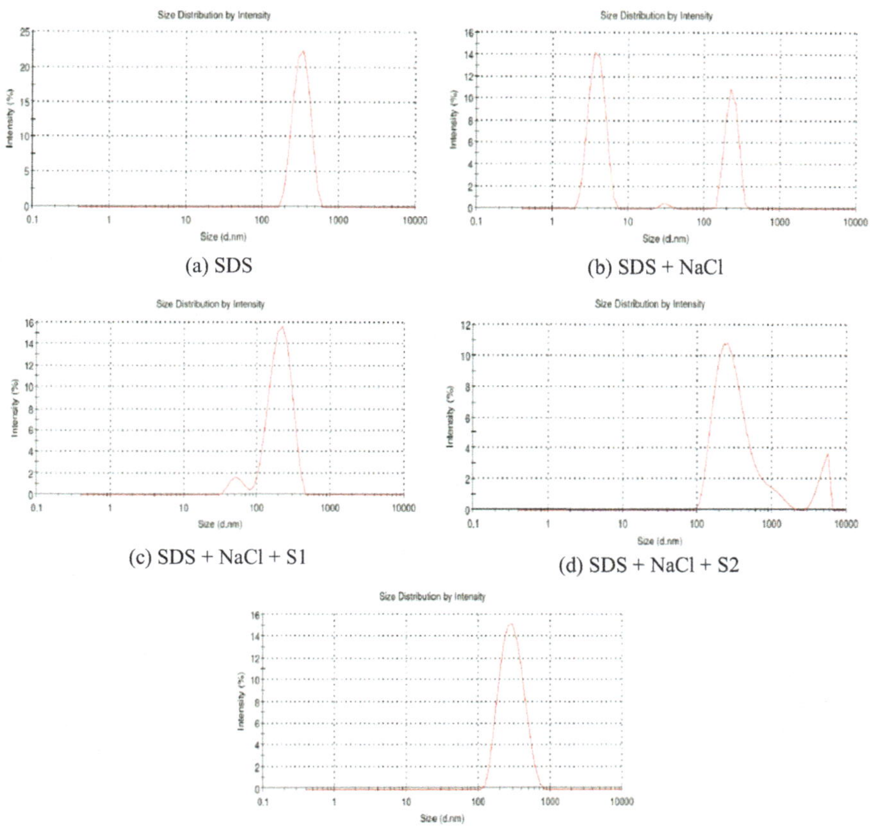

Fig. 3 Particle size distribution of aqueous solutions of **a** anionic surfactant SDS, **b** SDS + NaCl, **c** SDS + NaCl with crude S1, **d** SDS + NaCl with crude S2 and **e** SDS + NaCl with crude S3

Table 2 Particle size distribution (Z_{avg}) of different aqueous solutions

	Surfactant (nm)	Surf + NaCl (nm)	Surf + NaCl + S1 (nm)	Surf + NaCl + S2 (nm)	Surf + NaCl + S3 (nm)
SDS	407	239.8	205.6	312.13	260.8
CTAB	268.4	100.9	340.6	720.7	224.3
TX-100	10.57	12.83	291.3	790	391.2

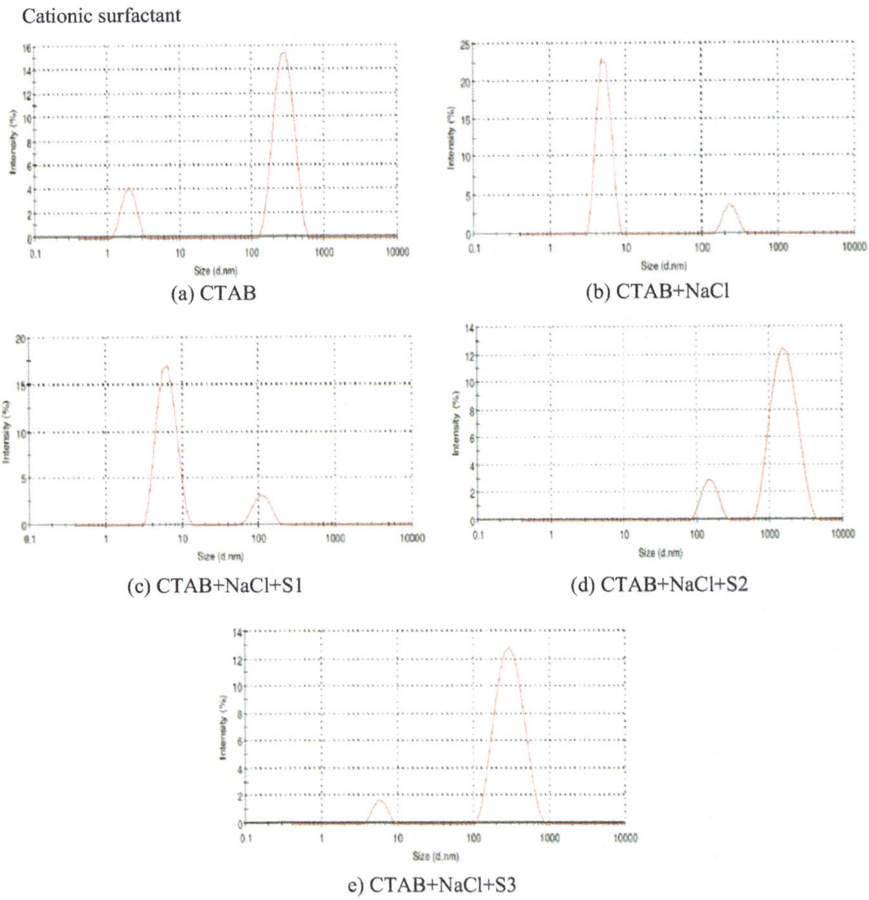

Fig. 4 Particle size distribution of aqueous solution of **a** CTAB, **b** CTAB + NaCl, **c** CTAB + NaCl with crude S1, **d** CTAB + NaCl with crude S2 and **e** CTAB + NaCl with crude S3

different crude oils. The micelle size of a non-ionic surfactant and non-ionic surfactant with NaCl almost shows the same size indicating that the micelle size of the non-ionic surfactant does not change on addition of salt. But, the average particle

Non-ionic surfactant:

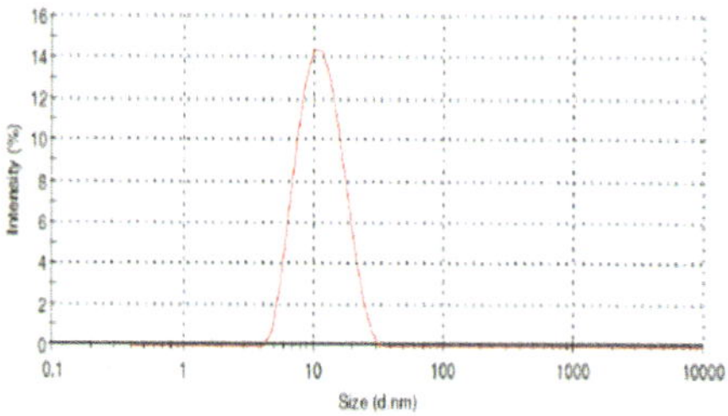

(a) TX100

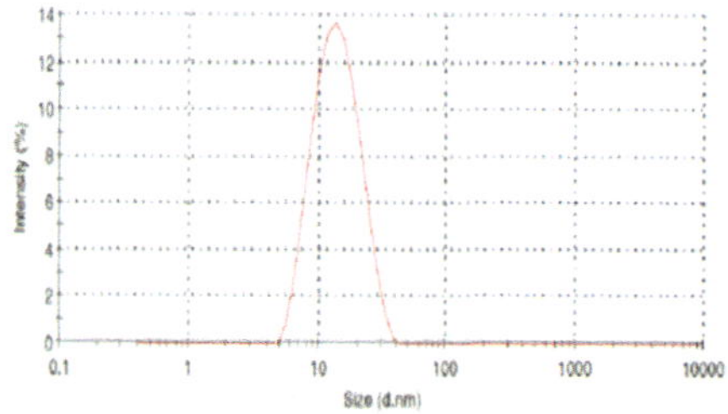

(b) TX100+NaCl

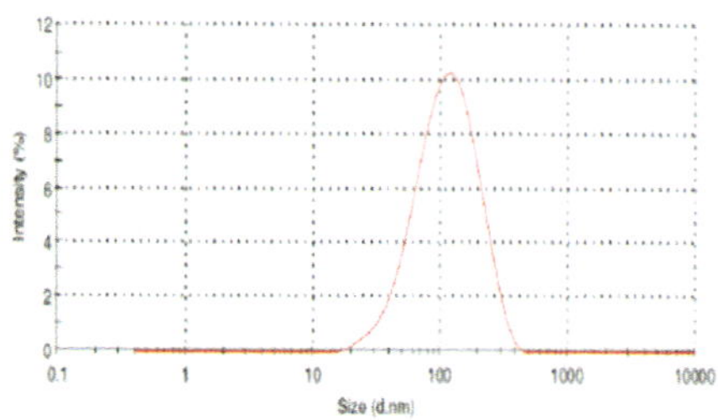

(c) TX100 +NaCl +S1

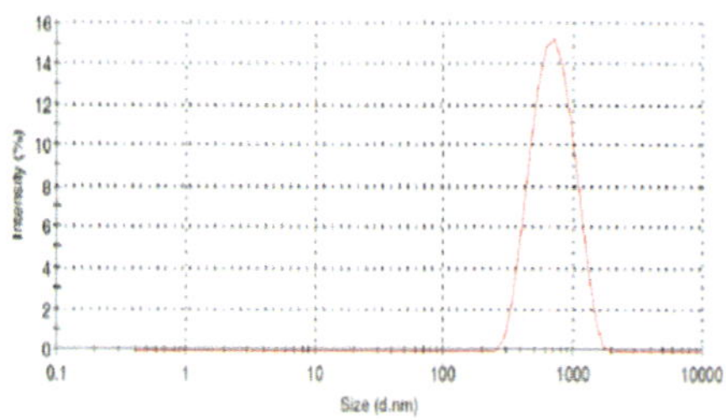

(d) TX100+NaCl+S2

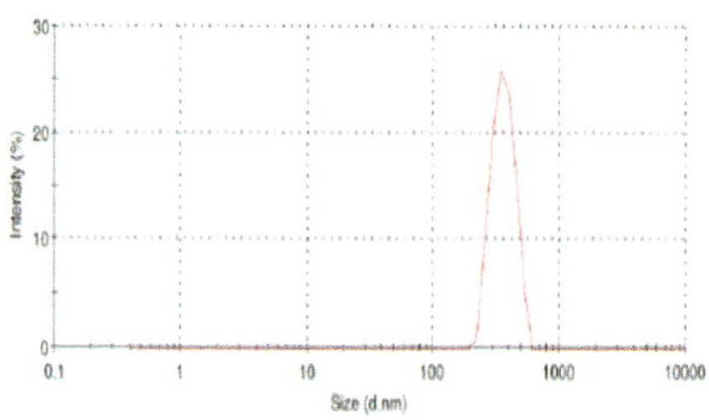

(e) TX100 +NaCl + S3

Fig. 5 Particle size distribution of aqueous solutions of **a** TX-100, **b** TX-100 + NaCl, **c** TX-100 + NaCl with crude S1, **d** TX-100 + NaCl with crude S2 and **e** TX-100 + NaCl with crude S3

size increases on the addition of crude oils with a non-ionic surfactant–brine solution as seen in Fig. 5. The size is the largest for S2 crude oil, intermediate with S3 and lowest for S1.

3.2.2 UV–Vis Absorption Spectroscopy

The absorbance spectra of crude oil with different surfactant solutions are shown in Fig. 6. The absorbance spectra of all three crude oil with CTAB and SDS are almost alike, whereas the absorbance spectra of all the crude oil samples with TX-100 seem to be shifted towards red wavelength and show absorbance in both UV and visible region. The UV–Vis spectrum of crude oil with CTAB and SDS

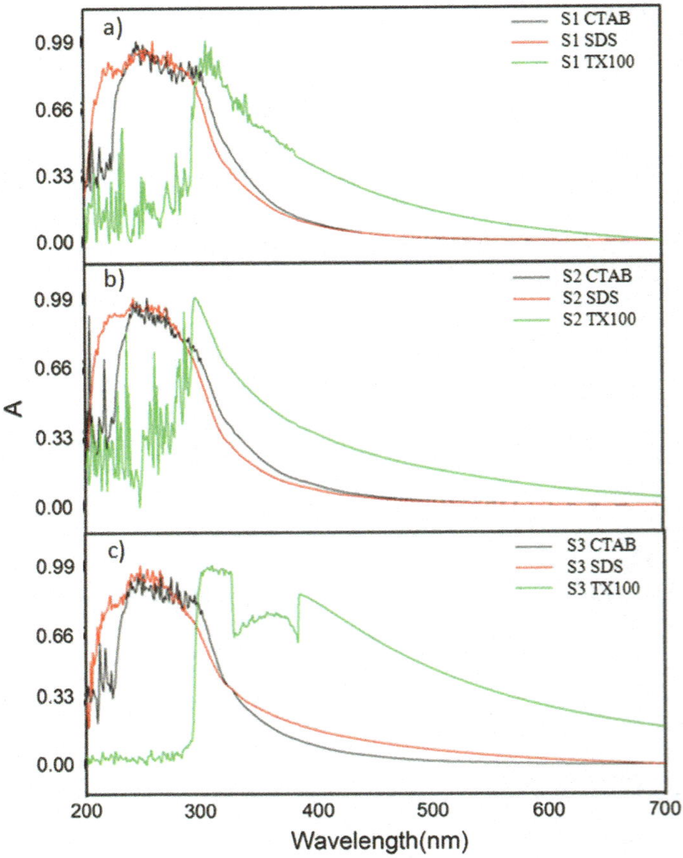

Fig. 6 UV–visible absorption spectra of different surfactant with **a** crude oil S1, **b** crude oil S2 and **c** crude oil S3

shows absorbance only in the UV region. This demonstrates the entrapment of different components into the micelle of surfactants of various charges like anionic SDS and cationic CTAB entrapping smaller aromatic moieties as compared to non-ionic TX-100. The spectrum of TX-100 shows a sharp peak near 300 nm with a little absorption in the visible region indicating the entrapment of larger aromatic moieties from the crude oil. Figure 7 shows the UV–Vis spectra of the interaction of individual anionic, cationic and non-ionic surfactant with different crude oil. As seen from Fig. 7, the spectra of cationic and anionic surfactant with different crude oil are similar indicating the similar behaviour of ionic surfactant with different crudes. The spectrum of TX-100 surfactant with S1 and S2 is similar but with S3 the spectra show absorption in both UV and visible region demonstrating entrapment of bigger size FAR in the micellar structures.

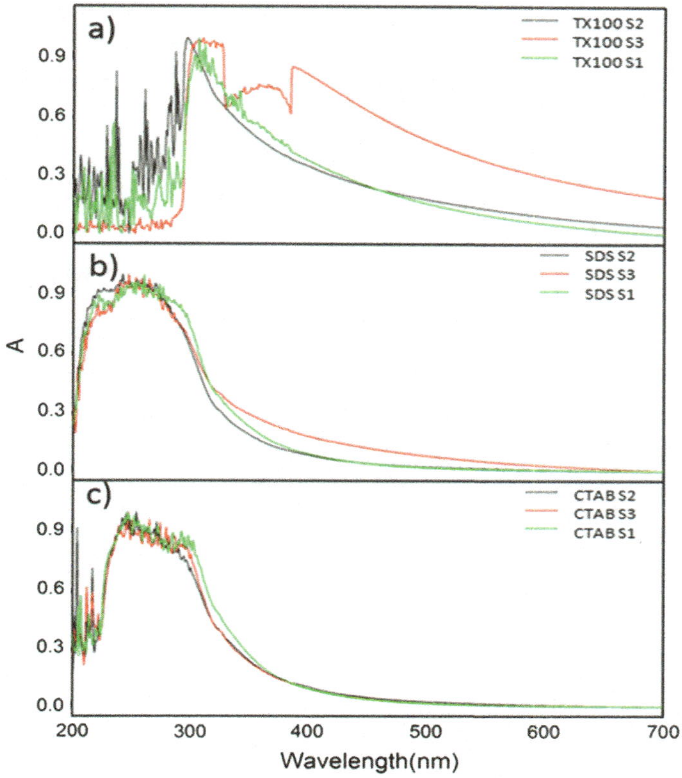

Fig. 7 UV–visible absorption spectra of different crude with **a** TX-100 **b** SDS and **c** CTAB

3.2.3 Fluorescence Spectroscopy

Fluorescence emission spectroscopy is a useful tool to study the entrapment of crude oil in the surfactant solution. The fluorescence emission spectrum of each crude oil with SDS, CTAB and TX-100 is shown in Fig. 8. The emission spectrum tends to shift towards the higher wavelength with the increase in aggregation of crude oil due to the increase in the number of fluorescent molecules present in crude oil [16]. The emission spectrum of S1 crude oil with non-ionic surfactant is red shifted indicating the entrapment of medium FAR (5–7) in the micelle while ionic surfactant (SDS and CTAB) entrapped (4–5 FAR) rings. For S2 crude oil sample, the spectra show entrapment of medium size FAR (4–6) rings with a non-ionic surfactant and cationic and 4–5 FAR with SDS. In case of S3 crude oil, medium size FAR (5–7 FAR) rings are entrapped with non-ionic (TX-100) surfactant (4–5 FAR) and (4–6 FAR) rings entrapped with SDS and CTAB, respectively. From this study, it can seen that the emission spectra of all three crude oil with non-ionic surfactant are red shifted and broader compared to other surfactants as seen in Fig. 9 indicating higher state of aggregation of crude oil molecules with entrapment

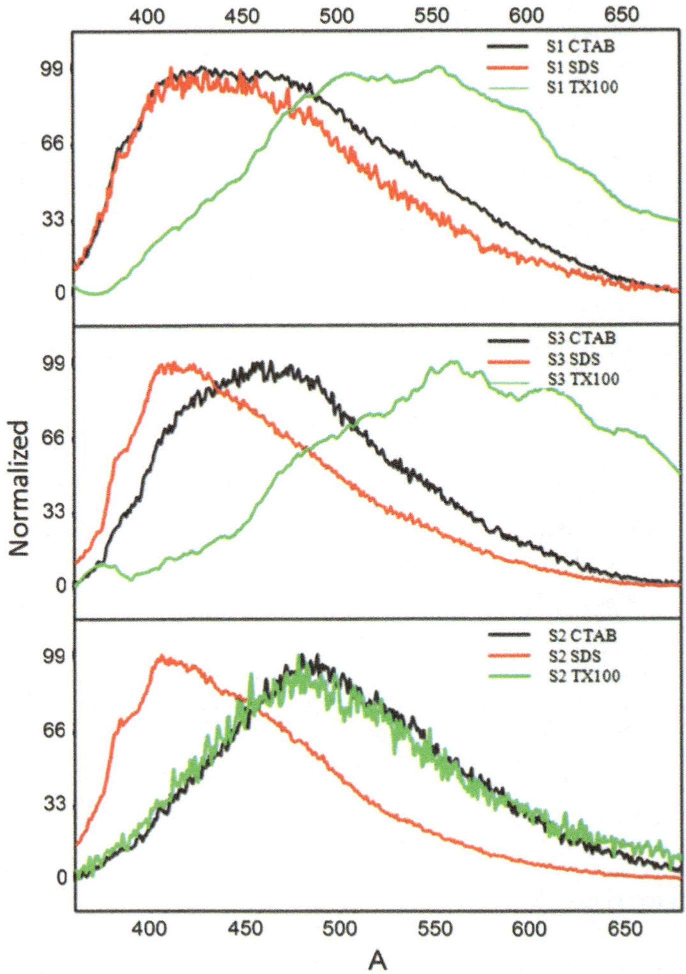

Fig. 8 Fluorescence emission spectra (at 350 nm excitation wavelength) of different surfactant with **a** crude oil S1, **b** crude oil S3 and **c** crude oil S2

of medium size FAR [16, 31]. This is also observed from the size distribution plot of a non-ionic surfactant that the size of micelle aggregates increases after addition of crude oil. As compared to TX-100, SDS and CTAB emission spectra are blue shifted and narrower which indicates lower aggregation state with SDS having the lowest state of aggregation.

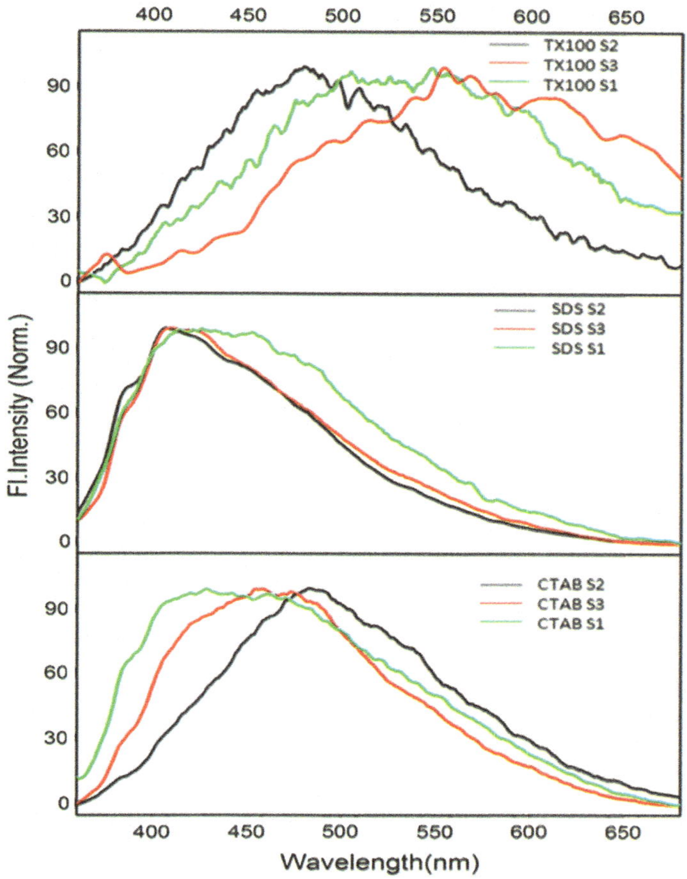

Fig. 9 Fluorescence emission spectra (at 350 nm excitation wavelength) of the study crude oils (S1, S2 and S3) with surfactant **a** TX-100, **b** SDS and **c** CTAB

4 Conclusion

Understanding the interaction between different type of surfactant and different crude oil is of the utmost importance to design the injection fluid for surfactant flooding. In this study, the interactions between three different crude oils with an anionic, cationic and non-ionic surfactant are studied using DLS, UV–Vis spectroscopy and fluorescence spectroscopy. The study demonstrates that different surfactant interacts in dissimilar ways with crudes. The crude oil S1 contains 4-7 FAR which entrapped medium size FAR 5–7 rings with non-ionic surfactant while (4–5 FAR) PAH rings were entrapped within the micelle of ionic surfactants (SDS and CTAB). Crude oil S2 having 3–7 FAR entrapped medium size FAR (4–6) with non-ionic and cationic surfactant while it entrapped (4–5 FAR) with SDS. S3 crude oil having (3–7 FAR) entrapped medium size FAR (5–7 FAR) rings with non-ionic (TX-100) surfactant and (4–5 FAR) and (4–6 FAR) rings with SDS and CTAB, respectively. For all the cases, it was observed that non-ionic surfactants have a higher state of aggregation with entrapment of comparatively larger size FAR while ionic surfactants (SDS and CTAB) have a lower state of aggregation with entrapment of smaller size FAR from the crude oil. For proper designing of injection fluid, it is important to understand the entrapment mechanism of FAR within the microenvironment of micelles structures of the surfactants with different charges.

Acknowledgements RA and SS both are thankfully acknowledging PDPU for the research fellowship. UKB acknowledges the support of GSPC for providing the crude oil samples for the work. Petroleum Engineering Laboratory and Reservoir Characterization Laboratory (RCL) of SPT and PDPU are thankfully acknowledged for the optical characterization data.

References

1. Wang Y, Zhao F, Bai B (2010) Optimized surfactant IFT and polymer viscosity for surfactant-polymer flooding in heterogeneous formations. In: SPE improved oil recovery symposium, Society of Petroleum Engineers
2. Al Adasani A, Bai B (2011) Analysis of EOR projects and updated screening criteria. J Petrol Sci Eng 79(1–2):10–24
3. Sharma K, Saxena VK, Kumar A, Ghildiyal HC, Anurada A, Sharma ND et al (1998) Pipeline transportation of heavy viscous crude oil as water continuous emulsion in North Cambay Basin (India). SPE 39537
4. Gillies RG, Sun R, Shook CA (2000) Laboratory investigation of inversion of heavy oil emulsions. Can J Chem Eng 78:757–763
5. Al-Sabagh AM, Zaki NN, Badawi AFM (1997) Effect of binary surfactant mixtures on the stability of asphalt emulsions. J Chem Technol Biotechnol 69:350–356
6. Plegue TH, Frank SG, Fruman DH, Zakin JL (1989) Studies of water-continuous emulsions of heavy crude oils prepared by alkali treatment. SPE Prod Eng 4(02):181–183
7. Plegue TH, Frank SG, Fruman DH, Zakin JL (1986) Viscosity and colloidal properties of concentrated crude oil-in-water emulsions. J Colloid Interface Sci 114(1):88–105
8. Kessick MA, Denis CES (1982) Alberta Research Council. Pipeline transportation of heavy crude oil. U.S. Patent 4,343,323
9. Verzaro F, Bourrel M, Garnier O, Zhou, HG, Argillier JF (2002) Heavy acidic oil transportation by emulsion in water. In: SPE international thermal operations and heavy oil symposium and international horizontal well technology conference. Society of Petroleum Engineers
10. Wegener DC, Zornes DR, Maloney DR, Vienot ME, Fraim ML (2001) Phillips Petroleum Co. Heavy oil viscosity reduction and production. U.S. Patent 6,279,653
11. Sanchez LE, Zakin JL (1994) Transport of viscous crudes as concentrated oil-in-water emulsions. Ind Eng Chem Res 33(12):3256–3261
12. Adewusi VA, Ogunsola AO (1993) Optimal formulation of caustic systems for emulsion transportation and dehydration of heavy oil. Chem Eng Res Des 71:62
13. Tanford C (1980) The hydrophobic effect: formation of micelles and biological membranes, 2nd edn. Wiley, New York
14. Israelachvili JN, Mitchell JJ, Ninham BN (1976) Amphiphilic association structures, shape factors. J Chem Sci Faraday II 72:1525–1538

15. Murillo-Hernández JA, García-Cruz I, López-Ramírez S, Duran-Valencia C, Domínguez JM, Aburto J (2009) Aggregation behavior of heavy crude oil–ionic liquids solutions by fluorescence spectroscopy. Energy Fuels 23(9):4584–4592
16. Vallejo-Cardona AA, Cerón-Camacho R, Karamath JR, Martínez-Palou R, Aburto J (2017) A study of the effect of surfactants on the aggregation behavior of crude oil aqueous dispersions through steady-state fluorescence spectrometry. Appl Spectrosc 71:1519–1529
17. Lake LW, Schmidt RL, Venuto PB (1992) A niche for enhanced oil recovery in the 1990s. Oilfield Rev 4(1):55–61
18. Ryder AG (2005) Analysis of crude petroleum oils using fluorescence spectroscopy. In: Geddes CD, Lakowicz JR (eds) Reviews in fluorescence, annual volumes. Springer, New York, pp 169–198
19. Mullins OC (1998) Optical interrogation of aromatic moieties in crude oils and asphaltenes. In: Mullins OC, Sheu EY (eds) Structure and dynamics of asphaltenes, Plenum Press, New York, Chap 2, pp 2–77
20. Patra D, Mishra AK (2001) Concentration dependent red shift: qualitative and quantitative investigation of motor oils by synchronous fluorescence scan. Talanta 53:783–790
21. Aske N, Kallevick H, Sjoblom J (2001) Determination of saturate, aromatic, resin and asphaltenic (SARA) components in crude oils by means of infrared and near-infrared spectroscopy. Energy Fuel 15(5):1304–1312
22. Falla FS, Larini C, Le Roux GAC, Quina FH, Moro LFL, Nascimento CAO (2006) Characterization of crude oil by NIR. J Petrol Sci Eng 51(1–2):127–137
23. Ralston CY, Wu X, Mullins OC (1996) Quantum yields of crude oils. Appl Spectrosc 50 (12):1563–1568
24. Smith GC, Sinski JF (1999) The red-shift cascade: investigations into the concentration-dependent wavelength shifts in three-dimensional fluorescence spectra of petroleum samples. Appl Spectrosc 53(11):1459–1469
25. Cui X, Mao S, Liu M, Yuan H, Du Y (2008) Mechanism of surfactant micelle formation. Langmuir 24(19):10771–10775
26. Paradies HH (1980) Shape and size of a nonionic surfactant micelle: Triton X-100 in aqueous solution. J Phys Chem 84(6):599–607
27. Berr SS (1987) Solvent isotope effects on alkytrimethylammonium bromide micelles as a function of alkyl chain length. J Phys Chem 91(18):4760–4765
28. Phillies GD, Yambert JE (1996) Solvent and solute effects on hydration and aggregation numbers of Triton X-100 micelles. Langmuir 12(14):3431–3436
29. Phillies GD, Hunt RH, Strang K, Sushkin N (1995) Aggregation number and hydrodynamic hydration levels of Brij-35 micelles from optical probe studies. Langmuir 11(9):3408–3416
30. José-Alberto MH, Jorge A (2011) Current knowledge and potential applications of ionic liquids in the petroleum industry. In: Kokorin A (ed) Ionic liquids: applications and perspectives, InTech, pp 439–456
31. Bhui UK, Sanyal S, Saha R, Rakshit S, Pal SK (2018) Steady-state and time-resolved fluorescence spectroscopic study of petroleum crudes in aqueous-surfactant solutions: Its implications for enhanced oil recovery (EOR) during surfactant flooding. Fuel 234: 1081–1088

Heavy Oil Characterization and Enhanced Oil Recovery: Challenges and Opportunities

Sujit Mitra, Sudheer Kumar Singh, Shikha Trivedi, and Omkar Nath Gyani

1 Introduction

The main challenge in the production of heavy oil is its low mobility due to high viscosity. This leads to low primary recovery and necessitates implementation of enhanced oil recovery (EOR) methods at very early stages of the field life cycle. EOR and heavy oil are intimately linked with each other, as globally most of the heavy oil is produced through EOR only. That too, thermal EOR techniques are the most popular and common methods for heavy oil production. Since any EOR process is technology and cost intensive, to decide on the best possible EOR techniques for any heavy oil reservoirs, it is crucial to have detail characterization of the oil. To understand thermal EOR methods for heavy oil, it is important to know the basic characteristics of heavy oil as well as different thermal EOR methods.

In India, ONGC is operating one of the world's longest running In situ combustion (ISC) projects in the heavy oil belts of Gujarat. One of the most important factors in designing and successful implementation of the ISC projects was the detail crude oil characterization.

S. Mitra (✉) · S. K. Singh · S. Trivedi · O. N. Gyani
Institute of Reservoir Studies, Oil and Natural Gas Corporation Ltd., Chandkheda Campus, Ahmedabad 380005, India
e-mail: Mitra_Sujit@ongc.co.in

S. K. Singh
e-mail: singh2_sudheer@ongc.co.in

S. Trivedi
e-mail: trivedi_shikha@ongc.co.in

O. N. Gyani
e-mail: gyani_omkarnath@ongc.co.in

U. K. Bhui (ed.), *Macromolecular Characterization of Hydrocarbons for Sustainable Future*, Green Energy and Technology,
https://doi.org/10.1007/978-981-33-6133-1_6

1.1 What Is Heavy Oil?

Although there is no such universally accepted definition of heavy oil, but the one which is most commonly used is that of 12th World Petroleum Congress (WPC), 1987. It defines heavy oil as the oil having API gravity in the range of 10–22.3° and viscosity 100–10,000 cp. And the oil having API gravity of less than 10° is further subdivided into two categories, extra heavy oil and bitumen/tar sands/oil sands. Extra heavy oil is having viscosity same as that of heavy oil, i.e. 100–10,000 cp but the bitumen is having viscosity of more than 10,000 cp.

Next, the common question that comes to mind that why the oil becomes heavy? There could be many reasons but the most common one is biodegradation and predominant occurrence in shallow reservoirs. Both of them lead to low content of dissolved gas in the oil, and the shallow depth is responsible for low temperature as well.

Crude oil consists primarily of hydrocarbons or compounds comprising hydrogen and carbon only. Some elements such as sulphur, nitrogen and oxygen are also present in small quantities and are generally combined with the carbon and hydrogen in complex molecules. Typically, crude oil components are classified into four types—saturate, aromatics, resins and asphaltenes (SARA) based on solubility classes. As oil gets heavier or more viscous, their composition becomes more complex with molecular weights in the thousands. As a result, an alternative characterization is often used based on solubility. **Saturates** are generally iso- and cyclo-paraffins, while aromatics, resins and asphaltenes form a continuum of molecules with increasing molecular weight, aromaticity and heteroatom contents. The saturate fraction consists of non-polar material including linear, branched and cyclic saturated hydrocarbons (paraffins). **Aromatics,** which contain one or more aromatic rings, are slightly more polarizable. **Resins** and **Asphaltenes** are not defined as specific molecular structures. In practice, these molecules have molecular weights in the thousands, are polar and contain elements such as sulphur, nitrogen, oxygen and heavy metals. They have no definite melting point (decompose between 300 and 400 °C), exist in a dispersed state in crude oils but can aggregate to form precipitates and often are composed of condensed aromatic rings in the form of a non-homogeneous flat sheet. Resins are usually defined as the propane-insoluble but pentane-soluble fraction, while asphaltenes are soluble in carbon disulphide but insoluble in petroleum ether or n-pentane. Heavy oils contain more resins and asphaltenes (2–10%) with high molecular weight.

1.2 The Challenges in Production of Heavy Oil

The high viscosity of heavy oil is responsible for low mobility leading to very low primary recovery. So, most of the heavy oil fields across the globe have short primary production history as during primary recovery oil production sharply declines leading to uneconomical rate. Hence, to address this challenge, EOR methods were found to be most effective.

EOR is the recovery of oil by injection of fluids not normally present in the reservoir but excludes pressure maintenance or water flooding and not necessarily a tertiary recovery method. EOR techniques are used to target the oil which otherwise is difficult or impossible to recover by conventional recovery methods. In other words, conventional recovery targets mobile oil in the reservoir and EOR targets immobile, i.e. the oil which cannot be produced due to capillary and viscous forces. EOR has its inherent risk, challenges and benefits. EOR projects are generally complex, technology heavy, capital and resource intensive. Most importantly, any successful EOR applications need to be tailored to each specific reservoir. The implementation of EOR is intimately tied to the price of oil and overall economics. Another challenge for EOR projects is the long lead time required for such projects. Typically, it may take anywhere between 5 and 10 years, from the start of the concept—generating laboratory data through multiple experiments and conducting simulation studies leading to pilot design. Subsequently the implementation of the pilot in the field to establish the technical feasibility and finally full field commercialization depending on the economic viability.

The most popular and effective EOR methods for heavy oil are thermal EOR techniques. In thermal EOR, primarily heat is being used to reduce the viscosity of the oil to increase mobility. Although, in the recent past, polymer flood has also gained popularity, there are few successful examples as well. The commercially successful thermal EOR processes can be broadly classified into steam-based processes (cyclic steam stimulation or CSS and steam flood) and in situ combustion (ISC), with steam being the more extensively applied process.

For heavy oil reservoirs, the recovery mechanism of both these processes targets the increase in productivity due to sharp reduction in oil viscosity with increase in temperature. Additionally, there is significant reduction in Sor (residual oil saturation) through both these displacing agents resulting into substantial increase in recovery factor. In case of steam-based processes, the heat is delivered through steam generated at surface and injected into the reservoir. Therefore, the depth of reservoir, reservoir pressure and reservoir thickness are critical for the effectiveness of this process in delivering heat into the reservoir and propagation of the steam front.

However, in case of in situ combustion, the heat is generated in situ within the reservoir with injection of air. This eliminates the loss of heat down the wellbore and thus relaxes the depth criteria. In the ISC process, the oxygen of the injected air in contact with the reservoir oil initiates oxidation reactions. Oxidation being an exothermic process liberates heat and that in turn reduces the viscosity of the oil and increases its mobility. The combustion front that is created, sustained and propagated is primarily responsible for the displacement of the oil. A small fraction of oil is used as fuel for propagation of the combustion front. Hence, the temperature of the reservoir and the oxidation characteristics of the oil are the crucial parameters for the successful application of this process.

High viscosity and low API gravity of crude oil make in situ combustion process the most effective EOR method to improve the oil recovery from the heavy oil fields of North Cambay Basin. Thermal Process Laboratory of Institute of Reservoir Studies of Ahmedabad is engaged in these characterization and conducting the combustion tube experiments to study the feasibility of thermal EOR processes for these heavy oil fields. Incorporation of crude oil analysis data into the thermal EOR simulation model can help optimize the thermal EOR process on laboratory scale and can help better predict the feasibility of its implementation on field scale.

As described earlier, thermal EOR methods are also technology and cost intensive. So, to design any thermal EOR pilots or commercial schemes, it is very important to characterize the heavy oil that is to be produced [1]. Parameters, viz. SARA composition, Asphaltene Maltene fractionation, API gravity, density, viscosity profiling and thermal characterization have been carried out. But in this paper, the thermal EOR processes shall be discussed as the oil characterization is the key to the success of thermal EORs.

2 Samples and Experimental Procedure

Lanwa, Balol and Santhal fields comprise the heavy oil belt in the Ahmedabad-Mehsana block of North Cambay Basin. Reservoir pressure and temperature are 90–100 kg/cm^2 and 70 °C, respectively. Crude oil density is around 0.95–0.975 g/cm^3, and API gravity is 13–17°. Viscosity ranges from 100 to 1500 cp. The lowest viscosity is in Santhal ($\sim$150 cp) and the highest in Lanwa (1000–1500 cp) and Balol is having oil of 150–1000 cp. Heavy oil of these fields presents high viscosity, high density, low API gravity and higher content of nitrogen, oxygen, sulphur and heavy metals along with higher content of high molecular weight hydrocarbons in comparison with conventional crude oil. These parameters directly impact the recovery of crude oil from the reservoir.

2.1 Crude Oil Samples and Sample Preparation

Reservoir fluid samples for the study were collected from wells of Santhal, Balol and Lanwa fields. Collected fluid samples were heated on water bath at 45–50 °C for 30–80 h to allow separation of aqueous layer from crude oil layer. After removal of separated water from the crude oil layer, water cut (%Volume) was determined for the crude oil samples using Dean–Stark method. Oil samples were then dehydrated using anhydrous calcium chloride.

2.2 *Crude Characterization*

2.2.1 Density Measurement

Density and API gravity parameters were measured at 15.56 °C (60 °F) for crude oil samples using density meter (DDM2910, Rudolph Research Analytical) [2] under atmospheric condition.

2.2.2 SARA Analysis

SARA analysis is used for many crude oil components. When a SARA test sample is heavy oil with minimal light ends, it is worth considering Iatroscan TLC-FID for which there is a standard method (IP-469). IP-469 determines all four compound classes by adsorption chromatography, so the asphaltenes content may differ numerically to that determined by IP-143. Typically, Iatroscan asphaltenes are about 80% of the IP-143 asphaltenes. Measuring asphaltenes by IP-143 and then determining SAR by IP-469 are possible. IP-469 does not involve a gravimetric finish, and light ends are still lost during the analysis, so this method works best with >300 °C distillates or residues. Asphaltenes can be pentane or heptane insoluble. Resins may be classified as a solubility fraction, such as heptane soluble, pentane insoluble, heptane soluble and acetone or ethyl acetate insoluble, or as the polar fraction eluted from a polar adsorbent with a polar solvent, such as pyridine, toluene or methanol [3].

Saturates and aromatics are determined by adsorption chromatography, typically from silica or silica/alumina. Saturates are eluted with a paraffinic solvent, such as pentane or heptane. Aromatics are eluted either with paraffinic or moderately polar solvents, such as toluene or DCM.

Procedure for the separation of asphaltene and maltene components of crude oil involved addition of n-pentane to crude oil samples at a ratio of 40 mL n-pentane to 1 g of oil followed by sonication for 45 min. Mixture was then allowed to settle for 24 h. Supernatant was poured through a preweighed Whatman#2 filter paper. Residue was again mixed with n-pentane at a ratio of 4:1 of n-pentane to original oil. After sonication for 30 min, mixture was left to settle for 16 h. Final mixture was filtered using the same filter paper. The precipitated asphaltene was washed on the filter paper with n-pentane three times per day for five days. After five days of washing, filtrate was almost colourless, and washed asphaltene samples were dried in the fume hood till constant weight was obtained. Similarly, the filtrates containing maltene component were also dried at room temperature till constant weight was obtained.

2.2.3 Rheological Study

Modular Compact Rheometer (MCR102, Anton Paar) was used to generate the viscosity profile of crude oil and maltene fractions as follows:

- Viscosity as a function of varying temperature (30–150 °C) at shear rate of 25 s^{-1}.
- Viscosity as function of varying shear rate from 1 to 100 s^{-1} at constant temperature (reservoir) of 70 °C.

2.2.4 Thermal Analysis

Thermal analysis is a popular and convenient tool in understanding the mechanism of oil-cracking kinetics, which is one of the fundamental mechanisms of in situ combustion [4]. Two thermal analysis methods are commonly used for the evaluation of oil samples under heating are thermogravimetry and differential scanning calorimetry.

(a) Thermogravimetry (TG/DTG)

TGA experiments were carried out at atmospheric pressure using oil samples of Santhal, Balol and Lanwa field in oxidative atmospheres to generate the TGA distillation data (weight loss with temperature), activation energy and log preexponential factor at different conversion level.

The size of the oil samples was in the range of 2–5 mg. The oil samples were heated from 30 to 700 °C at different heating rates of 1, 6 and 10 °C/min in oxidative environment (Zero Air). During analysis, zero air was flowing at rate of 20 mL/min.

(b) Differential Scanning Calorimetry (DSC)

Differential scanning calorimeter (DSC) measures the heat flow as a function of temperature. The technique provides information on the minimum ignition temperature of the crude to sustain the combustion. The DSC runs were carried out using DSC 2920 of TA instruments USA. Oil and oil mixed with rock mixtures were analysed using DSC in oxidative environment at heating rate of 5 °C/min with sample size of 5–30 mg. Zero air grade was used for oxidative environment.

3 Results

3.1 Physico-Chemical Characterization of Crudes

The SARA analysis of Santhal, Balol and Lanwa oil (Table 1) goes in line with their descending API gravity and ascending viscosity trend. API gravity and density of Santhal, Balol and Lanwa across the field are given in Table 2.

3.2 Rheological Behaviour of the Crudes

Viscosity data for the crude oil samples shows the variation of viscosity with temperature ranging from 30 to 150 °C at a constant shear rate of 25 s^{-1}. Decrease in viscosity with increase in temperature can be clearly observed (Fig. 1). Data for the variation of viscosity with shear rate at constant temperature and pressure shows that beyond shear rate of 20 s^{-1}, viscosity is independent of shear and thus sample shows Newtonian behaviour (Fig. 2). Viscosity data for the maltene samples shows decrease in viscosity with increase in temperature. However, the viscosity values for the maltene samples are lower than the corresponding viscosity values for the crude oil samples as evident from the viscosity versus temperature curves. Viscosity as a function of shear rate for the maltene samples follows Newtonian behaviour beyond 20 s^{-1}.

Table 1 Sara composition

Sample No.	Density (g/cc)	°API	Mol, wt	Sat, wt (%)	Aro, wt (%)	Res, wt (%)	Asph, wt (%)
Lanwa#I	0.9853	12.11	760.0	51.78	30.26	9.49	8.47
Lanwa#II	0.9827	12.49	784.0	47.46	33.62	9.90	9.01
Lanwa#III	0.9764	13.42	552.0	41.77	36.59	13.38	8.25
Balol#I	0.9751	13.61	584.0	40.07	23.89	30.62	5.42
Balol#II	0.9722	14.05	543.0	38.39	27.88	28.95	4.77
Balol#III	0.9679	14.69	512.0	35.33	27.50	33.08	4.09
Balol#IV	0.9612	15.71	492.0	40.15	25.84	28.80	5.21
Santhal#I	0.9624	15.53	555.0	40.82	24.05	31.46	3.67
Santhal#II	0.9567	16.40	515.0	42.76	24.58	30.51	2.15
Santhal#III	0.9556	16.57	601.0	39.82	24.37	30.84	4.97
Santhal#IV	0.9541	16.81	509.0	42.82	27.40	27.11	2.68
Santhal#V	0.9528	17.01	511.0	33.46	20.23	43.05	3.26

Table 2 API and density

Well name	API gravity (60 °F)	Density (60 °F)
Santhal#a	15.85	0.9603
Santhal#b	15.56	0.9622
Santhal#c	16.32	0.9572
Santhal#d	16.00	0.9593
Santhal#e	16.33	0.9572
Balol#a	14.74	0.9676
Balol#b	13.48	0.9760
Balol#c	13.34	0.9769
Balol#d	12.76	0.9809
Balol#e	12.92	0.9798
Lanwa#a	11.70	0.9881
Lanwa#b	11.67	0.9883
Lanwa#c	13.62	0.9741
Lanwa#d	13.18	0.9780
Lanwa#e	11.08	0.9924

3.3 *Thermal Behaviour of Crudes*

Experimental data for thermodynamic properties and reaction kinetics in the area of low-temperature oxidation, cracking, combustion and coking is required to make meaningful prediction of ISC process. These data are not provided by the combustion tube experiments. DSC and TGA have been used to generate parameter for chemical reaction such as energy of activation, heat capacity, preexponential factor, rate constant and heat of reaction [5]. The technique also provides information on the nature and qualitative composition of oil, the minimum ignition temperature of the crude to sustain the combustion and the residue left after heating.

(a) Thermogravimetry (TG/DTG)

Using thermogravimetry analysis (TGA), the dependence of the change in sample weight during programmed temperature change in a chosen gas atmosphere can be measured. The first derivative of the weight change with respect to time is called derivative thermogravimetry (DTG) and is a criterion for the reaction rate.

TG curve at heating rate of 1, 6, and 10 °C/min is shown in Fig. 3a–i for Santhal, Balol and Lanwa, respectively. The activation energy based on TG curve for Santhal, Balol and Lanwa field is shown in Fig. 4.

(b) Differential Scanning Calorimetry (DSC)

The thermograms (Fig. 5a–c) in oxidative environment show three distinct regions for the each oil category. First small peak corresponds to the peak due to low-temperature oxidation (LTO) wherein the oxygen is absorbed in the oil and second larger peak indicates to high-temperature oxidation (HTO) representing the

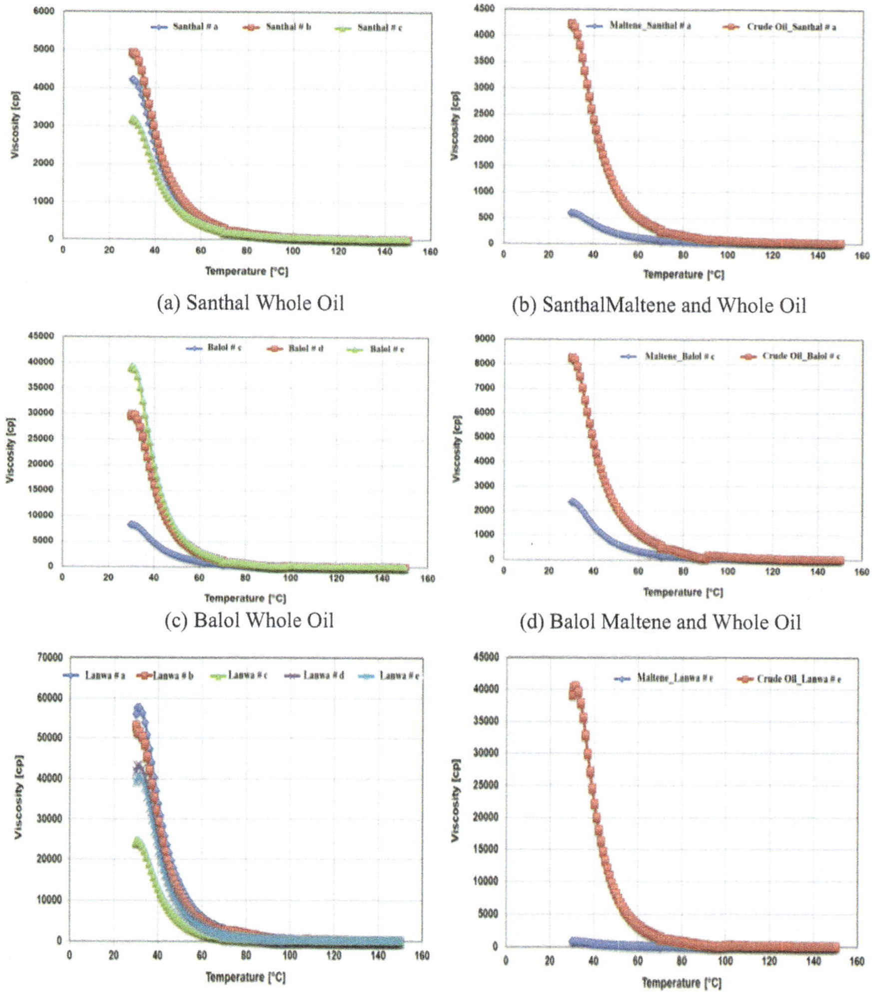

(a) Santhal Whole Oil

(b) SanthalMaltene and Whole Oil

(c) Balol Whole Oil

(d) Balol Maltene and Whole Oil

(e) Lanwa Whole Oil

(f) Lanwa Maltene and Whole Oil

Fig. 1 Viscosity versus temperature

combustion of fuel [6]. The oxidation of crude oil (LTO) starts around 220 °C and ends at 320 °C. The maxima of this peak lie around 280 °C, whereas in case of second peak, the oxidation starts around 400 °C and ends around 510 °C. The peak maxima of this are around 480 °C. The peak maximum is also a function of rate of heating. The third region represented by the valley between two peaks in the temperature range of 320–400 °C denotes fuel deposition.

Figure 5d represents the catalytic effect of certain minerals on coke-burning reaction. From the comparison of the thermograms of oil and rock mixture with that of oil alone, it was noticed that the combustion is facilitated in the oxidative

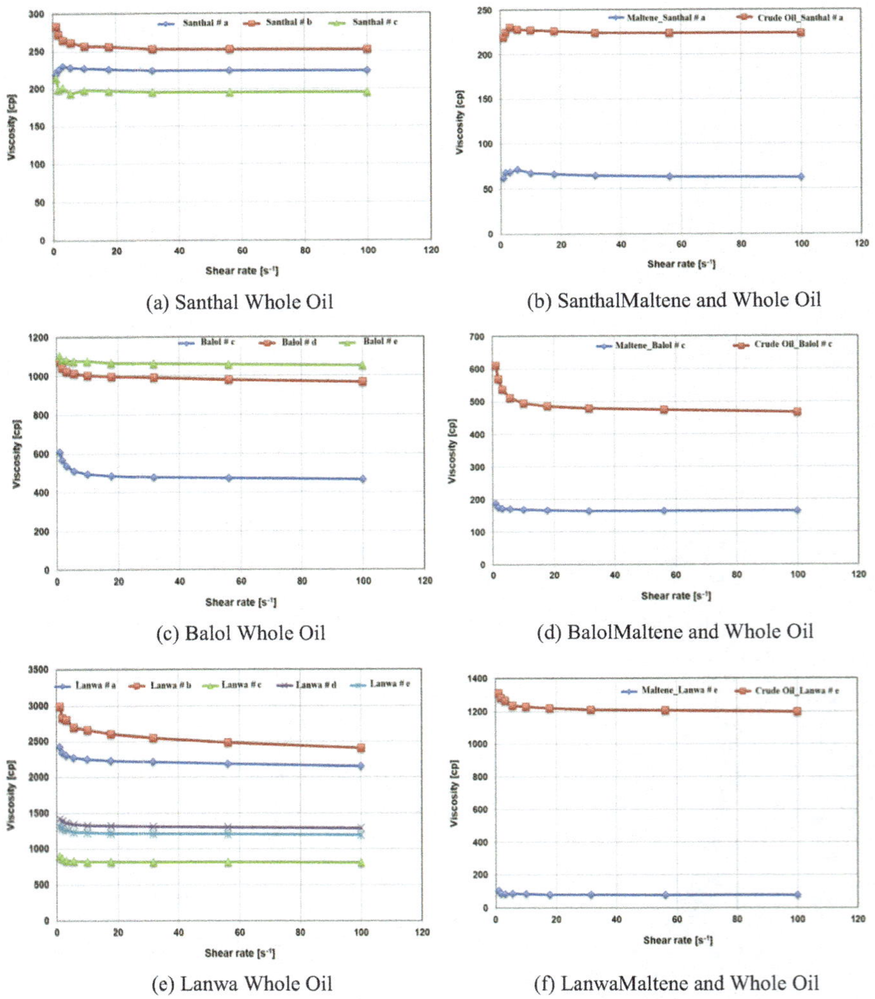

(a) Santhal Whole Oil

(b) SanthalMaltene and Whole Oil

(c) Balol Whole Oil

(d) BalolMaltene and Whole Oil

(e) Lanwa Whole Oil

(f) LanwaMaltene and Whole Oil

Fig. 2 Viscosity (70 °C) versus shear rate

atmosphere which is indicated by the shifting of the peak towards lesser temperature.

Comparative DSC thermogram in Fig. 5e for the oil Santhal, Balol and Lanwa in the order of increasing combined aromatic and resins content has been observed as widened HTO regimes, respectively.

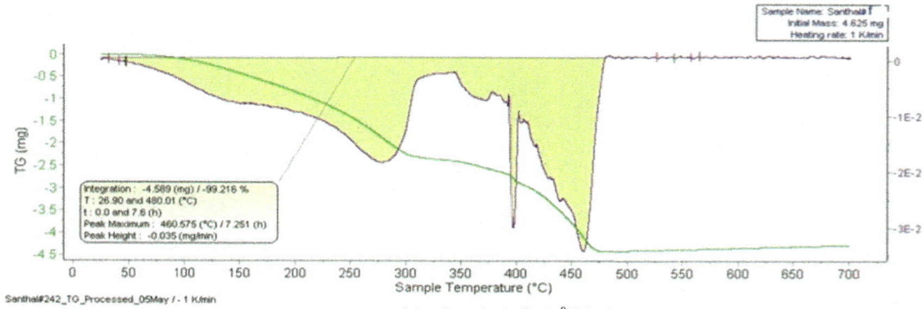

(a) Santhal @ 1 °C/min

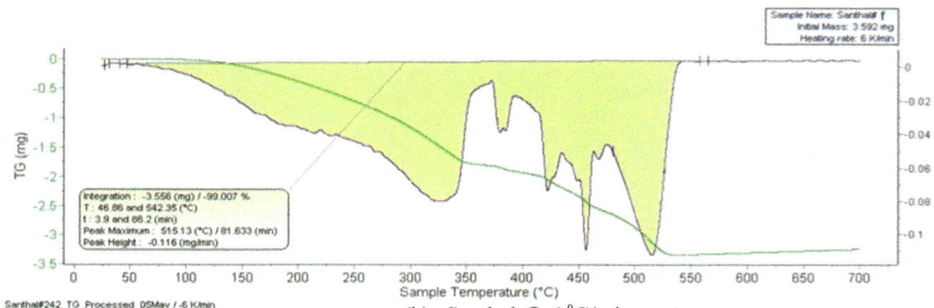

(b) Santhal @ 6 °C/min

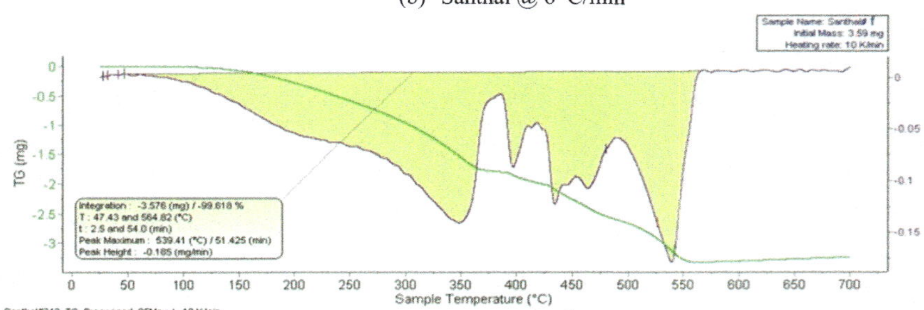

(c) Santhal @ 10 °C/min

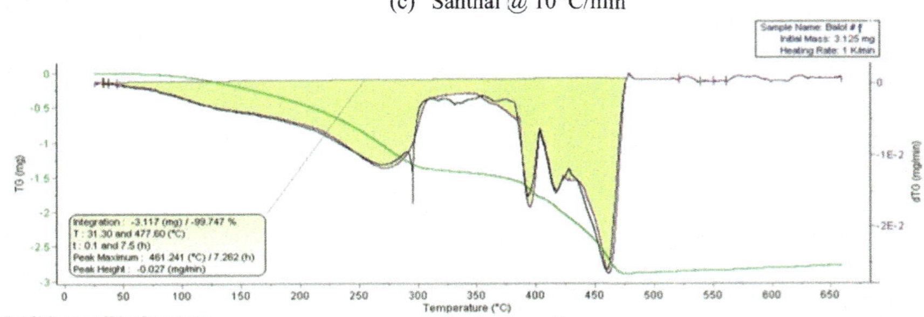

(d)Balol @ 1 °C/min

Fig. 3 TG curves with changing heating rate

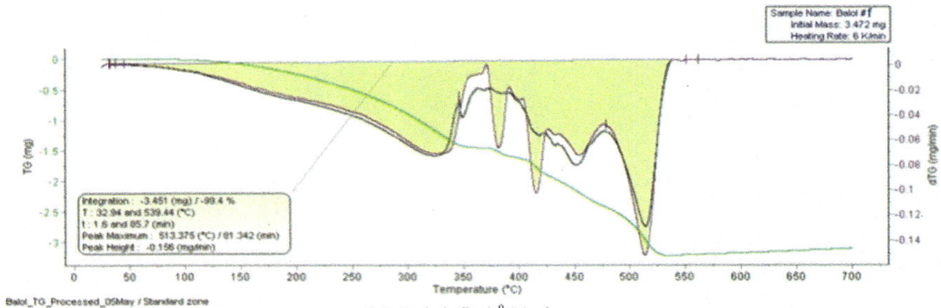

(e) Balol @ 6 °C/min

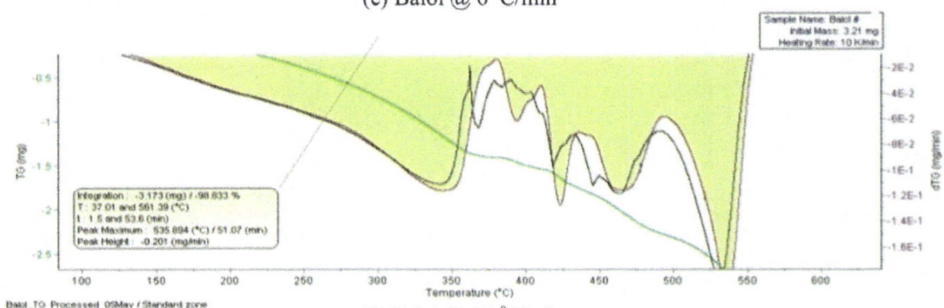

(f) Balol @ 10 °C/min

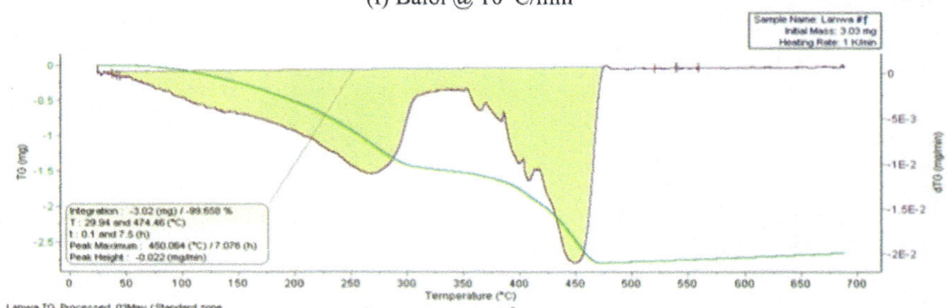

(g) Lanwa @ 1 °C/min

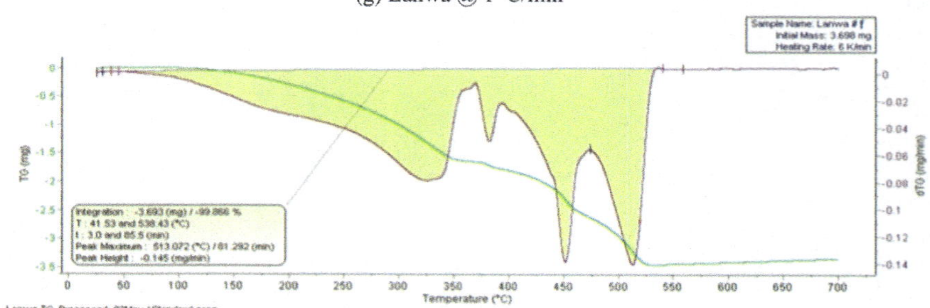

(h) Lanwa @ 6 °C/min

Fig. 3 (continued)

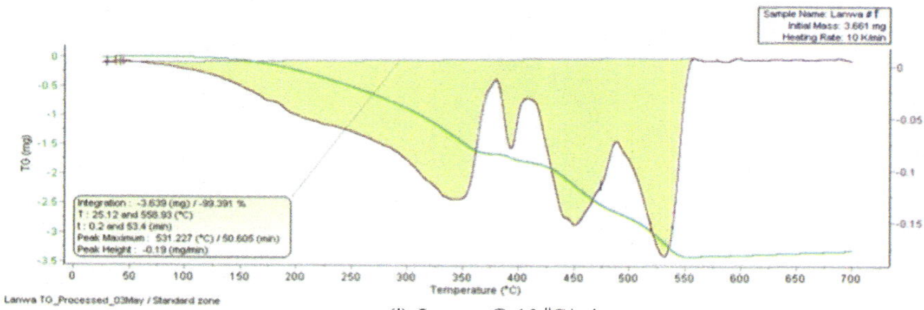

(i) Lanwa @ 10 °C/min

Fig. 3 (continued)

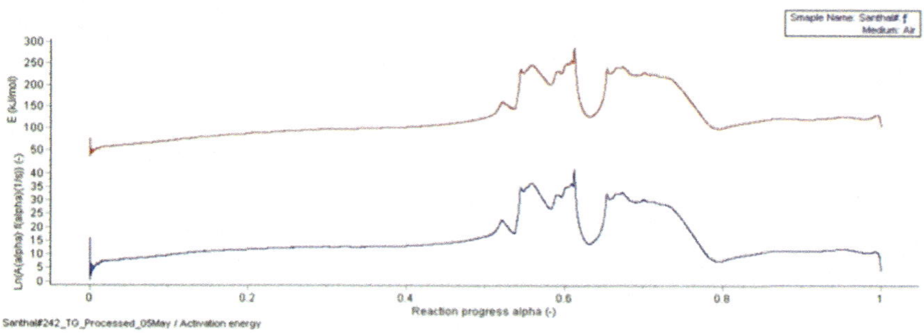

(a)Santhal

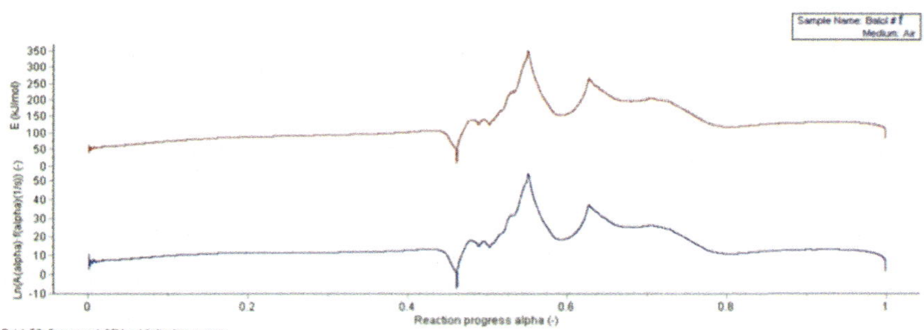

(b) Balol

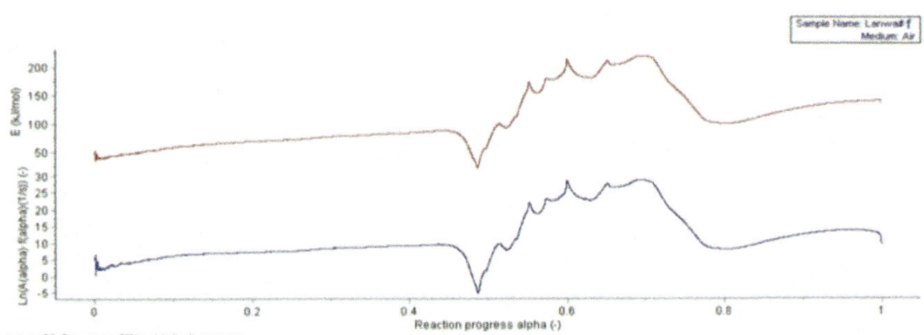

(c) Lanwa

Fig. 4 Activation energy and preexponential factor using TG curve

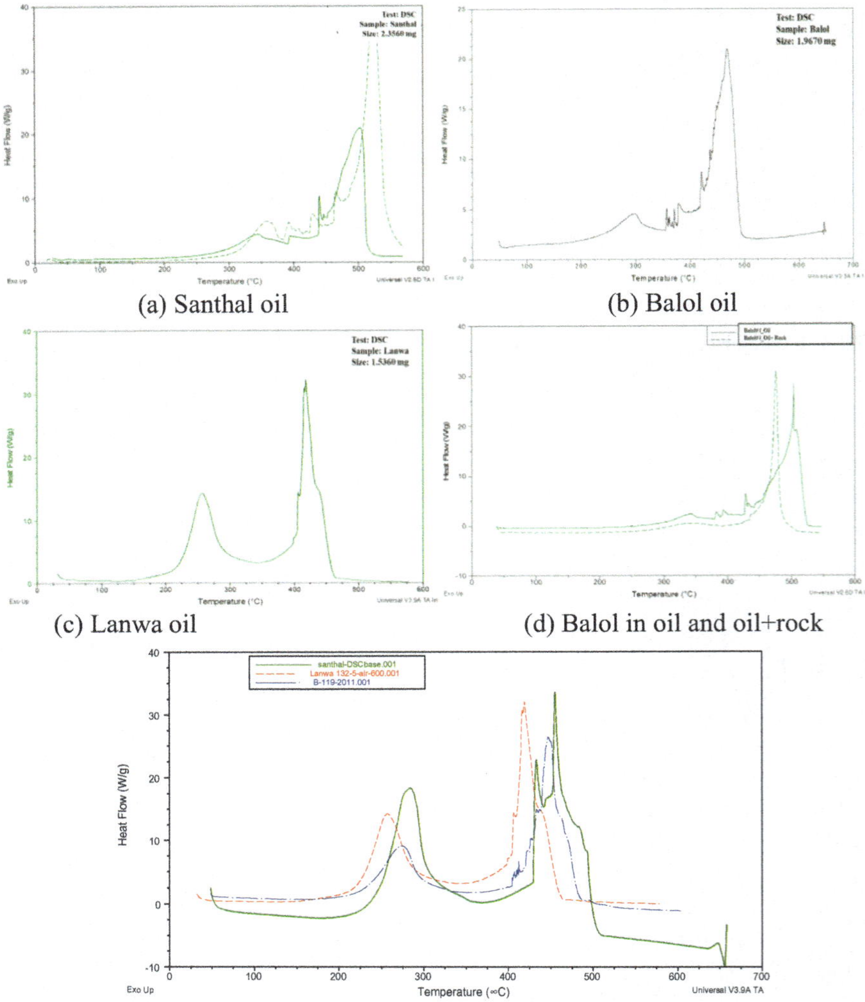

(a) Santhal oil (b) Balol oil

(c) Lanwa oil (d) Balol in oil and oil+rock

(e) Santhal, Balol & Lanwa comparison

Fig. 5 DSC thermograms

4 Discussion

4.1 Rheology and Production Planning

Production of heavy oil is directly linked with the rheological properties of the fluid. As evident from the result, increase in temperature exponentially reduces the viscosity and thereby greatly improving mobility of the oil. The important factor for heaviness of the crude is the presence of polar component Asphaltene in the crude oil. Higher percentage of Asphaltene increases the oil viscosity multifold and oil also becomes heavier. However, asphaltenes are also consumed as a fuel during in situ combustion process and result into peak temperature of ~500 °C which in turn reduces the viscosity of reservoir fluid due to heating of reservoir and augment in production of crude.

4.2 *Thermal Behaviour*

The difference in the slope of TGA plots first becomes noticeable at a temperature of around 350 °C when flattening of the curve occurs. This is due to the absorption of oxygen by the oil. The oxidation process starts only when more than 50% of oil has already evaporated. The fall in sample mass may be attributed to the combustion of oil in addition to cracking and evaporation.

The heat flow versus temperature plot shows three distinct peaks. The first small peak in the temperature ranges of 220–320 °C, an intermediate peak in the temperature range of 320–400 °C and a large peak above 450 °C. The first peak indicates low-temperature oxidation (LTO) wherein the oxygen is absorbed in the oil. The second peak denotes fuel deposition, and the third large peak represents the combustion of fuel or high-temperature oxidation (HTO) [7].

Earliest fraction to oxidize in LTO that triggers the oil combustion is the saturate fraction. Saturates are followed by resins and aromatics. Though asphaltenes are the most resistant part to oxidize, once the oxidation starts, the complete combustion/oxidation takes place around 450 °C. But the oxidation of aromatics and resins is spread in a wider range, and these fractions give larger and broader exotherms in HTO [8]. In case of aromatics and resins, the oxidation spreads around 550–560 °C range.

The catalytic effect of certain minerals on coke-burning reaction is well documented in the literature. In in situ combustion, native rock may contain certain minerals that may influence the combustion process [6]. From the comparison of the data of oil and rock mixture with that of oil alone, it was noticed that the combustion is facilitated in the oxidative atmosphere which is indicated by the shifting of the peak towards lesser temperature. This is further confirmed by the decrease in activation energy in oil and rock mixture as compared to oil alone in oxidative atmosphere [9].

5 Conclusions and Way Forward

Based on the measured parameters for the heavy oil of Lanwa, Balol and Santhal for the purpose to EOR designing and its monitoring, following pertinent observation has been made:

- Viscosity of whole oil and maltenes increases with decrease in API gravity exponentially and largely remains Newtonian across the shear regime.
- Fields where viscosity is higher (>500 cp) at reservoir temperature can be easily produced by either thermal process as viscosity at temperature of about 300 °C drops to less than 5 cp.
- With decrease in API gravity, molecular weight and percentage of asphaltenes increase and supply the fuel requirement to sustain the combustion process in case of ISC.
- One of the important parameters generated using DSC is likely temperature requirement for ignition to take place in HTO mode. Higher the asphaltene and resin content, ignition temperature requirement goes high.
- Also the presence of certain native core can catalyse the reaction at lower temperature hence ignition temperature can go lower.

Acknowledgements Authors express their sincere appreciation to Oil and Natural Gas Corporation Limited (ONGC) for the permission to publish this work. The authors thankfully acknowledge the assistance rendered by Thermal Laboratory of IRS. Also, cooperation extended by Mehsana Asset in several ways is hereby thankfully acknowledged.

References

1. Verkoczy B, Jha KN (1984) The role of thermal analysis techniques in the in-situ combustion process. SPE paper no. 12677. In: SPE/DOE forum symposium on enhanced oil recovery, Tulsa, 15–18 April 1984
2. Kok MV (1993) Use of thermal equipment to evaluate crude oils. Thermochim Acta 214:315
3. Kok MV, Karacan CO (1997) Behavior and effect of SARA fractions of oil during combustion. In: International thermal operations and heavy oil symposium. Society of Petroleum Engineers
4. Bae JH (1977) Characterization of crude oil for fireflooding using thermal analysis methods. Soc Petrol Eng J 17(03):211–218
5. Vossoughi S, Bartlett GW (1982) Development of a kinetic model for in-situ combustion and prediction of the process variables using TG/DSC techniques. Soc Pet Eng. AIME 1:12
6. Vossoughi S (1983) Study of the clay effect on crude oil combustion by thermogravimetry and differential scanning calorimeter. J Therm Anal 27:17
7. Ciajola A, Barbella R (1984) Pyrolysis and oxidation of heavy oils and their fractions in a TG apparatus. Fuel 63:657
8. Ranjbar M, Pusch G (1991) Pyrolysis and combustion kinetics of crude oils, asphaltenes and resins in relation to thermal recovery process. J Anal Appl Pyrolysis 20:185
9. Fassihi MR, Brigham WE, Ramey Jr HJ (1984) The reaction kinetics of in-situ combustion: part 2—modeling. Soc Pet Eng J 24(4):408–416

Feasibility of the Gas and Downhole Water Sink-Assisted Gravity Drainage (GDWS-AGD) Process to Enhance the Recovery of Oil in Reservoirs with Strong Aquifer

Watheq J. Al-Mudhafar, Dahlia A. Al-Obaidi, Dayanand Saini, Andrew K. Wojtanowicz, and Mohammed S. Al-Jawad

1 Introduction

To enhance oil recovery in both secondary and tertiary processes, the hybrid process of gas-assisted gravity drainage and downhole water sink (GDWS-AGD) has been developed for both miscible and immiscible injection modes [1]. The GDWS-AGD process incorporates the two processes: gas-assisted gravity drainage (GAGD) and downhole water sink (DWS). In the GAGD process, gas is injected at the top of the reservoir through a vertical injector in a gravity-stable mode. Due to the gravity drainage and the difference in density between gas and liquids, gas accumulates at the reservoir's top to formulate a gas cap, while oil and water drain downwards to the bottom of the reservoir [2]. For oil production, several horizontal producers are placed above the oil–water contact (OWC). To adopt the DWS technology, a series of horizontal wells are placed for water sink located below the OWC and exactly underneath the oil wells, as illustrated by Al-Mudhafar et al. [3]. The GDWS-AGD process includes two completions, one above OWC for oil

W. J. Al-Mudhafar (✉)
Basrah Oil Company, Basrah, Iraq
e-mail: watheq.almudhafar@utexas.edu

D. A. Al-Obaidi · M. S. Al-Jawad
University of Baghdad, Baghdad, Iraq
e-mail: dahliaalobedi@coeng.uobaghdad.edu.iq

M. S. Al-Jawad
e-mail: mjawad@coeng.uobaghdad.edu.iq

D. Saini
California State University, Bakersfield, CA, USA
e-mail: dsaini@csub.edu

A. K. Wojtanowicz
Louisiana State University, Baton Rouge, LA, USA
e-mail: awojtan@lsu.edu

U. K. Bhui (ed.), *Macromolecular Characterization of Hydrocarbons for Sustainable Future*, Green Energy and Technology,
https://doi.org/10.1007/978-981-33-6133-1_7

production, while the second interval for water production through the water zone [4]. These two completions are isolated by a packer; the water is produced from the bottom completion to the surface or reinjected to another water layer [5].

The advantages of the hybrid GDWS-AGD process are high oil recovery, better sweep efficiency, low water cresting/coning, low water cut and good gas injectivity, resulting from decreasing the reservoir pressure due to the water sink. There are severe limitations about the GAGD process implementations concerning the high water cut levels, mainly when applied on reservoirs bounded with infinite active aquifers [3]. The DWS technology, as described and illustrated by Wojtanowicz [4], has been integrated with the GAGD method to overcome these limitations to improve oil recovery and reduce water cut and water cresting/coning in the reservoir with high coning tendency. In the combined technologies, 7-in. production casing is dual-completed for two 2-3/8-in. horizontal tubings: one above the oil–water contact for oil production and one underneath for water sink. The two completions are hydraulically isolated inside the well by a packer. The bottom (water sink) completion employs a submersible pump and water-drainage perforations.

The performance of the GDWS-AGD process is affected by the reservoir property variations, especially heterogeneity and anisotropy, which influence sweep efficiency and recovery factor. Reservoirs with high heterogeneity make the flooding front unstable and cause early gas breakthrough and viscous fingering in CO_2 flooding performance. At the same time, anisotropy affects the vertical flow displacement of gas, oil and water. Dip angle assists gravity segregation of CO_2 and forms a stable flood front, resulting in delaying the breakthrough, and more hydrocarbons can be recovered [6].

In this study, CO_2 was used as an injected solvent through the GDWS-AGD process to improve oil recovery and reduce the water cut. GDWS-AGD process was evaluated by implementing the black oil flow simulation model to the PUNQ reservoir for ten years to assess its effectiveness in improving oil recovery and reducing cresting tendency and water cut. It is worth to mention here that the use of black oil simulation offers an opportunity to study the oil recovery mechanism of the GDWS-AGD process (i.e. compositionally independent pressure depletion and maintenance via immiscible CO_2 flooding) in an easy-to-implement and modelling procedure with shorter run time compared to compositional simulation. Additionally, the use of black oil simulation for simulating the immiscible CO_2 flooding scheme further reduces the data requirements such as composition and quality of the oil and the minimum miscibility pressure (MMP).

The pay zone of the PUNQ reservoir is a heterogeneous reservoir surrounded by the north and west with a strong edge water aquifer. Next, a sensitivity analysis was adopted using proxy modelling and the design of experiments (DoE) to determine the most influencing petrophysical parameters that impact the GDWS-AGD process flow performance. The parameters were porosity, horizontal and vertical permeability for each layer rock compressibility and radius of the aquifer.

2 Gas and Downhole Water Sink-Assisted Gravity Drainage (GDWS-AGD) Process

The hybrid integration of the GAGD process and DWS method has been introduced to improve oil recovery in reservoirs with infinite edge and/or bottom water drive [7]. Vertical injectors are located at the top of the reservoir to inject the gas and formulate a gas cap, while oil and water are drained by a series of horizontal wells located above and bottom of the oil–water contact (OWC) for oil and water production, respectively. These water and oil producers are the same length and diameter: the horizontal oil wells are located above OWC, while horizontal water producers are located below OWC underneath oil wells to reduce the water cresting tendency and eliminate water cut [4]. Two 7-in. production tubings are installed bilaterally and completed with two 2-3/8-in. horizontal tubings, the first completion is for oil production placed at the bottom of the oil pay zone, and the second one is for water production, which is placed underneath the oil producer at the bottom of the water zone. The two completions are hydraulically isolated inside the well by a packer. The schematic of the hybrid GDWS-AGD process is shown in Fig. 1.

In this study, the GDWS-AGD process is applied to the PUNQ reservoir to study its feasibility and efficacy to increase cumulative oil production and reduce water cresting and water cut. The reduced water cresting and water cut can help operators extract hydrocarbons more efficiently and greener manner.

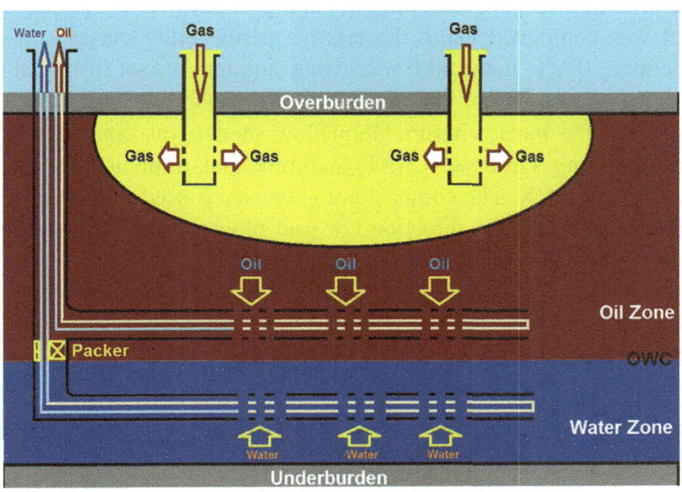

Fig. 1 Sketch of the GDWS-AGD process (Reused by permission from Copyright Clearance Center: Society of Petroleum Engineers, SPE Improved Oil Recovery Conference, Gas and downhole water sink-assisted gravity drainage GDWS-AGD EOR process: field-scale evaluation and recovery optimization, Al-Mudhafar WJ, Wojtanowicz AK, Rao DN © 2018)

3 PUNQ Reservoir Description and Modelling

The PUNQ reservoir model was generated during the PUNQ project, which stands for production forecasting with uncertainty quantification, has been taken from a reservoir engineering study on a real field by Elf Exploration Production [8, 9]. It was represented as a small-sized reservoir simulation flow model and used as a test model for the newly proposed methods. The PUNQ model consists of 19 × 28 × 5 grid cells (180 m by 180 m for each grid), which about two-thirds of grid cells (1761) are active. It is noted here that the PUNQ model is an open-source reservoir model for research, and its original geomodel is not accessible to the authors. Hence, the grid system was not adjusted further to have more subdivisions in the layers. Moreover, formulating a gas cap requires a thickness interval in the reservoir. Further, the use of the black oil flow simulation model enables the use of large grid sizes.

The geometry of the field has been modelled using corner-point geometry. The field's top structure map shows that the field is bounded to the north and west to an infinite aquifer and to the east and south by a fault. Also, a small gas cap is in the first layer and in the centre of the dome-shaped structure. The field initially contains six vertical oil production wells, which are situated around the gas–oil contact (GOC). Due to the active water drive, no water injection wells are needed [8, 10, 11]. The top structure map and location of the original production and injection wells in the PUNQ reservoir can be found in Floris et al. [8].

For the original reservoir model, the porosity and permeability (horizontal and vertical) distribution maps were built with the geostatistical technique. The reservoir model was completed using the relative permeability curves, pressure–volume–temperature (PVT) data and Carter-Tracy aquifer data set from real field data. There is no capillary pressure data in the original model [8–10].

In this paper, the black oil simulation flow model was constructed using the Computer Modelling Group's (CMG's) commercial implicit explicit black oil simulator, namely IMEX. The corner-point geometry is used to model the reservoir with 19 grid blocks in the x-direction, 28 grid blocks in the y-direction and five layers in the z-direction. The reservoir is a water-wet reservoir, and the petrophysical parameters and geological structure of the reservoir model were kept as the original model. A modified PUNQ reservoir model was regenerated using a set of

Table 1 Average porosity and horizontal and vertical permeabilities for each layer in the PUNQ reservoir (Reused by permission from copyright Clearance Center: Society of Petroleum Engineers, SPE Reservoir Simulation Symposium, Quantifying uncertainty for the PUNQ-S3 problem in a Bayesian setting with RML and EnKF, Gao G, Zafari M, ©2005)

Average of property	Layer 1	Layer 2	Layer 3	Layer 4	Layer 5
Porosity (%)	0.17	0.08	0.17	0.16	0.19
Horizontal permeability (md)	432	33	432	196	654
Vertical permeability (md)	137	13	137	64	205

relative permeability and capillary pressure curves. The average reservoir porosity and vertical and horizontal permeabilities used in the study are summarized in Table 1.

4 Simulation of the GDWS-AGD Process

The PUNQ black oil reservoir simulation model was adopted for the GDWS-AGD process evaluation with a 10-year prediction period. The original six horizontal oil-producing wells have been eliminated from the model. However, six new wells were placed

- two vertical CO_2 injectors (G-INJ1 and G-INJ2),
- two horizontal oil producers (HO1 and HO2) and
- two horizontal water producers (NEW WELL1 and NEW WELL2).

CO_2 was injected in the first layer of the reservoir in immiscible mode; the second and third layers were left to provide a gravity-stable column. The two horizontal oil wells were placed in the fourth layer for oil production. In contrast, the two horizontal water sinkers were in the fifth layer below OWC and exactly underneath the two horizontal oil producers. The perforations of oil and water producers were extended in the x-direction of the reservoir.

Figure 2 illustrates the location of the CO_2 injectors and the oil and water producers. Three regions of infinite active water drive aquifers have bounded the reservoir from the northern, western and bottom of the reservoir and modelled using the Carter-Tracy approach. The GDWS-AGD process was performed to calculate the cumulative oil production and water cut for both horizontal oil producers after a 10-year prediction period in comparison with the GAGD processes evaluation.

The operational constraints that were chosen for GAGD and GDWS-AGD processes were the maximum gas injection rate for the two injectors, the maximum surface rate and the minimum bottom hole flowing pressure for the two horizontal producers (oil and water). Table 2 shows these operational parameters constraints through the GAGD and GDWS-AGD processes.

Figure 3 shows the comparison between GAGD and GDWS-AGD in terms of cumulative oil production and oil production rate for ten years of production. As evident from the comparison shown in Fig. 3, cumulative oil production increased from 3.8×10^5 m^3 in the GAGD process to 4.7×10^5 m^3 for the GDWS-AGD process. The gain in cumulative oil production and growing gas injectivity resulted from reducing the average reservoir pressure due to water sink through the GDWS-AGD process.

The reduction in the reservoir pressure improved the gas injectivity and the cumulative oil production only and minimized the water cut and cresting tendency. Figure 4 shows a reduction in water cut for horizontal oil producers through ten

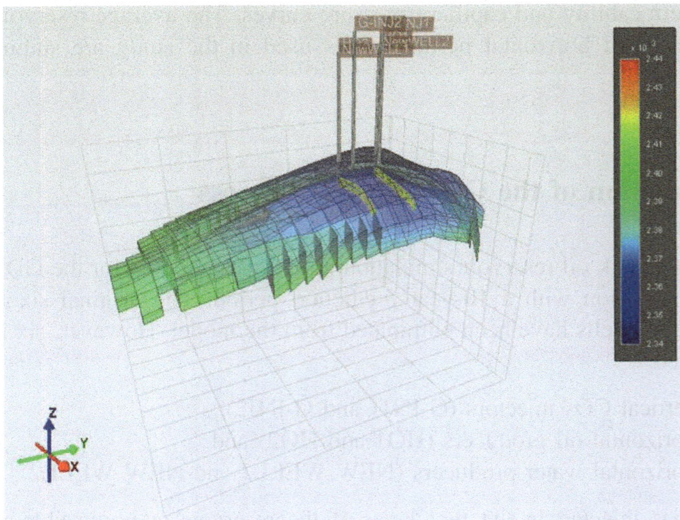

Fig. 2 Locations of the CO_2 injectors and the horizontal oil and water producers in the PUNQ reservoir

Table 2 Well constraints of the GDWS-AGD and GAGD processes

Wells	Constraints	GAGD	GDWS-AGD
Injectors	MaxSTG	280,000 m^3/d	280,000 m^3/d
Oil producers	MaxSTO	1,000 m^3/d	1,000 m^3/d
	Min BHP	12,000 kPa	12,000 kPa
Water producers	Max STW	–	3,500 m^3/d
	Min BHP	–	12,000 kPa

years of implementing the GDWS-AGD process. This reduction is about 5% less than compared to the same production period for the GAGD process.

5 GDWS-AGD Sensitivity Analysis

A sensitivity analysis is an essential step in detecting the most influencing petrophysical parameters on the reservoir flow performance and eliminating the non-influential petrophysical parameters in the integrated reservoir simulation studies [12]. On the other hand, the design of experiments (DoE) is a statistical method to create a suitable set of experiments for simulation purposes and identify the most sensitive parameters affecting the response. Additionally, the interaction between the parameters can be evaluated through the DoE approach to quantify the

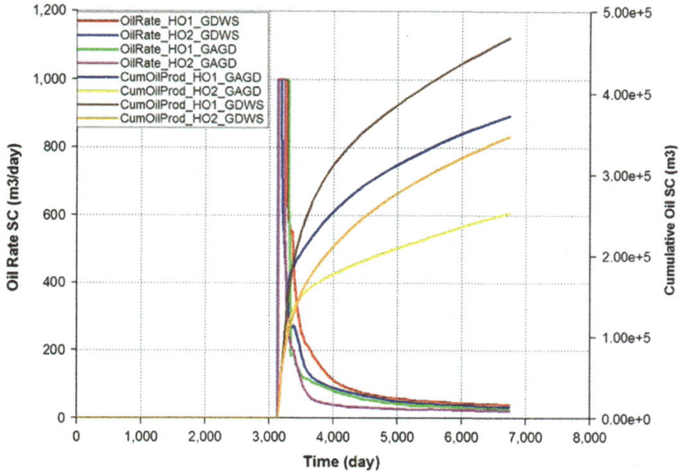

Fig. 3 Comparison of the cumulative oil production and rate through the GDWS-AGD and GAGD processes

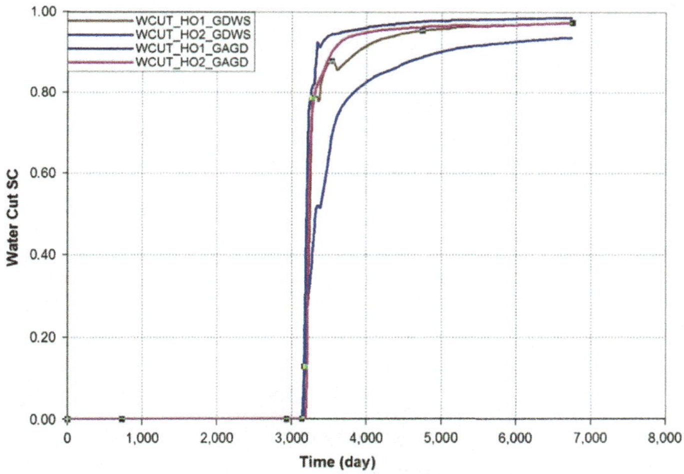

Fig. 4 Comparison of the water cut through the GDWS-AGD and GAGD processes

common effects on a process. Many computer experiments were created through the DoE approach, which incorporated a multilevel of each parameter and were evaluated by the black oil simulation model to find the flow response factor [13].

In this paper, a proxy model was built to be as an alternative to the black oil reservoir flow simulation model to get the optimal scenario in much less computational time for millions of runs through a studied process. The proxy model of either linear or polynomial regression is required to provide a sensitivity analysis formula that helps to determine the most influential parameters affecting the process. Sobol analysis is the way that provides a base to determine which factor or parameter is more influential than others.

This study's primary purpose is to detect the most significant parameters that impact the cumulative oil produced and the water cut level through the GDWS-AGD process. The Sobol analysis and response parameter cross-plots were used to achieve these purposes. The 45° line and the 95% confidence curve between the simulator-based flow response and proxy model also reflect the proxy model's accuracy.

For the GDWS-AGD process, sensitivity analysis of PUNQ reservoir properties on the cumulative oil production and water cut level was conducted to investigate these parameters' effects for each layer and not for the entire reservoir. Porosity, horizontal permeability, vertical permeability, rock compressibility and radius of the aquifer were selected. Figure 5 illuminates the original porosity distribution for the entire reservoir with well locations.

In the sensitivity analysis, the original porosity and vertical and horizontal permeability values were doubled for the maximum level and divided by two for the minimum level. Next, the generated multiple simulation runs (experiments) were evaluated by the black oil reservoir simulator to detect the most sensitive parameters on the cumulative oil production and water cut for ten years of the production period.

Figure 6 shows the sensitivity analysis of reservoir parameters on reservoir flow response. This figure shows that the different experiments that combine various levels of parameters led to assorted values of the reservoir flow responses (cumulative oil production and water cut). This fact can further be illuminated in Fig. 7 that shows the ranges of field cumulative oil production and water cut values given all these experiments (i.e. simulation jobs).

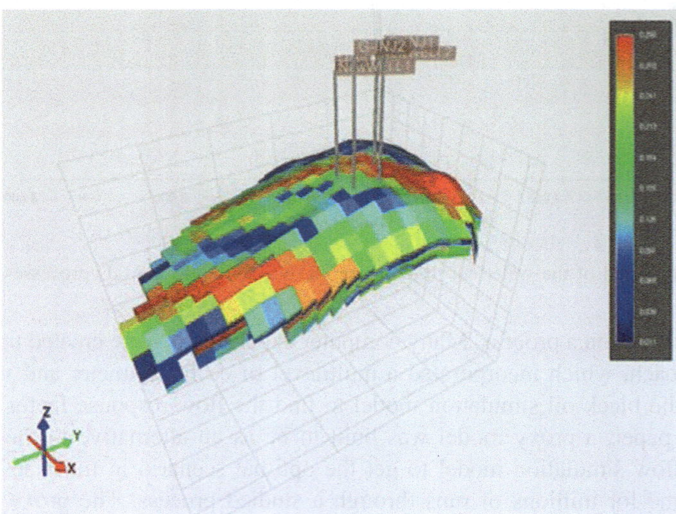

Fig. 5 Well locations through the porosity map

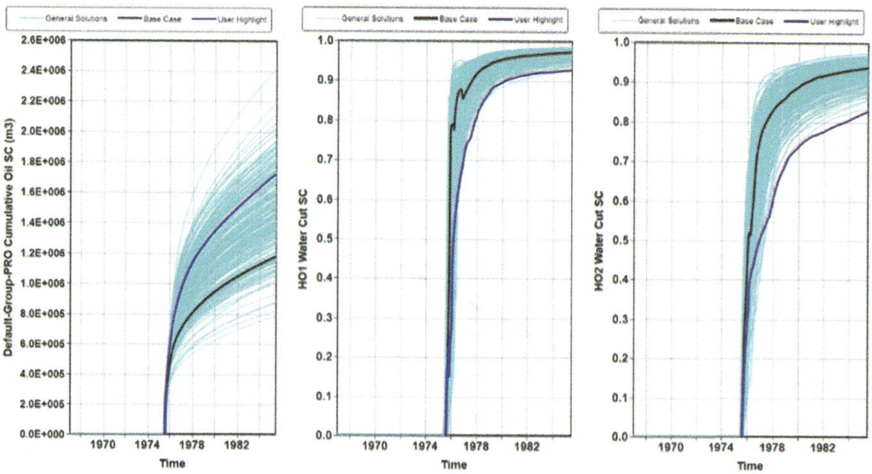

Fig. 6 General solutions of the cumulative oil production and water cut in the two horizontal oil producers

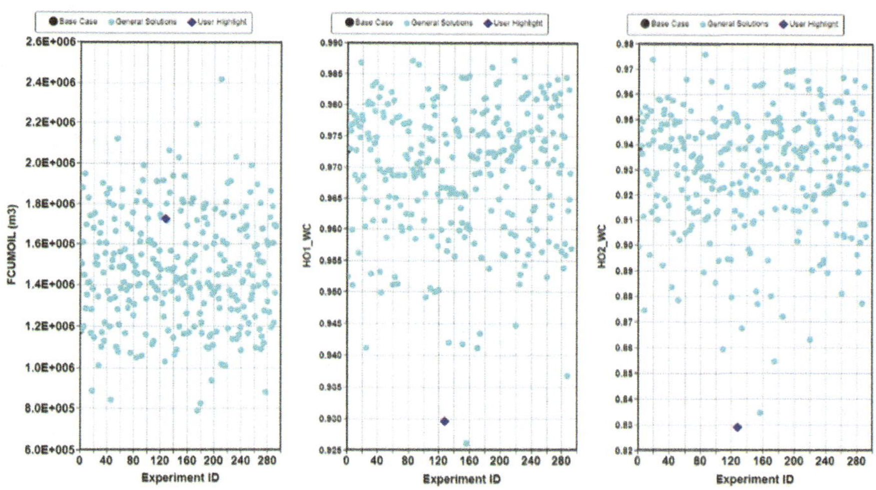

Fig. 7 Sensitivity space of the cumulative oil production and the water cut in the horizontal oil producers

The simulator and proxy model mismatch for any response has been quantified to validate the proxy model. Figure 8 shows the 45° line and the 95% confidence curve between the simulator and the proxy model for cumulative oil through the GDWS-AGD process. Figures 9 and 10 are for water cut in both horizontal oil producers HO1 and HO2 through the same process.

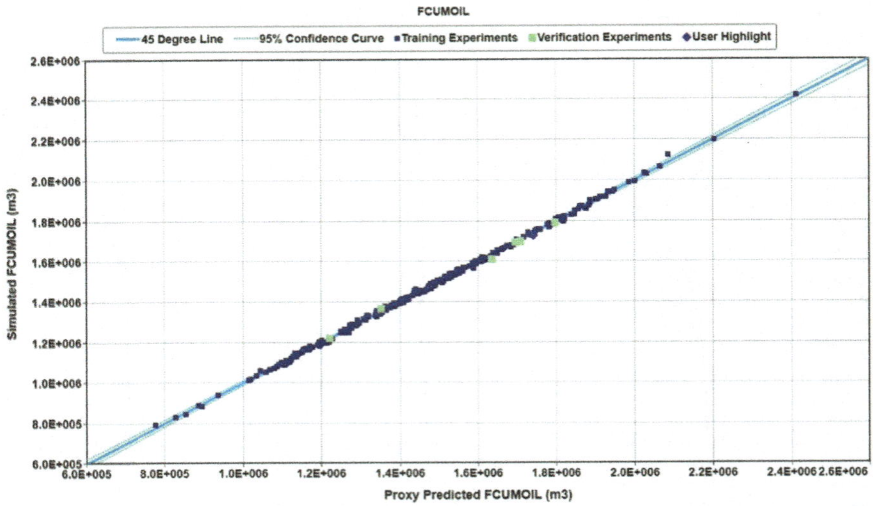

Fig. 8 Cross-plot between the simulator- and proxy-based cumulative oil production through the GDWS-AGD process

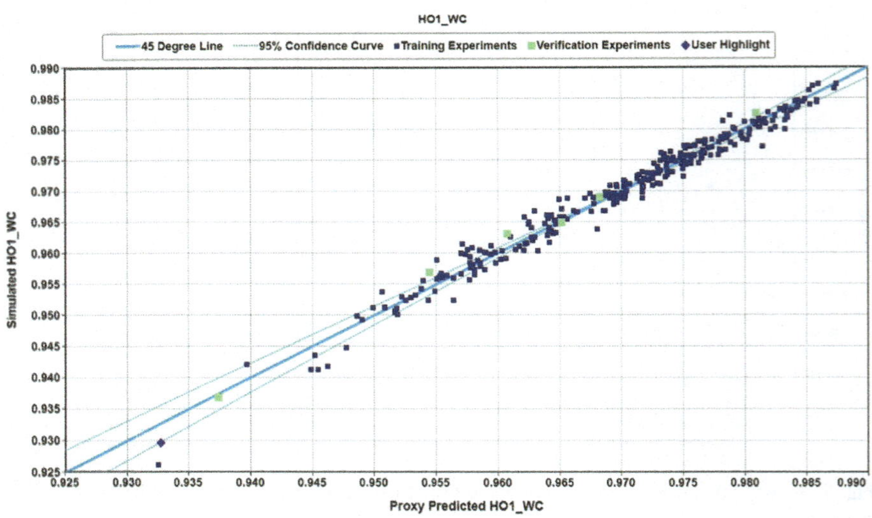

Fig. 9 Cross-plot between the simulator- and proxy-based water cut in the HO1 Oil producer

The Sobol analysis and the cross-plot of the response parameters were adopted to detect the most influential parameters, which affect the flow performance through the GDWS-AGD process. As shown in Fig. 11, which represents the Sobol results analysis for cumulative oil, the most influential parameters are the porosity for layers 5, 4 and 3, with 31, 27 and 19%, respectively.

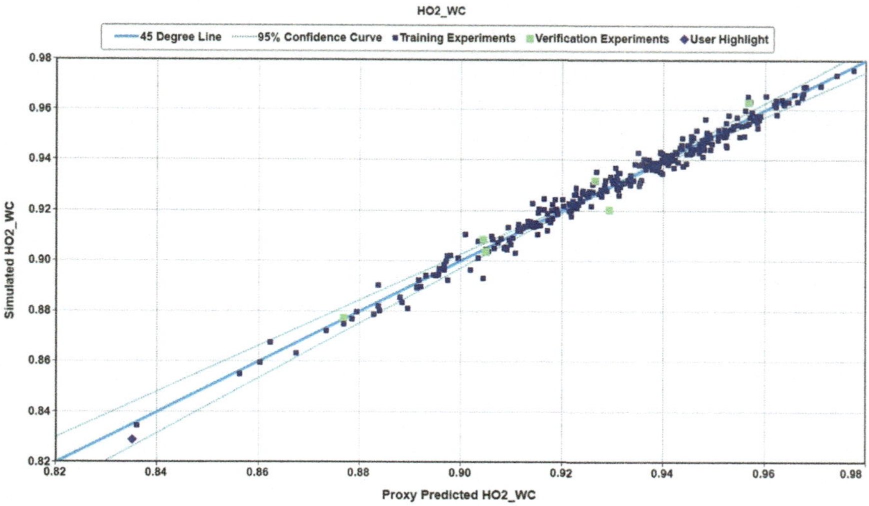

Fig. 10 Cross-plot between the simulator- and proxy-based water cut in the HO2 Oil producer

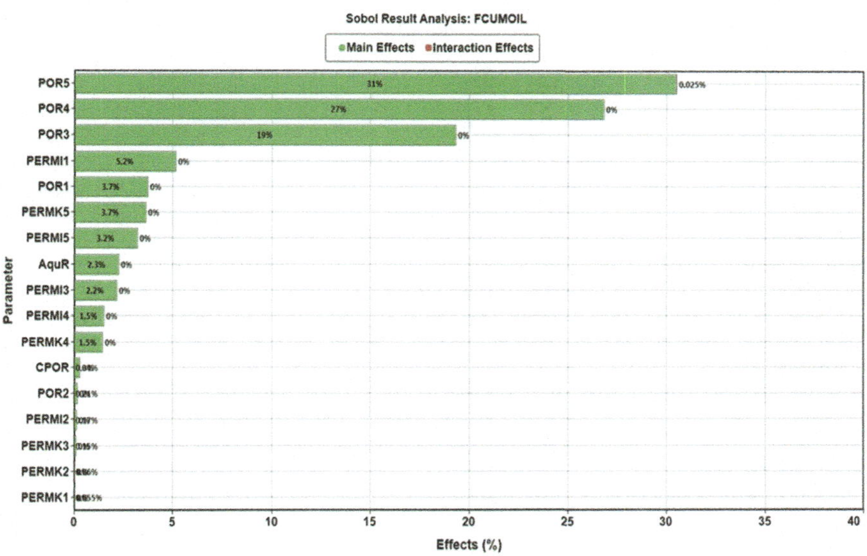

Fig. 11 Sobol analysis with respect to the cumulative oil production objective function

These parameters and the other influential parameters on cumulative oil through the GDWS-AGD process are summarized in Table 3. For the water cut through the two horizontal oil producers, the Sobol analysis shows that the horizontal permeability for layer 4 is the most influential parameter (i.e. 3.3% interaction) with the

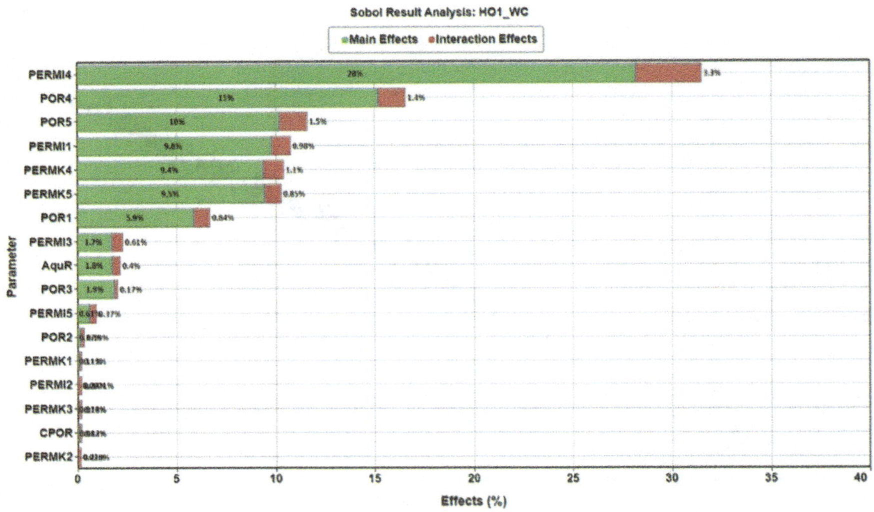

Fig. 12 Sobol analysis of the water cut in HO1 oil producer through the GDWS-AGD process

other parameters, as shown in Figs. 12 and 13. The other influential parameters on the water cut for the two horizontal oil producers are summarized in Table 4. Tables 3 and 4.

Table 3 Most influential parameters affecting the GDWS-AGD process

Parameter	Main effects (%)	Interaction effects (%)
Porosity/layer 5	31	0.025
Porosity/layer 4	27	0
Porosity/layer 3	19	0
Horizontal permeability/layer 1	5.2	0
Porosity/layer 1	3.7	0
Vertical permeability/layer 5	3.7	0
Horizontal permeability/layer 5	3.2	0
Radius of aquifer	2.3	0
Horizontal permeability/layer 3	2.2	0
Horizontal permeability/layer 4	1.5	0
Vertical permeability/layer 4	1.5	0
Rock compressibility	<1%	0
Porosity/layer 2	<1%	0
Horizontal permeability/layer 2	<1%	0
Vertical permeability/layer 3	<1%	0
Vertical permeability/layer 2	<1%	0
Vertical permeability/layer 1	<1%	0

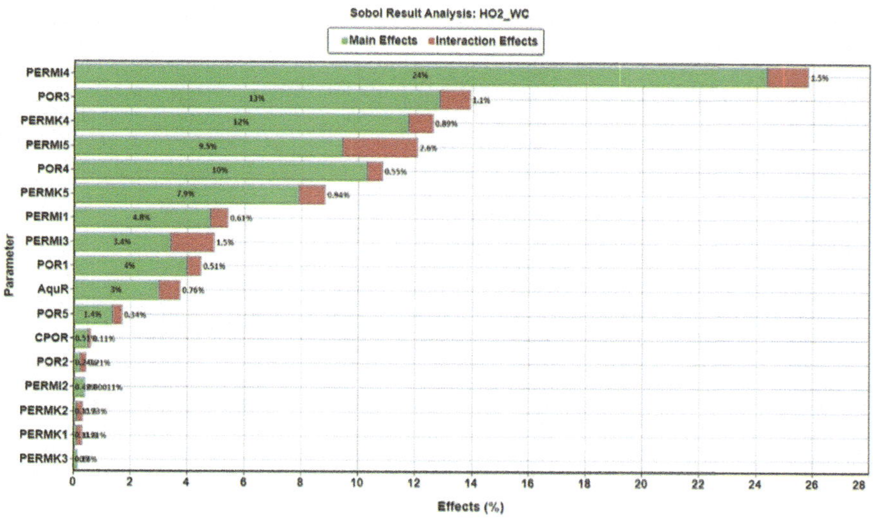

Fig. 13 Sobol analysis of the water cut in HO2 oil producer through the GDWS-AGD process

It is noted here that in the present study, the focus was on the evaluation of gas and downhole water sink-assisted gravity drainage (GDWS-AGD) process involving an immiscible CO_2 flooding scheme. Hence, it was deemed unnecessary that the impact of the most influential parameters affecting the GDWS-AGD process is also evaluated for the GAGD process as well. Similarly, sensitivity runs for CO_2 versus other hydrocarbon gases for the GAGD-AGD process were deemed beyond the present study's scope.

6 Summary and Conclusions

A hybrid process of gas and downhole water sink-assisted gravity drainage (GDWS-AGD) was developed to improve the cumulative oil production and reduce water cut and cresting tendency in the reservoir with infinite edge and/or bottom aquifer. The main advantage of GDWS-AGD process is that it enables the operators to extract the hydrocarbons in a more efficient and greener manner.

The black oil reservoir flow simulation model was conducted for the GDWS-AGD process in the PUNQ reservoir for a 10-year production period. Two vertical CO_2 injectors, two horizontal oil producers and two horizontal water sinkers were constituted to simulate the process. For sensitivity analysis, a proxy model was built to be an alternative to the black oil reservoir flow simulation model using the design of experiments (DoE) to identify the most influential reservoir parameters on the GDWS-AGD process performance for each layer of PUNQ

Table 4 Influential parameters on water cut for HO1 and HO2 oil producers through the GDWS-AGD process

HO1 oil producer			HO2 oil producer		
Parameter	Main effects (%)	Interaction effects (%)	Parameter	Main effects (%)	Interaction effects (%)
Horizontal permeability/layer 4	28	3.3	Horizontal permeability/layer 4	24	1.5
Porosity/layer 4	15	1.4	Porosity/layer 3	13	1.1
Porosity/layer 5	10	1.5	Vertical permeability/layer 4	12	0.89
Horizontal permeability/layer 1	9.8	0.98	Horizontal permeability/layer 5	9.8	2.6
Vertical permeability/layer 4	9.4	1.1	Porosity/layer 4	10	0.55
Vertical permeability/layer 5	9.5	0.85	Vertical permeability/layer 5	7.9	0.94
Porosity/layer 1	5.9	0.84	Horizontal permeability/layer 1	4.8	0.61
Horizontal permeability/layer 3	1.7	0.6	Horizontal permeability/layer 3	3.4	0.51
Radius of aquifer	1.8	0.41	Porosity/layer 1	4	0.76
Porosity/layer 3	1.9	0.17	Radius of aquifer	3	0.34
Horizontal permeability/layer 5	<1%	0.37	Porosity/layer5	1.4	0.11
Porosity/layer2	<1%	<0.1	Rock compressibility	<1%	0.21
Vertical permeability/layer 1	<1%	<0.1	Porosity/layer 2	<1%	0
Horizontal permeability/layer 2	<1%	<0.1	Horizontal permeability/layer 2	<1%	0.1
Vertical permeability/layer 3	<1%	<0.1	Vertical permeability/layer 2	<1%	0.23

(continued)

Table 4 (continued)

HO1 oil producer			HO2 oil producer		
Parameter	Main effects (%)	Interaction effects (%)	Parameter	Main effects (%)	Interaction effects (%)
Rock compressibility	<1%	<0.1	Vertical permeability/ layer 1	<1%	0.2
Vertical permeability/ layer 2	<1%	<0.1	Vertical permeability/ layer 3	<1%	0

reservoir. Porosity, horizontal permeability, vertical permeability, rock compressibility and radius of the aquifer were selected.

A comparison between the GAGD and GDWS-AGD processes showed an increase in cumulative oil production (about 10^5 m^3) for the GDWS-AGD process compared with the GAGD process. Besides, GDWS-AGD caused decreasing in the tendency of cresting and water cut for both horizontal oil producers from 97 to 92% compared with the water cut levels for the GAGD process.

For the proxy model, it is clear from the Sobol analysis that the porosity for layer 5 was a more influential parameter than others on cumulative oil through the GDWS-AGD process, with 31% main effects and 0.025% interaction effects. On the other hand, horizontal permeability for layer 4 was the most influential parameter, with 24% main effects and 1.5% interaction effects through the water cut objective functions.

Acknowledgements The authors thank the Society of Petroleum Engineers (SPE) for granting reuse permission to use material from their previous work (SPE-195332-MS). The financial support for Dr. Saini's contribution to the work reported here was provided by the National Science Foundation (NSF) sponsored Center for Research Excellence in Science and Technology (CREST) Phase II grant (NSF Award#1547784) at California State University, Bakersfield, CA, USA.

References

1. Al-Mudhafar WJ, Wojtanowicz AK, Rao DN (2017) Hybrid process of gas and downhole water sink-assisted gravity drainage (GDWS-AGD) to enhance oil recovery in reservoirs with water coning. Soc Pet Eng. https://doi.org/10.7122/502487-MS
2. Rao DN, Ayirala SC, Kulkarni MM et al (2004) Development of gas assisted gravity drainage (GAGD) process for improved light oil recovery. Soc Pet Eng. https://doi.org/10.2118/89357-MS
3. Al-Mudhafar WJ, Rao DN, Srinivasan S (2018) Reservoir sensitivity analysis for heterogeneity and anisotropy effects quantification through the cyclic CO_2-assisted gravity drainage EOR process—a case study from South Rumaila Oil Field. Fuel 221:455–468. https://doi.org/10.1016/j.fuel.2018.02.121

4. Wojtanowicz AK (2006) Down-hole water sink technology for water coning control in wells. Wiertnictwo Nafta Gaz: Tom 23(1):575–586
5. Wojtanowicz AK, Xu H (1995) Downhole water loop—a new completion method to minimize oil-well production watercut in bottom-water-drive reservoirs. Pet Soc Can. https://doi.org/10.2118/95-08-06
6. Li D, Lake LW (2013) Scaling Fluid flow through heterogeneous permeable media. Soc Pet Eng. https://doi.org/10.2118/26648-PA
7. Al-Mudhafar WJ, Wojtanowicz AK, Rao DN (2018) Gas and downhole water sink-assisted gravity drainage GDWS-AGD EOR process: field-scale evaluation and recovery optimization. Soc Pet Eng. https://doi.org/10.2118/190163-MS
8. Floris FJT, Bush MD, Cuypers M et al (2001) Methods for quantifying the uncertainty of production forecasts: a comparative study. Pet Geosci 7(S):S87–96. https://doi.org/10.1144/petgeo.7.s.s87
9. Barker JW, Cuypers M, Holden L (2000) Quantifying uncertainty in production forecasts: another look at the PUNQ-S3 problem. Soc Pet Eng. https://doi.org/10.2118/62925-MS
10. Gao G, Zafari M (2005) Quantifying uncertainty for the PUNQ-S3 problem in a Bayesian setting with RML and EnKF. Soc Pet Eng. https://doi.org/10.2118/93324-MS
11. Gao G, Zafari M, Reynolds AC (2006) Quantifying uncertainty for the PUNQ-S3 problem in a Bayesian setting with RML and EnKF. Soc Pet Eng. https://doi.org/10.2118/93324-PA
12. White CD, Royer S (2003) Experimental design as a framework for reservoir studies. Soc Pet Eng. https://doi.org/10.2118/79676-MS
13. Box GEP, Hunter JS, Hunter WG (2009) Statistics experimenters: design, innovation, and discovery. Wiley. ISBN: 978-0-471-71813-010. https://doi.org/10.1080/00401706.1979.10489788

Hydrocarbon Processing: Futuristic Issues and Challenges

Kunal Mehta

1 Introduction

The hydrocarbon industry is inherently embedded with a host of complexities and challenges on account of various reasons and characteristic of the inputs involved. The biggest challenge in upstream oil and gas industry is to assert where to locate oil; how deep and how far to drill to reach the reservoir; and how to design, construct, operate, develop and manage a field to deliver the greatest possible return on investment with the lightest, safest and smallest operational footprint.

The oil industry is increasingly focusing on unconventional hydrocarbon recovery due to the continuous growth in energy demand. Nowadays, exploitation of unconventional oils is significantly increasing but the problems associated with production of these crude oils are much more severe & complex due to their specific chemical composition. The operating physico-chemical condition varies widely from subsurface reservoir to refineries through surface installations and flow lines. The safe and efficient delivery of hydrocarbon crude oil from reservoir to refinery is the most important component of flow assurance. Asphaltene, being the most polar component, exhibits self-associating feature which promotes aggregation with subsequent increase in viscosity of the crude oil [1]. Experimental results clearly demonstrated that there is exponential relation between asphaltene and the viscosity of oil [2–5]. Flow assurance is one of the main challenges experienced due to high viscosity and pour point of the crude emulsion having relatively higher wax, resins, asphaltene, etc. The solution for the flow assurance of crude oil is provided by addition of certain chemicals such as solvent, pour point depressant and other specialized chemicals/nanomaterials. Finding the remedial solution for flow

K. Mehta (✉)
Head Surface Chemistry, Oil and Natural Gas Corporation Ltd., C-407, Vasudhara Bhawan, ONGC, Bandra (East), Mumbai 400051, India
e-mail: mehta_kunal@ongc.co.in; kunalmehtac5@gmail.com

U. K. Bhui (ed.), *Macromolecular Characterization of Hydrocarbons for Sustainable Future*, Green Energy and Technology,
https://doi.org/10.1007/978-981-33-6133-1_8

assurance of crude oil will help in optimizing production and leads to proper production development plan.

Brine generally accompanies crude oil during its recovery from subsurface reservoir and forms emulsion on the way to surface facilities. Most water-in-oil (W/O) emulsion is formed during crude oil production from subsurface reservoirs. The stringent water content requirement of processing units demands very effective demulsification due to technical and commercial implications. The presence of surface-active molecules particularly asphaltene, resins and inorganic solids in the crude oil at the oil–water interface produces a mechanically strong rigid viscoelastic stagnant film that resist droplet coalescence. The important factors that affect the stability of emulsion include solids, heavy fraction present in crude oil, temperature, pH, & salinity. W/O emulsions are complex, multi-component & multiphase system. The effective viscosity of these mixtures is a function of many parameters such as the presence of solids, the presence of emulsifiers, the viscosity of the individual phases, the droplet size and distribution, the density of constituents, the volume fraction of the constituents, the age of the emulsion as well as the temperature of the emulsion [6, 7]. Heavy oils contain a large amount of asphaltene and more generally of amphiphilic molecules. As mentioned, these molecules interact and reorganize at the interface, forming strong films that give very stable emulsions. Hence, crude oil demulsification can be very difficult and inefficient and it may require high residence times. Therefore, it is one of the most frequently discussed and studied subjects in order to find efficient industrial methods to easily and economically break emulsions and maintain crude quality for supply to refinery.

The H_2S mitigation has emerged as one of the most critical operational challenges and requires multipronged approach to circumvent the problem. As most of the hydrocarbon-producing fields are very old, effective effluent management and maintaining reservoir integrity have become burning issues. Moreover, maintaining the health and integrity of pipelines and operating system have become a perennial source of headache for operational team.

Each step from bringing the oil to surface, taking it to process installations and demulsification and meeting the norms of downstream processing while having an effective effluent management, maintaining the integrity of the reservoir and pipelines and also taking care of health, safety and environment norms and regulatory guidelines, require a constant effort under a precise strategy by experienced professionals. This contribution discusses an overview of the major challenges in hydrocarbon processing with reference to the Ahmedabad Asset, ONGC, including the recent advancement in the field and futuristic approaches.

2 Hydrocarbon Processing

At present, Ahmedabad Asset is carrying out exploration and production activities in 22 fields located within 70 km from Ahmedabad city. The asset produces approximately 3700 MT crude [8] and processes approximately 13000 m^3 of crude

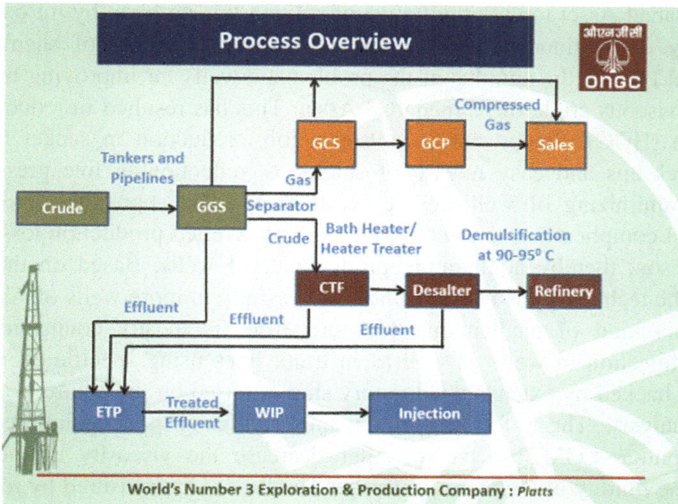

Fig. 1 Schematic representation of the basic overview of hydrocarbon processing in Ahmedabad Asset, ONGC

oil emulsion daily (including emulsion of other assets). The average water cut of the crude in Ahmedabad Asset is 50–70%. This means that the asset handles approximately a large quantity of effluent every day. Besides, the asset also has approximately 4.0 LCMD production of associated gas which is either utilized for internal purposes or sold to the local vendors. The process overview is presented in Fig. 1.

3 Operational Challenges and Technological Intervention

3.1 Flow Assurance

The crude oil being produced from different fields of Ahmedabad Asset poses the problem of poor fluidity, especially during winters and monsoon season due to its relatively high wax, resin and asphaltene content resulting in high pour point and high viscosity. On account of high pour point and high viscosity coupled with waxy depositions, the transportation of the crudes poses tremendous challenges for operational purpose. Since most of the wells are on artificial lift, there is pressure build up in the flow lines especially during winter/monsoon seasons with a resultant loss of production of crude oil. This problem also leads to many complications requiring remedial measures involving lot of downtime. Ahmedabad Asset is using pour point depressant (PPD)/flow improver (FI) for its operational need to improve the fluidity of highly waxy crudes of Viraj, Limbodra, Gamij, etc., especially in winter season for more than 10 years.

Ahmedabad Asset took the initiative of solving this problem by introducing the technology of continuous injection of an optimized mixture of chemicals like xylene and PPD in the annulus of the problematic wells for improving the fluidity of highly viscous crude of Ahmedabad Asset. This has resulted in reduced hot oil circulation (HOC)/effluent circulation/steam jobs, reduction in sucker rod pump (SRP) stuck-ups and easy release of stuck-ups, reduction in line pressures and avoiding/minimizing of well closure for flushing. The appreciable reduction in operational complications and remedial jobs have averted production losses during winter season, thereby augmenting productivity of wells. Based on the accrued benefits, the technology is being applied successfully in more wells of Ahmedabad Asset. This kind of challenging fluid properties is mostly countered through chemical injection in wells as well as in trunk lines using specific chemical formulations backed by extensive laboratory studies based on the nature of crude and SARA analysis. These formulations mainly utilising pour point depressants, xylene/naphtha, EGMBE, & demulsifier, decrease the viscosity of the oil and increase the fluidity of the emulsion/crude. The flow is also ensured by regular line flushing/cleaning jobs in wells and trunk lines, HOC/hot effluent circulation (HEC) jobs and regular pigging to remove any kind of deposition. The dispatch through RNW trunk line carrying the crude of Gamij, Ramol, Nandej and Wasna is maintained throughout the year to ensure the uninterrupted supply to desalter plant (Fig. 2).

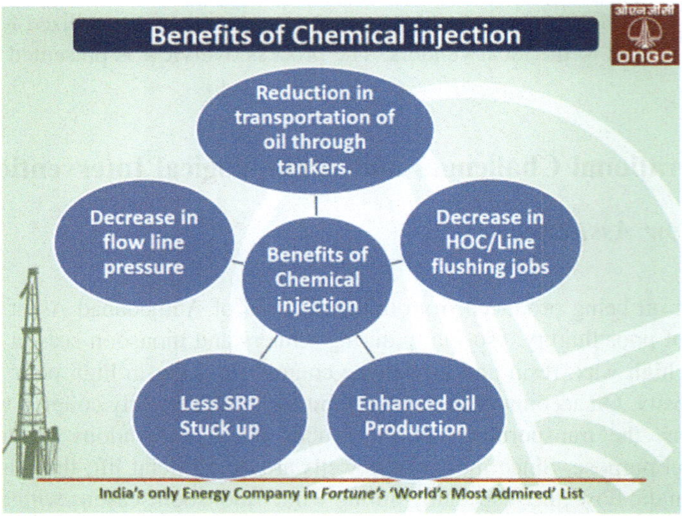

Fig. 2 Chemical injection and the benefits during hydrocarbon processing in Ahmedabad Asset, ONGC

Flow assurance in Gamij–Ramol Trunk line during winter (2016–17) using in-house formulation

Recommended laboratory formulation of PPD (500–1500 ppm), xylene (1500–4500 ppm) and demulsifier (100–150 ppm)

- Observed viscosity reduction of 55–75% of the blank.
- Observed pour point reduction of 3–6 °C.
- Average pressure range—Before treatment: 21.0–23.5 kg/cm^2. After treatment: 16.5–23.0 kg/cm^2.
- Significant reduction in back pressure after chemical injection line flushing and uninterrupted flow of crude oil to GGS.

3.2 *Demulsification*

The crude oils from different wells of different fields are sent via pipeline to common tank farm (CTF) in the mixed form. Crude oils from all the producing fields of Mehsana Asset and Ahmedabad Asset (Kalol, Nawagam, Ramol–Nandej and Gamij field) are stored in the storage tank of CTF Nawagam before processing. The stored oils from both the assets are pumped in a single line (without any specific rate or pattern) for treatment mainly desalting and demulsification. The water content is around 10–20% at initial stage. Desalter plant, Nawagam, processes about 13000 m^3 of combined crude emulsion with heterogeneous nature from Ahmedabad, Mehsana and Cambay Assets to achieve COSA specifications of quality of BS&W <0.2% prior to the dispatch to refinery. At CTF, the water content and salinity are reduced to below permissible level *(organic chloride <5 ppm, water cut (WC) <0.2%)* before being sent to refineries. Desalter, at times, encounters process upsets, especially in winters due to the sudden change in nature of crude as the proportion of the crude received from various fields changes, governed by numerous factors including change in production pattern. The main problems associated in Desalter Plant / CTF are the following:

1. Difficulty in reducing the water content of crude oil to below 0.2%.
2. Due to varying water content and nature of emulsion in the initial mixture of the crudes, appropriate dosing of the demulsifiers could not be properly done.

Long residence time (16–18 h) in the storage tank and higher water content (more than the permissible limit >0.2%) in the final processed crude oil lead to severe technical and financial impact to the organization. Moreover, reductions of the residence time of the treated crude oil may help reducing many operational, management and financial issues. The asset is constantly striving to achieve more effective demulsification at Group Gathering Station (GGS) and Central Tank Farm (CTF) to reduce processing load at desalter plant and to design a system that reduces the settling time and reprocessing of crude/emulsion.

3.3 H_2S Mitigation

Recently, high H_2S content is observed in some wells of Wasna, S/Kadi, Paliyad, Wadu and Nandej fields. The origin of H_2S in Ahmedabad fields may be biogenic, as opined by IRS, Ahmedabad, which may be due to activity of sulphate-reducing bacteria (SRB) in injection water [9]. This study steered the H_2S control strategy of the asset towards controlling SRB being one of the most effective ways to contain H_2S. The asset is constantly working towards the process of

- Regular dose of bactericide in injection water.
- Regular backwashing of all injectors and overnight soak with bactericides for better control of GAB/SRB.
- Regular backwashing of all injection lines.
- Regular microbiological analysis of injection water.
- Employing different types of bactericides to avoid development of resistance to particular type of bactericide.
- Identification of more effective bactericides.

At wells and GGSs, the H_2S is mainly contained by using caustic dosing being economical and effective as well. However, caustic may have long-term effect on the health of the wells and cause corrosion in flow lines and processing systems. Asset uses amine, aldehyde and non-amine non-aldehyde-type bactericide alternately to curb the microbial growth. Bactericide glutaraldehyde and THPS [9] showed very encouraging results in laboratory as per studies carried out by IRS, and procurement is underway for field application. Field trial of H_2S scavenger—non-triazine/triazine-based chemical—is under process at GGS Nandej. Other chemical systems and technologies for H_2S scavenging are being explored.

3.4 Effective Effluent Management

The effluent management in the asset is guided by the principle of quality enrichment, recycling and re-utilization. Ahmedabad Asset handles a large quantity of effluent per day that is treated in the effluent treatment plant (ETP). The treated effluent is utilized in water injection to maintain reservoir pressure or is disposed in the ED wells as per the requirement of the particular field and availability of infrastructure.

Before initiating water injection, compatibility studies are carried out at IRS to ensure that the health of reservoir is not affected. Water quality parameters like turbidity, TSS, DO, O&G are also maintained within permissible limits to preserve the reservoir characteristics and maintain the health of injection wells, thereby reducing the maintenance jobs like CTU interventions, acid jobs, BHA, etc.

The quality parameters of ETP and water injection plant (WIP) are under strict vigilance and are counter checked at ETP outlets, WIP inlets and outlets and well heads to ensure there is no deterioration in the quality of injected water during storage and in the flow lines. The water, at WIP (effluent/effluent and bore/bore) before being injected, is treated with oxygen scavenger, bactericide, corrosion inhibitor and scale inhibitor to maintain the health of the pipelines, well and reservoir as well.

The storage water tanks and the pipeline are subjected to shock treatment of bactericides and Sodium Hypochlorite to prevent any SRB generation and chocking due to biological deposits.

3.5 Maintaining Pipeline Integrity

Monitoring and assessment of the health of the old and ageing pipelines and operating systems to prevent accidents and leakages and sustain pipeline operations is one of the major focus areas of asset. This is also a regulatory requirement as per SOP for integrity assessment of cross-country pipeline issued by OISD, GOI, MOPNG, etc.

The pipeline health is monitored closely by the asset. The pig residue is analysed on quarterly basis by RGL including analysis of Fe_2O_3 and sulphur content along with sulphur content in crude in major feed lines. The results are analysed and studied by IEOT being the premier institute for corrosion monitoring and mitigation, and measures suggested are taken up on priority basis. There is also regular dosing of oil corrosion inhibitor/gas corrosion inhibitor in pipelines as per the recommendations of IEOT.

4 Conclusions and Way Forward

The Ahmedabad Asset is contributing to the energy needs of the nation from last 57 years. The major challenges facing the asset today are reversing the decline and enhancing production and augmenting surface facilities for flow assurance and processing of tough crude emulsions with higher water content from the mature fields. Moreover, issues like H_2S mitigation, effluent and reservoir management, maintaining health and integrity of pipeline and operational systems, are being addressed with priority. The asset is under process of implementing its first full-scale EOR project for augmenting production with the likelihood of generation of tougher emulsion in future—a challenge that can only be handled by skill development and technology upgradation.

For the issues in flow assurance, the asset is looking into the possibility of using chemicals such as wax and asphaltene dispersant/dissolvers for highly wax/asphaltene crudes and is also in contact with the academia for exploring an effective solution specific to the crudes of Ahmedabad Asset. Besides, a greener substitute of xylene is also being explored. The asset is carrying out field trial of new broad spectrum demulsifier for desalter as per the specification suggested by Regional Geochemical Laboratory (RGL), Vadodara [10]. An extensive R&D study is required, including crude characterization, development of a broad spectrum demulsifier suitable for wide variance in the nature of and development of low-temperature demulsifiers to suit the field conditions and operational constraints at GGS and CTFs. Possibility for cleaning of water injection lines with nanotechnology-based novel chemical is being explored. For effluent treatment, hiring of mini-ETP for the treatment of Gamij field effluent generated at Gamij GGS I is under process. The effluent otherwise used for the flushing purpose will be utilized in water injection in Gamij field.

These issues facing the industry are mainly associated with the different fractions of the crude oil components like wax and asphaltene. So, in present scenario, greater thrust on R&D is required for detailed molecular-level characterization of crude oil along with in-depth analysis of these issues to address the challenges and mitigate the problems. Also, the collaboration between industries and academia is vital to develop novel chemicals and technologies which may help to contain the problems effectively.

Acknowledgements The author is thankful to Mr. Debasish Basu, Ex. Executive Director, ONGC; Mr. Ravi Mathur, Ex. GGM(P); Mr. G.M. Das CGM(Chemistry); and all the members of Chemistry and Production Section of ONGC, Ahmedabad, for their support and encouragement. Thanks are also due to Mrs. Vartika Roy, Suptdg. Chemist, ONGC, Ahmedabad Asset, for extending all support and assistance in preparing this work.

References

1. Guzmán JD, Betancur S, Carrasco-Marín F, Franco CA, Nassar NN, Cortés FB (2016) Importance of the adsorption method used for obtaining the nanoparticle dosage for asphaltene-related treatments. Energy Fuels 30(3):2052–2059
2. Pierre C, Barré L, Pina A, Moan M (2004) Composition and heavy oil rheology. Oil Gas Sci. Technol. 59(5):489–501
3. Argillier JF, Coustet C, Henaut I (2002) Heavy oil rheology as a function of asphaltene and resin content and temperature. In: SPE international thermal operations and heavy oil symposium and international horizontal well technology conference. Society of Petroleum Engineers
4. Hénaut I, Barré L, Argillier JF, Brucy F, Bouchard R (2001) Rheological and structural properties of heavy crude oils in relation with their asphaltenes content. In: SPE international symposium on oilfield chemistry. society of petroleum engineers

5. Ghanavati M, Shojaei MJ, SA AR (2013) Effects of asphaltene content and temperature on viscosity of Iranian heavy crude oil: experimental and modeling study. Energy Fuels 27(12): 7217–7232
6. Farah MA, Oliveira RC, Navaes-Caldas J, Rajagopal K (2005) Viscosity of water-in-oil emulsions: variation of temperature and water volume fraction. J Petrol Sci Eng 48:169–184
7. Sefton E, Sinton D (2010) Evaluation of selected viscosity prediction models for water in bitumen emulsions. J Petrol Sci Eng 72(1–2):128–133
8. Annual Report and DPR, Ahmedabad Asset
9. Kukreti V, Rana DP, Kiran K (2017) Biocide optimization for south Kadi and Wasna fields of Ahmedabad Asset' MEOR Lab. IRS, Ahmedabad
10. RGL Recommendation report BDA/RGL/FFL/Desalter Plant/2018-19 dt 05/09/18

Macromolecular Characterization of Petroleum Crude for Deeper Insights into Acute Flow Problems: A Case Study from Lakwa Oil Field, Assam, India

Partha P. Das, Tapan K. Paul, Shyamalee Gogoi, and Farhat Arifi

1 Introduction

Bioenergy or renewable energy derived from biological sources had been the staple source of power and heat prior to the Industrial Revolution. Since then, economic developments have largely relied on fossil fuels and global energy demand has been increasing in lockstep with environmental damages due to fossil fuel use. Needless to say, continued dependence on fossil fuels put a crimp in the global efforts for reducing greenhouse gas emissions that contribute to global warming. Major fillip for the development of bioenergy had therefore come from the search for alternatives to fossil fuels, particularly those used in transportation, and there has been an unprecedented scaling up of the production of biofuels over the last two decades [1]. In spite of an ambitious energiewende worldwide to a future based on renewable, sustainable and low carbon economies, there is no prospect of a complete substitution of fossil fuels on the horizon. It would therefore be prudent to fall back on innovative solutions to the existing problems in the upstream as well as downstream hydrocarbon industries in a much greener way.

P. P. Das
Oil and Natural Gas Corporation Ltd., Assam Asset, Nazira 785686, Assam, India
e-mail: ppd0203@gmail.com

T. K. Paul · S. Gogoi · F. Arifi
Surface Chemistry, Oil and Natural Gas Corporation Ltd., Assam Asset, Sivasagar 785640, Assam, India
e-mail: gogoi_shyamalee@ongc.co.in

F. Arifi
e-mail: arifi_farhat@ongc.co.in

T. K. Paul (✉)
A-303 Sopan Apartments, Opposite ICICI Bank & Central Bank of India, New C G Road, Chandkheda, Ahmedabad 382424, Gujarat, India
e-mail: drtpall@gmail.com

U. K. Bhui (ed.), *Macromolecular Characterization of Hydrocarbons for Sustainable Future*, Green Energy and Technology,
https://doi.org/10.1007/978-981-33-6133-1_9

Petroleum crude is a complex mixture of several constituents like the non-polar saturates/paraffins, aromatics and polar organic compounds viz. asphaltenes and resins along with some metallic elements (porphyrins). Its nature and composition vary widely from reservoir to reservoir. Among the indigenous surfactants contained in the crude oils, asphaltenes and resins are known to play important roles in the formation and stability of emulsions. Asphaltenes are polar species formed by condensed polyaromatic structures containing alkyl chains and heteroatoms (such as O, S and N), while resins are mainly naphthenic aromatic hydrocarbons or aromatic ring systems with acyclic chains (Fig. 1). Chemically, asphaltenes are disc-shaped polycyclic molecules having a tendency to form stacked aggregates or micelles that sets them apart from other constituents in the crude. Asphaltene aggregation is the cause of complex nonlinear effects in several phenomena such as adsorption at solid surfaces, precipitation, fluid's rheology, emulsion stability, etc. Asphaltenes, as natural emulsifiers, stabilize crude oil emulsions through the formation of a viscoelastic, physically cross-linked network of asphaltenic aggregates at the oil–water interface [2].

The currently accepted picture of crude oil (Fig. 2) shows the asphaltene molecules as being surrounded by resins through their polar end groups with their paraffinic ends acting as tails to make the transition to the relatively non-polar bulk of the oil containing aromatics and saturates [3–6]. Within the complex mixture in a crude oil, the high molecular weight polar components particularly the asphaltenes cause serious problems during its production and refining. Both upstream and downstream processes involve variation of temperature and flow conditions, which cause the precipitation of waxes and asphaltenes leading to many problems. Asphaltenes are now the current focus of scientists for their role in building up of viscosity from subsurface reservoir condition to surface flow lines and storage conditions.

The asphaltene molecules remain dispersed in oil as a colloidal system with asphaltenes and resins as the dispersed phase and saturates and aromatics as the continuous phase, forming a micellar structure in which the core of the micelle is formed by one or several asphaltene molecules. This core is encircled by the resin molecules which in turn are surrounded by aromatics ensuring a progressive

Fig. 1 Examples of molecular structures of some components present in crude oil: **a** Asphaltenes **b** Resins and **c** Naphthenic acids

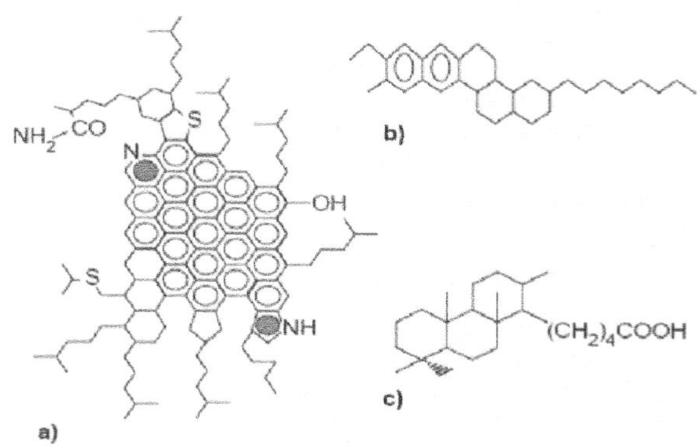

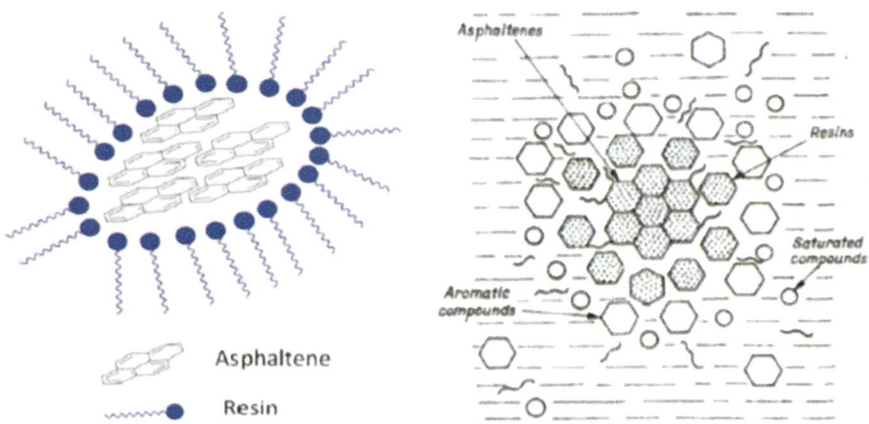

Fig. 2 Schematic view of the orientation of resins around asphaltene molecules in the emulsion

transition to the bulk where saturated hydrocarbons are usually predominant [7]. Resins have a strong tendency to associate with asphaltenes and are crucial in holding the asphaltene micelles in suspension. On the other hand, aromatics act as a bridge between the mixed micelle and saturates, making the mixed micelles effectively dispersed in saturates [8]. Resins thus form a protective shield around the asphaltenes [9], removal of which results into their precipitation [10]. Disturbing the equilibrium of attractive forces between resins and asphaltenes occasioned by factors like change in pressure due to higher liquid withdrawal, among others, may lead to precipitation of asphaltenes at the well bore. The result may be the formation of organic skin choking the well bore and reduced liquid influx.

The study of the crude oil composition (saturates, aromatics, resins and asphaltenes) demonstrates that the stability of asphaltenes in crude oils is a phenomenon that is related to the fractions of the components and the influence of each component on the other. To recognize the asphaltene deposition potential of petroleum systems, the concept of colloidal instability index (CII) was introduced [11]. It is defined as

$$\mathrm{CII} = \frac{\mathrm{Saturates} + \mathrm{Asphaltenes}}{\mathrm{Aromatics} + \mathrm{Resins}}$$

Crude oil with a CII value below 0.7 is considered to be stable while that with a value higher than 0.9 is unstable. Organic skins from wellbores can be removed by stimulation jobs with solvent mixtures of high aromaticity. Formulations are designed with aromatic compounds like toluene, xylene and naphthalene to make CII value close to 0.7. The formulation is squeezed into the formation and kept for soaking for about 12 h under pressure. Solvent stimulation jobs executed in Assam Asset have been showing encouraging results.

Seamless flow of well fluid from the reservoir to the group gathering stations (GGS), as well as transportation of processed crude from the central tank farms (CTF) to refineries through pipelines, is the most important aspects of flow assurance. When it comes to flow properties of any crude, pour point and cloud point/ wax appearance temperature (WAT) are two very important crude oil parameters that have to be taken into consideration. Pour point is defined as the temperature below which the crude oil ceases to flow, whereas cloud point or WAT is the temperature at which the first wax crystal appears on cooling the crude.

Solid deposition is a common problem in crude oil production, transportation and storage operations causing substantial economic losses for the petroleum industry. Much as the high molecular weight n-alkanes (n-paraffins) are the main components in wax deposits, long iso- and cycloalkanes and high molecular weight polyaromatics (asphaltenes/resins) are also often found in these deposits. This is corroborated by complex agglomerates of asphaltenes and waxes found in the storage tank bottom sediments [12].

At a thermodynamically suitable temperature (WAT), the paraffins crystallize out of the solution and start to build a 3D network with a complex morphology which entraps the rest of the fluid inside. This is the so-called wax-oil gel which becomes harder with time. The gelling significantly increases the fluid viscosity and changes the rheological behaviour to non-Newtonian [13]. The wax crystals separating out from the crude deposit on the inner walls of pipelines choking the flow lines that may need line flushing/mechanical removal, or, in time, lead to shut down of well operations. Moreover, production tubings also get choked due to wax deposition on inner walls of the tubulars. The most commonly used technique for clearing the choked flow lines is by flushing of the lines at regular intervals. On the other hand, wax deposits on the tubulars can be cleared by either hot oil circulation (HOC) or by scraping operations. Many a time the scraper is found to be left in the well resulting into expensive fishing operations. The outcome is loss of production and, by extension, loss of revenue. Another useful technique for mitigation of wax deposition problems in pipelines is by electrical heating of flow lines at regular intervals. The same technology can be gainfully utilized for elimination of wax/ paraffin deposits in tubulars by downhole heating cables. Another major problem with waxy crudes is the reusing of the pipelines after prolonged shut-down where the cooled oil, eventually, develops a strong gel structure with high yield stress.

The role of asphaltene during wax crystallization is very complex, with involvement of several parameters some of which are yet to be clearly understood. The polar asphaltenes remain dispersed rather than dissolved in the crude oil matrix. However, as shown in Fig. 1, their association with resins maintains them in a single phase. Once the asphaltene concentration reaches the critical value, where the complete dispersion cannot be maintained by the rest of the oil, the fragile equilibrium is broken, and asphaltene molecules tend to flocculate among themselves, crystallizing out of the crude in the fullness of time. The flocculated asphaltenes may provide the crystallization sites for paraffins, and they themselves may co-precipitate with the waxes as well [14]. Under such circumstances, it is very difficult for waxes to build a proper wax network within this highly flocculated system because of spatial interference from asphaltenes. This makes waxes crystallize on the asphaltene particles producing an unorganized asphaltene-paraffin composite rather than a proper wax network. The upshot is depressed yield stress and WAT. Asphaltenes can thus act as natural flow improvers.

Several remedies are suggested to mitigate the problem, some of which are (a) subjecting the crude to a special heating–cooling treatment cycle to modify the wax crystal structure and controlled thermal conditioning to take advantage of the natural pour point depressing effect of resin and asphaltene present in oil [15–17], (b) application of microwave and ultrasound irradiation [18], (c) application of magnetic field [19], (d) lining and coating pipelines with fibre-reinforced plastics to reduce the wettability of paraffins with walls [20] and (e) covering inner wall surfaces of pipelines with polypropylene to inhibit wax deposition [21]. However, these are not widely accepted in the oil industry due to some drawback or the other in each of these methods and chemical treatment with polymeric additives by modification of the wax crystal habit is often found to be the most economically viable solution.

Pour point depressants (PPDs) are polymeric additives having a non-polar paraffin part along with a polar component in the form of acrylate, methacrylate, acetate, etc., and are completely soluble in crudes. These are present in a comb-like shape in which the pendant chains act as the paraffin part and co-crystallize with wax, while the backbone and polar end groups sterically hinder such possibilities of crystallization. If the crystalline parts of the additives are comparable in structure to the paraffin waxes present in a given crude oil, it serves as a nucleating site for them. This results into modifying the habit of wax crystals, which is generally orthorhombic, to a compact pyramidal form. In consequence, the crystals are prevented from agglomerating and forming a gel-like structure [22]. With modified shapes, the wax crystals cannot intergrow and interlock with each other and remain dispersed in the system. The outcome is lowering of the pour point to make the flow of crudes easier.

The higher viscosity of crude is occasioned by the presence of H-bonding between asphaltene and resin molecules as well as by overlapping of sheets of aromatic rings of resins and asphaltene molecules. Crude viscosity can therefore be reduced by using suitable additives like PPD/flow improvers that can break these H-bonds. These additives compete for making stronger H-bonds with resins or asphaltenes replacing existing H-bonds between asphaltene and resin molecules themselves. Moreover, these polymeric substances sterically hinder mutual overlapping of the aromatic ring planes of resins and asphaltenes and hence reduce viscosity.

In keeping with the increasing needs for migration towards green chemistry, presently chemical treatment of crude in ONGC, Assam Asset, is done with PPD/flow improvers in a way to minimize the generation of by-products or wastes that are hazardous or not eco-friendly. The PPD/flow improvers are therefore used in liquid formulations as liquid flow improvers (LFI) to make them completely miscible with the crude without any change in composition in transit through pipelines to the downstream industries.

Table 1 Characteristics of crudes of the major fields of ONGC, Assam Asset

Parameters	Field			
	Rudrasagar	Geleky	Lakwa	Lakhmani
API gravity	22.6–33.7	27–31	25.9–28.8	29.1–29.4
Pour Point (°C)	21–30	27–33	18–42	21–36
Asphaltene (% w/w)	0.4–1.3	0.6–0.7	0.2–1.0	0.2–0.7
Resin (% w/w)	10–20	11–14	9–15	9–15
Wax (% w/w)	7.8–13.9	9–19	7–17	7–11

1.1 Crude Oil Characteristics of Assam Asset

The Assam Asset comprises four major production areas viz. Lakwa, Lakhmani, Rudrasagar and Geleky fields. Crudes of the Asset, with high wax and asphaltene contents of 7% to as high as 20% and 0.2–1.5%, respectively (Table 1), are very viscous in nature causing serious flow problems. The situation gets aggravated in monsoon and more in winter months due to high pour points (30–42 °C) along with low temperature (30–35 °C) of crudes at the well head.

Production from oil wells is associated with variation in flow conditions like liquid withdrawal rate, temperature, pressure, etc., that may lead to precipitation of organic components of the crude like waxes and asphaltenes. These solids may get deposited at the well bore, in production tubings and flow lines causing constriction and, by extension, reduced throughput. Tubing deposits are generally cleared by mechanical scraping, while choked flow lines need to be flushed. These cleaning operations, however, turn out to be quite expensive as they involve loss in production due to interruptions in well operations. Treatment of crude with PPD/Flow improvers is found to be the most economically viable option.

High wax and asphaltene contents have been the bane of production operations in Lakwa and Geleky fields and, in part, in Charali area of Rudrasagar and Kuwargaon area of Lakhmani fields of Assam Asset. There is significant loss of production, particularly in winter months every year due to frequent tubing scraping and flow line flushing operations. To mitigate the Asset's unrelenting flow problems, chemical dosing pumps have been installed at various well sites, in which well-specific flow assurance formulations with Xylol and PPD designed at the field laboratories are currently being dosed in the well annulus/flow lines.

1.2 Case study Well of Lakwa Field (Safrai# 10)

To mitigate the Asset's recurring flow problems, ONGC, Assam Asset has implemented her own flow assurance formulations with the PPD (Deva flow 009) and Xylol in a number of wells with a history of frequent flushing/scraping operations. These, however, were not found to be effective enough in reducing the pour point or improving flow properties. In this backdrop, the exploratory well Safrai# 10 of Lakwa field of Assam Asset with persistent flow problems was chosen for pilot studies for identification of low temperature dosed PPD/flow improvers.

2 Materials and Methods

(i) Collection of crude oil sample:

Approximately 10 litres of untreated crude oil sample from Safrai# 10 well head was collected and was homogenized and conditioned at 60 ± 2 °C for 1 h to dissolve any wax particles.

(ii) Physico-chemical characterization of the Safrai# 10 crude:

Detailed physico-chemical characteristics of the emulsion were determined adopting tests and analytical techniques as per standard IP/ASTM methods as mentioned below:

(a) The water content of the crude sample was determined using Dean and Stark method using toluene as refluxing solvent as per ASTM D 4006.
(b) The density/API gravity and pour point were determined as per ASTM D 1298 and ASTM D 5853, respectively.
(c) The viscosity profile of the crude was studied using Wells-Brookefield Cone and Plate Rheometer (Model: RV DV-III Ultra, with CPE-41 cone spindle) at a shear rate of 25 s^{-1}. The Rheometer and a Brookefield refrigerated external circulating water bath (Model: TC 502P) are connected to a PC, and both are operated and controlled in unison by proprietary Brookfield 'RheoCalc' software.
(d) ARW contents of the crude were determined gravimetrically at the Formation Fluid Lab, Regional Geoscience Laboratories, ONGC, Sivasagar, Assam.

(iii) Dose optimization studies:

Chemicals required: Commercially available LR grade Xylene, Merck chemicals, was used as the solvent. The five different variants of LFIs were received from three vendors namely, M/s Krishna Antioxidants Pvt Ltd, Mumbai, M/s Devadrill Tech (I) Ltd, India and Nalco-Champion Dai-ichi India Pvt Ltd, Pune.

(a) Preparation of stock solution: 2% (w/v) solution of LFI was prepared in Xylol and stored as the stock solution in a volumetric flask.
(b) The untreated crude was doped with LFI at doses ranging from 500 to 3000 ppm at a doping temperature of 40 °C.

3 Results

I. The primary oil analysis as well as dose optimization studies at a dosing temperature of 60 °C with Devaflow 009 was done at the field laboratory at Lakwa CTF, while ARW (asphaltene-resin-wax) analysis was done at ONGC RGL, Sivasagar. The physico-chemical characteristics of Safrai# 10 crude are tabulated in Table 2.

II. The second part of the studies viz. identification of suitable low temperature dosed LFIs to find a solution to the nagging flow problems at Safrai# 10 was carried out by the R & D group of Oil India Ltd., at Duliajan, Assam, India [23] on Assam Asset's request.

Five different LFIs viz. Devaflow 009, Devaflow 108 LA Cristol SGT 07, Cristol LFI 12 and Cristol SGT 12 were used for the studies on Safrai# 10 crude. Initial experiments were done with Devaflow 009 and Devaflow 108 LA available with OIL Duliajan, but both were found to be ineffective/marginally effective on Safrai# 10 crude. On the contrary, the remaining three LFIs, namely Cristol SGT 07, Cristol LFI 12 and Cristol SGT 12, were found to improve the flow properties of Safrai# 10 crude significantly. Even at a much lower doping temperature of 40 °C, there was a reduction in pour point by 15 °C along with a significant lowering of crude viscosity. The efficacy of the LFIs was found to improve further when doped at 50 °C. But keeping the low well head temperature at Safrai# 10 in view, particularly during winters, the lower doping temperature of 40 °C is considered to be more worthwhile.

As can be seen from Table 3, Cristol SGT 07 at a dose of 2000 ppm lowers the pour point by ~50% to 18 °C, even as its apparent viscosity remains much higher than that of the other two LFIs (Table 4). On top of it, its congealing as well as melting points are also somewhat higher. All in all, Cristol SGT 07 is not a good choice as an LFI. On the other hand, Cristol LFI 12 as well as Cristol SGT 12 reduce the pour point by ~40% to 21 °C at a dose of 2000 ppm and is deemed to be better LFIs for Safrai# 10 crude. But Cristol SGT 12 has an added advantage of lower melting as well as congealing points by 4 °C vis-à-vis Cristol LFI 12 and would therefore be more convenient in handling and application at well head, a fortiori during the winter months.

Table 2 Physico-chemical characteristics of Safrai # 10 crude

Parameters	Water content	API gravity	Pour Point (°C)	Asphaltene (% w/w)	Resin (% w/w)	Wax (% w/w)
	6%	32.2	33–36	2.6	20	14.2

Table 3 Rheology and pour point studies on Safrai# 10 crude with LFI

Name of LFI	Dosing rate (ppm)	Pour Point (°C)	Apparent Viscosity (cP) at 25 s^{-1}						
			33 °C	30 °C	27 °C	24 °C	21 °C	18 °C	15 °C
Doping temperature: 40 °C									
Untreated Crude	0	33	526	NM	–	–	–	–	–
Cristol SGT 07	500	27	42	156	244	624	NM	–	–
	1000	24	61	70	101	173	430	NM	–
	1500	18	63	71	92	156	408	NM	–
	2000	18	65	69	100	141	307	931	NM
	3000	18	39	42	67	91	168	320	858
Cristol LFI 12	500	30	52	96	218	538	NM	–	–
	1000	21	49	59	71	109	182	414	NM
	2000	21	51	57	80	105	160	279	504
	3000	18	48	57	80	100	159	285	599
Cristol SGT 12	500	30	59	105	248	573	NM	–	–
	1000	24	79	84	107	136	220	422	NM
	2000	21	68	79	104	130	190	317	563
	3000	24	84	99	126	178	285	538	NM

*NM (not measurable) implies apparent viscosity (AV) > 982 cP (out of instrument range)

Table 4 Physical properties of LFI chemicals used in the study

Sl. No.	Brand name of the LFI	Congealing point (°C)	Melting Point (°C)	Apparent viscosity (cP) of the LFI product at 20 °C/25 s^{-1}
1	Cristol SGT 07	16	26	59
2	Cristol LFI 12	12	22	20
3	Cristol SGT 12	08	18	26

4 Discussions and Conclusions

The nature and content of wax in crudes vary widely from reservoir to reservoir, there cannot be any one-size-fits-all flow assurance solution, and a deeper insight into the nature and intensity of the flow problems is very much warranted. Flow assurance formulations need therefore be tailor-made for each type of crude based on its macromolecular characterization followed by design and synthesis of PPD/ flow improver based on Similia Similibus Curantur principle. The crystalline parts of the synthesized additives would then be analogous to the actual structure of the paraffin waxes present in a given crude oil and would serve as nucleating sites. The result would be modification of wax crystal habit preventing agglomeration of wax crystals and formation of tenacious gels with high yield points and significant lowering of pour points. Of the three LFIs studied, Cristol SGT 07 shows best results on the score of pour point reduction, but on the whole, it is not a suitable LFI for Safrai# 10 crude due to its much higher apparent viscosity. In contrast, Cristol LFI 12 and Cristol SGT 12 have performed reasonably well on Safrai# 10 crude with a high wax content of 14% on both counts—pour point and apparent viscosity—and at a much lower dosing temperature of 40 °C. That said, extensive work on synthesis and evaluation of performance-based polymeric wax crystal growth inhibitors based on macromolecular characterization of Assam Asset crudes is the way forward for an overarching solution to the persistent flow issues plaguing the Asset.

The PPD/flow improver, for the most part, is formulations with solvents/ chemicals with low flash-points and accordingly, low temperature dosed additives have several advantages over high temperature dosed ones. A higher dosing temperature of 60 °C needs additional heating facilities and hence additional fire safety precautions for heating the crude prior to dosing. To be sure, low temperature dosed ($\sim$40 °C) LFIs can do away with these extra burdens and are ready to be dosed directly to the flow lines.

To mitigate the tenacious flow problems encountered in Safrai# 10 and a large number of other wells, ONGC Assam Asset is eagerly looking for suitable chemical solutions to constrained flow arising out of asphaltenes and wax deposition.

Acknowledgements The authors are grateful to the ED-Asset Manager, ONGC Assam Asset, Nazira, India, for his encouragement for taking up the pilot studies on the exploratory well Safrai# 10 and for his valuable advices in the course of the work. Thanks are due to Mr S. Sinha, GM, R & D Department, Oil India Ltd., Duliajan, Assam, and his team for carrying out the useful studies on Safrai# 10 crude.

References

1. Popp J, Lakner Z, Harangi-Rakos M, Fari M (2014) The effect of bioenergy expansion: food, energy and environment. Renew Sustain Energy Rev 32:559–578
2. McLean JD, Kilpatrick PK (1997) Effects of asphaltene aggregation in model heptane–toluene mixtures on stability of water-in-oil emulsions. J Colloid Interface Sci 196(1):23–34
3. Abdel-Raouf MES (2012) Factors affecting the stability of crude oil emulsions. In Crude oil emulsions-Composition stability and characterization, Intech
4. Hammami A, Phelps CH, Monger-McClure T, Little TM (2000) Asphaltene precipitation from live oils: an experimental investigation of onset conditions and reversibility. Energy & Fuels 14(1):14–18
5. Hammami A, Ferworn KA, Nighswander JA, Over SVERRE, Stange E (1998) Asphaltenic crude oil characterization: An experimental investigation of the effect of resins on the stability of asphaltenes. Pet Sci Technol 16(3–4):227–249
6. Haskett CE, Tartera M (1965) A practical solution to the problem of asphaltene deposits-hassi messaoud field algeria. J Petrol Technol 17(04):387–391
7. Tissot BP, Welte DH (1984) Petroleum formation and occurrence, 2nd edn. Springer-Verlag, Berlin

8. Fan T, Wang J, Buckley JS (2002) Proceedings of the SPE/DOE improved oil recovery symposium, SPE 75228. Tulsa, Okla, USA
9. Andersen SI, Speight JG (2001) Pet Sci Technol 19(1):1–34
10. Al-Kafeef SF, Al-Medhadi F, Al-Shammari AD (2005) SPE Prod Oper 20(2):126–132
11. Amro MM (2005) J Petrol Sci Eng 46(4):243–252
12. Carbognani L, Orea M (1999) Studies on large crude oil alkanes. I. High temperature liquid chromatography. Pet Sci Technol 17:165–187
13. Singh P, Fogler HS, Nagarajan N (1999) J Rheol 43:1437–1459
14. Kriz P, Andersen SI (2005) Effect of Asphaltenes on crude oil wax crystallization. Energy & Fuels 19:948–953
15. Ramchandran V, Ostlund JA, Chawla H, Wattana P, Magnus N, Fogler HS (2003) The effect of asphaltenes on the gelation of waxy oils. Energy & Fuels 17:1630–1640
16. Thompson DG, Taylor AS, Graham DE (1985) Emulsification and demulsification related to crude oil. Colloids Surf 15:175–189
17. Chanda D, Sarmaha A, Borthakur A, Rao KV, Subrahmanyam R, Das HC (1998) Combined effect of asphaltenes and flow improvers on the rheological behavior of Indian waxy crude oil. Fuel 77:1163–1167
18. Bjorndalen N, Mustafiz S, Islam MR (2003) Numerical modelling of petroleum fluids under microwave irradiation for improved horizontal well performance. Int Commun Heat Mass Transfer 30:765–774
19. Rocha NO, Gonzalez G, Vaitsman DS (1998) Magnetic field effect on paraffin deposition. Quin Nova 12:11–17
20. Slack M (1992) Polyethylene liners for internal rehabilitation of oil pipelines. Mater Perform 31:49–52
21. Quintella CM, Musse APS, Martha TPO, Castro S, Mikelsons L, Watanabe YN (2006) Polymeric surfaces for heavy oil pipelines to inhibit wax deposition: PP, EVA 28 and HDPE. Energy & Fuels 20:620–624
22. Soni HP, Kiranbala, Bharambe DP (2008) Performance-based designing of wax crystal growth inhibitors. Energy & Fuels 22:3930–3938
23. Panigrahi R (2019) Flow assurance study of "Safrai-10" crude oil: the final report. R & D 26/355/2019

Feasibility Study of Crude Oil Asphaltenes as Light-Harvesting Materials for Organic Photovoltaics: Light Absorption Characteristics of the Thin Film with P3HT

Uttam K. Bhui, Abhijit Ray, and Madhav Pravinbhai Joshi

1 Introduction

A photovoltaic solar cell provides electrical energy from the solar energy which is clean, environment friendly in terms of sustainable future. Organic photovoltaic (OPV) devices are comprised of light absorbers that consist of organic semiconductors which are used for light absorption and charge transport to produce electricity from sunlight. Organic solar cell comprised of an organic absorber sandwiched between a hole collecting interlayer in one side with high work function where in opposite side it is in contact with an electron-collecting interlayer with low work function. The organic absorber must have the following functions: (1) it harvests a fraction of sunlight and converts optical radiation into mobile electron carriers; (2) it stores holes and electrons in spatially separated layers or domains, to yield a photovoltage; and (3) it provides charge transport pathways for carriers towards the hole- and electron-collecting interlayers to yield a photocurrent. Under illumination, the organic semiconductor layers absorb light, creating excitons that dissociate at donor–acceptor interfaces to form mobile holes and electrons

U. K. Bhui (✉) · M. P. Joshi
School of Petroleum Technology, Panddit Deendayal Energy University (PDEU, Formerly Pandit Deendayal Petroleum University-PDPU), Raisan, Gandhinagar, Gujarat 382007, India
e-mail: Uttam.bhui@spt.pdpu.ac.in

M. P. Joshi
e-mail: Madhav.joshi151@gmail.com

A. Ray
Department of Solar Energy, School of Technology, Pandit Deendayal Energy University (PDEU, Formerly Pandit Deendayal Petroleum University-PDPU), Raisan, Gandhinagar, Gujarat 382007, India
e-mail: Abhijir.Ray@sse.pdpu.ac.in

U. K. Bhui (ed.), *Macromolecular Characterization of Hydrocarbons for Sustainable Future*, Green Energy and Technology,
https://doi.org/10.1007/978-981-33-6133-1_10

which lead to the formation of an intermolecular charge-transfer complex [1–3] that simultaneously influences the photogeneration of mobile charge carriers, i.e. the current.

Organic photovoltaics (OPV) are a relatively new technology that has attracted tremendous research interest due to its strong potential, inexpensive and abundant supply on the earth. The OPV materials are usually electron acceptors which essentially transport the electrons after the exciton is generated by the absorption of light. Due to the high optical absorption coefficient of organic molecules, large amount of light could be absorbed with a small amount of materials where their low conductivity leads to an overall low electrical resistance for efficient device operation. New materials with optimized optical and electrical properties are synthesized for the most efficient solar cells that are built from building blocks with varying degree of electron-rich and electron poor moieties with wider optical absorption spectrum. Many electron-acceptors organic materials have been developed so far, including conjugated polymers as well as small molecules [4]. Conjugated molecular species such as polymer, dyes which are capable of absorbing light and conducting charge currently are of great interest as photovoltaic materials [5, 6]. Attraction of such materials primarily lies on their natural availability, low-cost bulk solution processing with less challenging manufacturing environment compared with crystalline silicon wafer production. Only a few electron acceptor materials out of them have shown reasonable performance, such as fullerenes and their derivatives. However, most of such compounds are expensive. In all organic solar cells, the degree of electronic coupling between the donor-like material used for hole transport, and acceptor-like material used for electron transport, plays a critical role. Therefore, search for the new inexpensive electron-acceptor materials is of primary interest in the OPV field.

Over the years, the cost of electricity produced by PV reduced significantly, but it is still notably higher compared to electricity produced by conventional power plants. This means for making a significant contribution to the global energy production, the cost of solar electricity needs to undergo substantial further cost reduction. There has been remarkable progress in the field of organic solar cell showing double digit efficiencies with reasonable lifetime. Bulk heterojunction concept by Yu et al. [7] with blending of donor–acceptor molecules provides many folds increase in conversion efficiency of organic photovoltaic. Bulk heterojunctions are commonly created by forming a solution containing two components in general with one belongs as a donor material and another acceptor. Well-defined morphologies of the blended donor (D) acceptor (A) moieties could provide suitable interfaces for excitons dissociation followed by preferred pathways connected to the corresponding electrodes. Combination of components in bulk heterojunction can work properly if the donor and acceptor have a proper HOMO-LUMO. An annealing step allows separation of the mixed two phases where the two components will self-assemble into an interpenetrating network connecting the two electrodes. Manipulation of the morphology of the blended film could be achieved through thermal annealing, varying the composition of the processing solvent mixture, chemical structure variation of the active materials, for improvement of the performance. There is still no full consensus how the ideal nanomorphology of the bulk heterojunction should look like when it comes to the efficiency optimization of a given material combination. In most of the models, the power conversion efficiency of organic solar cells not only depends on the optical blending but also in difference of the electron affinities of the donor–acceptor materials. However, before large-scale commercialization, further improvements, especially in the power conversion efficiency, are required.

Asphaltenes are an organic moiety having high molecular weight polycyclic aromatic hydrocarbon (PAH) with fused aromatic ring having heteroatoms and metal ions in different structural position [8, 9]. This type of PAH allows for diversifying the structural configuration with reactivity and electronic properties [10–12]. π–π interaction phenomenon for self-assembly make PAH an efficient molecular material for opto-electronic applications particularly for organic field effect transistors (OFETs) or organic solar cells (OSCs) [13–15]. Organic semiconductors can be doped intrinsically to be used as light absorber in dye-sensitized solar cell (DSSC). Abujnah and Chianelli [6] worked with different fraction of asphaltenes for dye-sensitized solar cell (DSCC) and proved that for at least one part of the fraction or asphaltene in whole can be efficiently utilized as good sensitizer in DSCC with a total overall efficiency of 4.55–7.0%. Asphaltenes are materials that could fulfil the requirement of absorbing a broad range of visible and near-infrared region light [16, 17] capable of conducting charge [18], stable and naturally abundant in crude oil and coal. Asphaltenes cause a lot of problems right from exploitation from the reservoirs, transportation through pipelines as well as in refining, and abundantly available and undesirable part of natural crude oils. With an aim for effective utilization of undesirable and problematic asphaltene and to understand the applicability of asphaltenes as light absorber for OPV, the present work has been undertaken which clearly demonstrates suitable optical characteristics of thin film prepared asphaltene blended with P3HT for organic photovoltaics.

2 Materials and Methods

2.1 Materials

Crude oil samples are collected from well head of different producing fields of Ahmedabad Asset of ONGC for separation of asphaltenes. One tar sample (Sample no 1) is also used in this study which actually contained high amount of asphaltenes. P3HT is purchased from market. Different solvents and chemicals of research grade are used in the study. High-quality glass slides are used as substrate to prepare thin film

2.2 Methods

Separation of Asphaltenes from crude oils and tar

The ASTM recommended procedure (ASTM D2007-80) is used for separating asphaltenes from crude oils and tar samples. The ASTM procedure specifies adding a volume of alkanes that is 40 volume of the aliquot of oil. For separating further into various polyaromatic hydrocarbons of different fused aromatic ring (FAR: C5+, C6+, C7+ and C8+), asphaltene are further added with solvent n-Pentane, n-Hexane, n-Heptane and n-Octane, respectively. In each case, solvent and crude oil is mixed and equilibrate at 40:1 ratio and after ageing for 24 h, a funnel filter assembly with filter paper is used to separate precipitated asphaltenes from the mixture (Fig. 1 and Table 1).

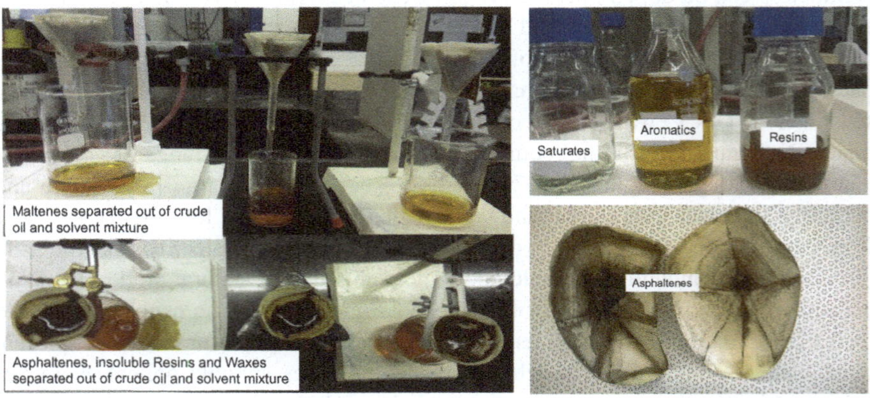

Fig. 1 Laboratory set-up for separating asphaltenes and other fractions from the crude oil and tar samples

Table 1 Different asphaltene fractions separated from crude oil samples using different solvents

Sample no (1 gm Sample)	C5+ in gm (With n-Pentane)	C6+ in gm (With n-Hexane)	C7+ in gm (With n-Heptane)	C8+ in gm (With n-Octane)
173	0.0671	–	0.02724	0.00388
1	0.46001	0.34569	0.2741	0.0002
229	–	–	0.02685	–
45	–	–	0.02273	–
112	–	–	0.01574	–
185	–	–	0.05243	–
64	–	–	0.00014	–
08	–	–	0.02772	–

Table 2 Thin film prepared from the following combinations of asphaltenes and P3HT

Sample No	Sample name	C5+ (gm)	C6+ (gm)	P3HT (gm)	Chlorobenzene (ml)
1	C5P3	0.05	–	0.05	10
2	C6P3	–	0.05	0.05	10
3	C56P3	0.025	0.025	0.05	10
4	C56	0.05	0.05	–	10

2.3 *Thin Film Preparation and Thermal Annealing*

Prior to the experiment, the substrate (glass slide) was ultrasonically cleaned with acetone, isopropanol and deionized water separately for 30 min and then dried. The slide was held firmly in position by atmospheric pressure. In a clean environment, one drop 5 μml of organic solution was allowed to fall with the help of a clean syringe onto the centre of a stationary glass slide. The speed of rotation was selected to 2500 rpm. The drop spreads out across the glass slide under the influence of the centrifugal force. As it moves, the solvent evaporates from the solution leaving a smooth continuous film. Sixteen films of four samples is prepared with four of each sample. After deposition, the thin film was placed in a high-temperature furnace under atmospheric conditions. A 10 °C/min temperature heating rate was used to raise the temperature to 200, 250 and 300 °C for thermal annealing. A dwelling time of 2 h was adopted in the experiment. After thermal annealing of thin film, efficiency is considerably increased due to crystallization of material (Table 2).

3 Results and Discussion

Asphaltenes may be regarded as condensed polycyclic aromatic hydrocarbon (PAH) compounds, which possess a semiconducting property like n-type semiconductors with an energy gap, called HOMO-LUMO gap (henceforth H-L gap), between the highest occupied molecular orbitals (HOMO) and the lowest unoccupied molecular orbitals (LUMO) in the visible photon energy ranges [19, 20]. The H-L gap is used as a direct indicator of kinetic stability where a large H-L gap implies high kinetic stability with low chemical reactivity since it is energetically unfavourable to add electron to a high lying LUMO [21–24]. Schematic representation of H-L gap ($\Delta E_{H\text{-}L}$) versus number of fused aromatic ring (nFAR) in Fig. 2 is clearly shows that various PAHs with different condensation (cata-condensate or peri-condensate) systems have different H-L gaps [21, 24]. Asphaltenes having PAH with four or fewer fused ring are mostly colourless, while asphaltene are highly absorptive in the visible to the near-infrared region indicating their larger (>4 ring) fused aromatic ring structure. It is very much clear that asphaltenes certainly possess larger PAH which have 5–10 fused aromatic ring (FAR) in the fused ring core. Larger PAH absorb at larger wavelength, as because the delocalization area in a π-system of an electron in larger aromatic box increasing the wavelength of the electron wavelength, thereby decreasing its transition frequencies. The population of larger PAH exponentially declines in asphaltenes with an electronic absorption edge at ~650 nm.

The interpretations of the optical absorption spectra in the hydrocarbons utilize an 'amorphous semiconductor' model [25] in which the slope of the spectra is related to a distribution of electronic states in similar type of chromophore molecules. In the case of petroleum samples, the slopes of the optical absorption spectra of crude oils relate the population distributions of aromatic molecules. Figure 3 shows the measured UVVis spectra of asphaltene fractions (C5+, and C6+) separated from crude oil samples using different alkanes as solvents. As solubility of different asphaltenes in different alkane solvents is related to chemical structure, pentane-insoluble asphaltenes (C5+) are smaller PAH, while n-heptane-insoluble (C6+) is larger in size. The absorption spectra of C5+ asphaltene is in lower wavelength compared to C6+ in the UV range of energy while the absorption is poor in the visible photon energies in both the cases. Previous studies with films of Tatarstan heavy crude oil indicate a possible existence of UV peaks [26] which are frequently observed in absorbance spectra of nano-sized carbon species. It appears from the result that C5+ and C6+ asphaltene has practically little absorption characteristics in the visible light energy of the solar spectrum.

P3HT has a molecular weight of 65.5 g/mol and a high hole mobility of $3.8 - 3.9 \times 10^{-4}$ cm^2/vs [27]. P3HT has a HOMO of 3 eV, with absorbance in the range of 450–600 nm in the visible light range and can self-assemble to well-defined nanostructure. Figure 4 shows the absorption spectra of various combinations of asphaltene fraction and P3HT. The broad absorption spectra of C5+P3, C6+P3 and C56P3 are due to structural ordering of asphaltene and P3HT. The absorption of lower wavelength (300–450 nm) is denoted to the absorption of asphaltenes. The absorption from 450 to 600 nm was attributed to the P3HT absorption peaks due to the vibronic splitting of π–π^* transition and their degree of

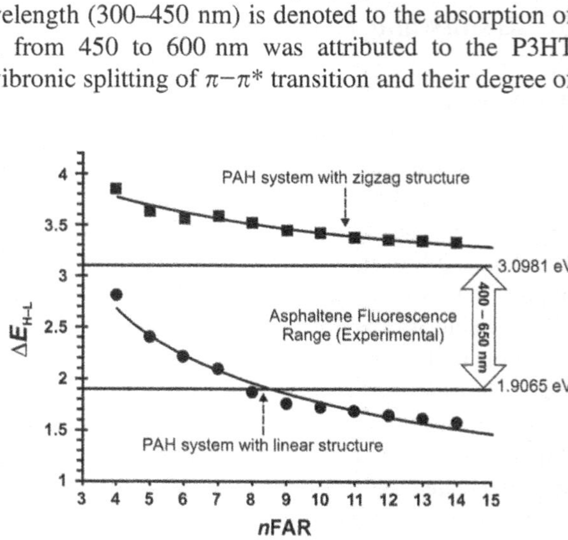

Fig. 2 $\Delta E_{H\text{-}L}$ versus number of fused aromatic ring (nFAR) of polycyclic aromatic hydrocarbon (PAH) with different structural systems. Experimental asphaltene fluorescence emission range is between 400 and 650 nm as indicated between two horizontal bar. Modified after Ruiz-Morales [21, 24]

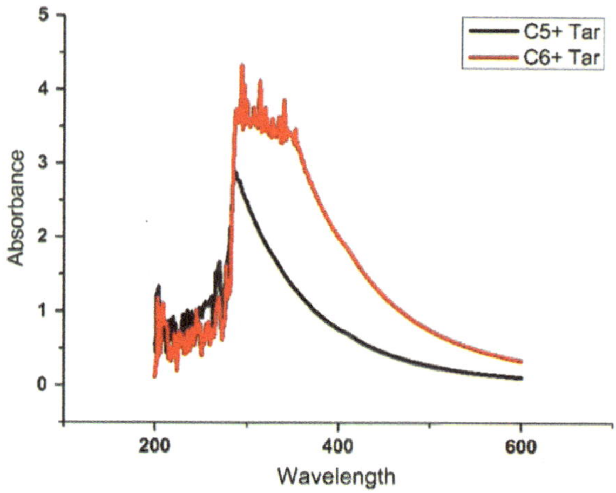

Fig. 3 UV-Vis spectra of asphaltenes of crude oil (tar) samples C5+ and C6+

polymer ordering. After forming the thin films on mixing with P3HT, broad optical absorption peak was noticed in the visible photon energies as shown in Fig. 4. More interestingly, C5+-based film shows higher optical absorption as compared to C6+ and their mixture. The strong and featureless UV absorption observed in the crude (Fig. 2) indicates the absorption by petroleum porphyrin [28] which largely modifies upon the formation of thin films towards the visible wavelengths (Fig. 4). Although asphaltene by its own shows an UV-absorption edge, its blend with P3HT has shown the same way absorbing visible photons most likely due to the extended defect states (Urbach states) in the HOMO-LUMO gap. Edokimov et al. [26] suggested this region corresponds to the longer wavelength range of absorbance spectra, which exhibits high structural sensitivity to phenomena of asphaltene aggregation They also suggested, an increase of this visible absorption at high dilutions signifies an appearance of some dangling bonds in non-associated asphaltene monomers explaining their high propensity to aggregation. In this aspect, the n-pentane-based blend (C5 and C56) tends to show higher visible absorption. Rogel showed in a molecular dynamic simulation that solvent plays important role in the interaction and dissociation of asphaltene aggregates [29]. In this study, the aggregates were shown to get dissociated by the interaction with different types of solvents. However, the dissociation not just depends on the solvent interaction but also the strength of Van der Waals interactions of the asphaltene and resins. Still the lower values of molecular weight of the solvent, number of aromatic carbons to total carbons, as well as number of bridging aromatic carbon to the number of peripheric non-bridging carbon (also known as aromatic condensation degree) would play key role here (Fig. 5).

Masson et al. [30] described that solid asphaltenes lack a crystalline phase, but they may have an ordered amorphous or mesoporous phase They identified two overlapping glass transitions in the amorphous domains of asphaltenes, at 70 °C which assume an oligo-aromatic structures. Specifically, one fraction of the asphaltenes undergoes melting or a glass transition (or endotherm, at around −13 °C), while

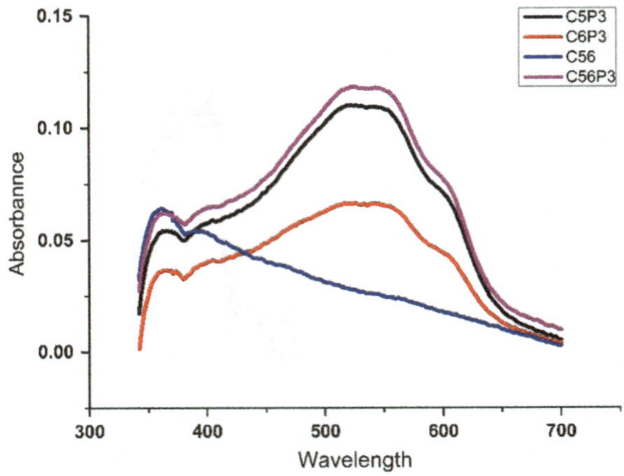

Fig. 4 UV-V is spectra of thin films of asphaltene specimen (C5+, C6+ and mixture C56) with P3HT

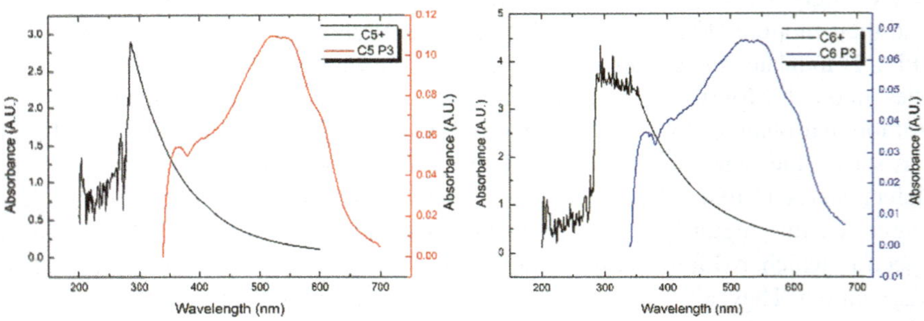

Fig. 5 Comparison between absorption spectra of asphaltene and thin films

another fraction dissolves into it at a higher temperature (as exotherm at 77 °C). In this regard, the compound shows its structure-dependent optical and electrical properties which could be beneficial to the use of opto-electronic application. Previous studies on the thin film blend of P3HT and 6,6-phenyl-C61-butyric acid methyl ester (PCBM) has revealed that solution heat treatment leads to a structural ordering of P3HT polymer which in turn modifies structural (including vibrational), morphological and optical properties of the matrix [31]. The author showed in their study that an enhanced crystallinity of P3HT leads to a red shift of the optical absorption edge of the blend. The temperature dependencies were thus studied further with the prepared thin film to elucidate the possible effects of structural transformations in the asphaltenes to absorb solar light and its probable photovoltaic conversion. The broad absorption of the blended thin film reflects significant degree of structural ordering (Fig. 4) while higher temperature ($\sim$250 °C) thermal annealing provide better absorption

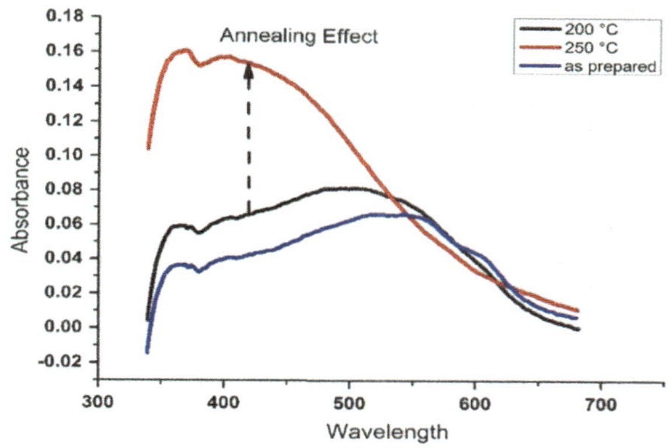

Fig. 6 UV-Vis spectra of thin film C6P3 annealed at different temperatures

characteristics of the prepared thin film (Fig. 6). As mentioned, structural phase transitions in the asphaltene at higher temperature can be a direct cause of the optical absorption characteristics of the prepared thin film with P3HT.

4 Conclusions

Petroleum crude contains undesirable and problematic asphaltene of different nature and structural types which show absorption in different wavelengths in the UV–visible range. The present study with different solubility fractions of asphaltene is to demonstrate their potential as solar cell component. The purpose of using P3HT as a donor and crude oil asphaltenes as the acceptor for fabricating solar cell is partly studied here with the help of optical characterization of asphaltenes and blended thin film. Though the used asphaltene fractions are showing absorption in the UV wavelength range, a significant visible light absorption is observed while blending it with hole transporting counterpart, P3HT. The broad absorption of the blended thin film reflects significant degree of structural ordering with formation of self-organized nanostructure forming bulk heterojunction. Annealing of the film up to 200 °C as well facilitate larger visible light absorption. The hole membrane's structural change leads to such absorption enhancement. The asphaltene thus proved to be a potential candidate for bulk heterojunction solar cells. A more detailed study of their photovoltaic properties including electronic transports while non-conducting resin component could be one of the limitations is required and underway.

Acknowledgments The authors are thankfully acknowledging PDPU for the research facilities. UKB acknowledges the support of GSPC for providing the crude oil samples for the work. Petroleum Engineering Laboratory and Reservoir Characterization Laboratory (RCL) of SPT and SRDC of PDPU are thankfully acknowledged for the optical characterization data and thin film preparation and characterization data, respectively.

References

1. Perez MD, Borek C, Forrest SR, Thompson ME (2009) Molecular and morphological influences on the open circuit voltages of organic photovoltaic devices. J Am Chem Soc 131:9281–9286
2. Potscavage WJ, Yoo S, Kippelen B (2008) Origin of the open-circuit voltage in multilayer heterojunction organic solar cells. Appl Phys Lett 93:193308
3. Vandewal K, Tvingstedt K, Gadisa A et al (2009) On the origin of the open-circuit voltage of polymer-fullerene solar cells. Nat Mater 8:904–909
4. Hains AW, Liang Z, Woodhouse MA, Gregg BA (2010) Molecular semiconductors in organic photovoltaic cells. Chem Rev 110(11):6689–6735
5. Troshin PA, Lyubovskaya RN, Razunov VF (2008) Organic solar cells: structure, critical characteristics, and outlook. Nanotechnol Russia 3(5–6):242–271
6. Abujnah RE, Chianelli RR (2014) Asphaltene as light-harvesting materials in dye-sensitized solar cell. In: Dinner et al. (eds) Progress in exergy, energy, and the environment. Springer International Publishing, Switzerland, https://doi.org/10.1007/978-3-319-04681-5_91
7. Yu G, Gao J, Hummelen JC, Wudi F, Heeger AJ (1995) Polymer photovoltaic cells-enhanced efficiencies via a network of internal donor-acceptor heterojunctions. Science 70:1989–1991
8. Groenzin H, Mullins C (2007) Asphaltene molecular size and weight by time-resolved fluorescence depolarization, asphaltenes, heavy oils, and petroleomics. In: Mullins OC, Sheu EY, hammami A, Marshall AG (eds), Springer, pp 17–62
9. Mullins OC, Sheu EY, Hammami A, Marshall A (2007) Asphaltenes, heavy oils, and petroleomics. Springer, New York
10. He B, Dai J, Zherebetskyy D, Chen TL, Zhang BA, Teat SJ, Zhang Q, Wang L, Liu Y (2015) Chem Sci 6:3180
11. Escande A, Ingleson MJ (2015) Chem Commun 51:2188
12. Matsuo K, Saito S, Yamaguchi S (2014) J Am Chem Soc 136:12580
13. Watson MD, Fechtenkotter A, Mullen K (2001) Chem Rev 101:1267
14. Wu J, Pisula W, Mullen K (2007) Chem Rev 107:718
15. Schmidt-Mende L, Fechtenkotter A, Mullen K, Moons E, Friend RH, MacKenzie JD (2001) Science 2001(293):1119
16. Ruiz-Morales Y, Mullins OC (2007) Electronic absorption edge of crude oils and asphaltens analysed by molecular orbital calculations with optical spectroscopy. Energy Fuels 21:944–952
17. Salmanova CK, Akhmedbekova SF, Mamedov AP, Kyazimov SM, Abdulova SN (2007) Transformation of resin and asphaltenes in photoirradiation. Chem Technol Fuels Oils 43 (5):415–421
18. Jiqian W, Chuan L, Longli Z, Guohe Q, Zhaomin L (2009) The properties of asphaltenes and their interaction with amphiphiles. Energy Fuels 23(7):3625–3631
19. Garrett CGB (1960) Organic semiconductors. Radiat Res 2(Supplement):340–348
20. Forster EO (1964) On the activation energy for conductance in aromatic hydrocarbons. Electrochim Acta 9(10):1319–1327
21. Ruiz-Morales Y (2002) J Phys Chem A 106:11283
22. Clar E (1964) Polycyclic hydrocarbons, vol 1. Academic, London
23. Randic M (2003) Chem Rev 103:3449
24. Ruiz-Morales Y (2007) Molecular orbital calculations and optical transitions of PAHs and asphaltenes In: Mullins OC, Sheu EY, Hammami A, Marshall A (ed) Asphaltenes, heavy oils, and petroleomics, Springer, pp 95–137
25. Mott NF, Davis EA, Street RA (1975) States in the gap and recombination in amorphous semiconductors. Phil Mag 32(5):961–996
26. Evdokimov IN, Eliseev NY, Akhmetov BR (2003) Assembly of asphaltene molecular aggregates as studied by near-UV/visible spectroscopy: I. Structure of the absorbance spectrum. J Petrol Sci Engg 37(3–4), 135–143

27. Supriyanto A, Mustaqim A, Agustin M, Ramelan AH, Suyitno Rosa S, Yofentina ES, Nurosyid F (2016) IOP Conf Ser Mater Sci Eng 107, 012050
28. Miller JT, Fisher RB, van der Eerden AMJ, Koningsberger DC (1999) Structural determination by xafs spectroscopy of non-porphyrin nickel and vanadium in maya residuum, hydrocracked residuum, and toluene-insoluble solid. Energy Fuels 13:719–727
29. Rogel E (2000) Simulation of interactions in asphaltene aggregates. Energy Fuels 14(3):566–574
30. Masson J-F, Polomark GM, Collins P (2002) Time-dependent microstructure of bitumen and its fractions by modulated differential scanning calorimetry. Energy Fuels 16(2):470–476
31. Otieno F, Mutuma BK, Airo M, Ranganathan K, Erasmus R, Coville N, Wamwangi D (2017) Enhancement of organic photovoltaic device performance via P3HT: PCBM solution heat treatment. Thin Solid Films 625:62–69

Coal

Petrographic and Organic Geochemical Characterizations of Early Eocene Lignites, Cambay Basin, Western India

Monalisa Mallick, Suryendu Dutta, Bhagwan D. Singh, Sharmila Bhattacharya, and Alpana Singh

1 Background

Early Eocene was a crucial time in the palaeogeographic and climatic history of Indian subcontinent. The Indian subcontinent occupied equatorial position during that time [1] on the course of its journey from southern hemisphere to the present location after the break-up of Gondwanaland. By and large, early Palaeogene was a period of extreme temperature and humidity, causing widespread mangrove swamp and coal formation [2]. Hence, the early Palaeogene era in western coastal margin of India is characterized by the occurrence of numerous lignite-bearing sequences. Cambay Basin, being one of the three peri-continental rift basins that opened up between Early Jurassic and Palaeogene during the northward drift of Indian subcontinent, also comprises several lignite sequences. Some of the major oil and gas fields bearing Eocene–Miocene reservoirs have been discovered in Cambay Basin. The main source rock, Cambay Shale that deposited during Palaeocene–Early

M. Mallick (✉)
CSIR-National Geophysical Research Institute, Uppal Road, Hyderabad 500 007, India
e-mail: m.mallick@ngri.res.in

S. Dutta
Department of Earth Sciences, IIT Bombay, Powai 400076, Mumbai, India
e-mail: s.dutta@iitb.ac.in

B. D. Singh · A. Singh
Birbal Sahni Institute of Palaeosciences, 53 University Road, Lucknow 226007, India
e-mail: bdsinghbsip@yahoo.co.in

A. Singh
e-mail: alpanas_in@yahoo.com

S. Bhattacharya
Department of Earth and Environmental Sciences, IISER Mohali, Punjab 140306, India
e-mail: sbhattacharya@iisermohali.ac.in

U. K. Bhui (ed.), *Macromolecular Characterization of Hydrocarbons for Sustainable Future*, Green Energy and Technology,
https://doi.org/10.1007/978-981-33-6133-1_11

Eocene period, is considered to be transmuted into lignite-bearing shale/clay sequences in the basin marginal parts, e.g. Vastan lignite sequence investigated in the present study [3].

Coal has unquestionably been regarded as an efficient source rock for gas generation. Although unequivocal evidence of oil generation from coal is limited, some coals have been proved to be oil-prone and contributed to oil fields, viz. Gippsland and Sydney basins, Australia [4–6], Taranaki Basin, New Zealand [7–9], Indonesian basins [10–12], Norwegian Sea [13], Danish North Sea [14–16], and Turpan Basin, NW China [17]. In fact, some lignite sequences in northern Cambay Basin are believed to contribute to nearby oil fields [18].

The lignite sequence of present study area, Vastan lignite mine, has long been a hotspot for palaeontological studies [19, 20]. However, the geochemical characteristics of Cambay Shale sequence from this area have not been investigated previously. The well-exposed mine section of Vastan lignite sequence has provided an excellent opportunity for geochemical studies.

The present study therefore constitutes petrographical, geochemical and biomarker analyses to establish source rock characteristics. While the maceral composition and geochemical parameters determine the source rock potential, the biomarker assemblage reveals source of organic matter and depositional environment.

2 Study Area and Age

2.1 Study Area

The study area, Vastan lignite mine (21°25′ N:73°07′ E), is part of NNW–SSE trending Cambay Basin located in the north-western margin of Indian shield on Indo-Arabian platform, in the western Indian state of Gujarat (Fig. 1).

2.2 Age

The age of this lignite sequence has been assigned as Early Eocene, determined by palaeontological, palynological and organic carbon isotopic studies. An Early Eocene age was inferred based on palynological investigation of this sequence [21]. The lower part of Vastan lignite sequence has been assigned as Late Thanetian/Sparnacian (~55 Ma) and the upper part as basal Ypresian (~54 Ma) age based on dinoflagellate cysts [22]. The second Eocene Thermal Maxima or ETM2 (ca. 53.7 Ma) has been reported from organic carbon isotopic values from a zone at the upper part of the sequence that corresponds to the dinoflagellate-bearing horizon [23]. Based on the large benthic foraminiferal assemblage (*Nummulites burdigalensisburdigalensis*) from the foraminifera-dominated horizon at the middle part

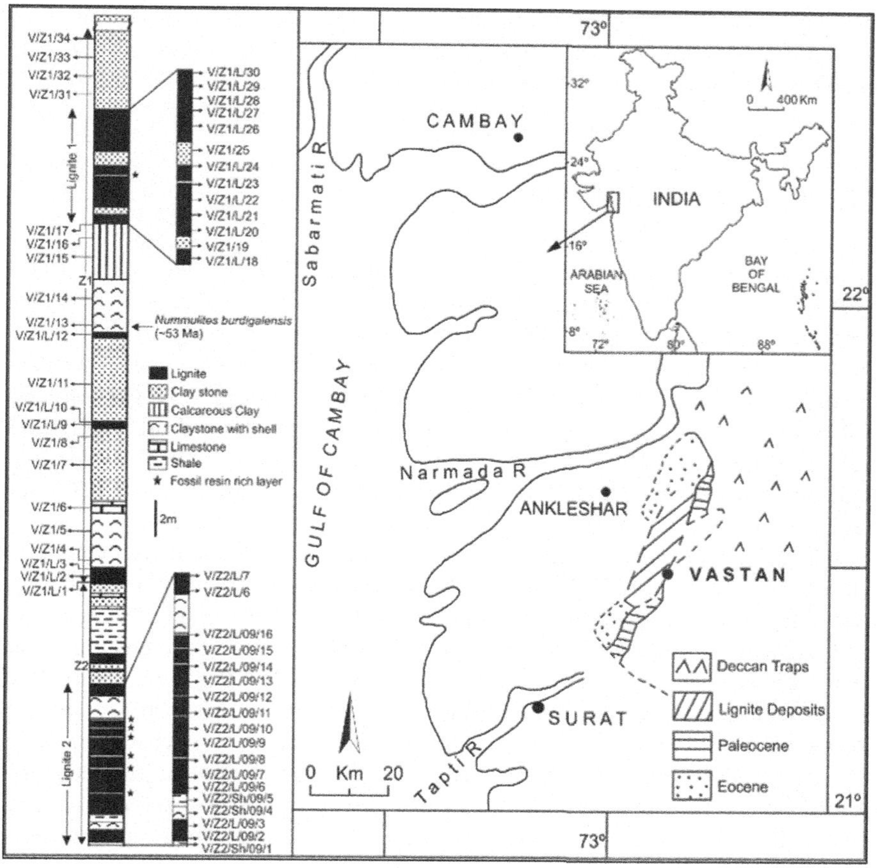

Fig. 1 Geological map showing the location of Vastan lignite mine

of the lignite sequence, a basal Cuisian (early Eocene) age for Vastan lignites has been suggested [24]. Palaeocene Eocene Thermal Maxima (PETM) has been reported from the lower part of the mine section [25].

3 Samples and Methods

3.1 Sampling

The subsurface lignite-bearing sequence of Vastan lignite mine has been referred to as Cambay Shale Formation [3]. The thickness of the Cambay Shale in Vastan mine section varies from 20 to 145 m, and the Cambay Shale here is characterized by alternating sequence of lignite beds, dark grey to black and greenish grey shales and clay–marls [3]. Samples were collected vertically at ½ m interval from about 44 m lignite-bearing section, containing two main lignite seams—upper seam (thickness 3–8 m) and lower seam (thickness 4–6 m) (Fig. 1).

3.2 Petrographical Studies

Representative 6 lignite samples from upper seam and 5 from lower seam sections were examined for petrographical study. Polished particulate lignite pellets were prepared, as per the specifications of International Committee for Coal and Organic Petrology (ICCP), by embedding ±1–2 mm grain particles in a homogeneous mixture of Buehler's epoxy resin and hardener in a ratio of 5:1. The hardened pellets were ground and polished successively on finer grades of CarbiMet (silicon carbide) papers and on microcloths with polishing alumina (oxide). The descriptions and terminology for the study of macerals were followed as per ICCP and published literatures [26–28].

The reflectance values were measured on huminite constituents (preferably on ulminite) on Leica DM4500P microscope equipped with immersion oil (refractive index: 1.518) and 50x objective lens using sapphire (0.594) as a reflectance standard, along with yttrium–aluminium–garnet (0.904) and gadolinium–gallium–garnet (1.725) standards. Microscope-photometry system (PMT III) and Software MSP 200 were used for the random reflectance measurements, calculations and data collection. The maceral analysis was carried out on the same microscope, simultaneously under normal reflectance (incident light) and fluorescence (blue light excitation) modes adopting single-scan method, using oil immersion objective. Quantitative estimation of macerals was made through automatic computerized point counter using 2.35 version of PetrogLite software. Software tool—Leica Application Suite (LAS)—was used for maceral images.

3.3 Rock-Eval Pyrolysis

For geochemical investigations, a suite of lignite and claystone samples was selected. The samples were pulverized in a brass mortar followed by drying in an oven at 50 °C for 6 h prior to further analysis. Pyrolysis experiments were conducted using a Rock-Eval 6 (standard) analyser manufactured by Vinci Technologies®. The full description of the method has been given in the published literature [29]. Briefly, samples were first pyrolysed under an inert N_2 atmosphere and the residual carbon was subsequently burnt in an oxidation oven. The amount of hydrocarbons (HC) released during pyrolysis was detected with a flame ionization detector (FID), while infrared detector measures continuously the released CO and CO_2. The samples were first pyrolysed from 300 °C, held for 5 min, then to

650 °C at a rate of 25 °C/min. The oxidation phase started with an isothermal stage at 300 °C, again held for 5 min, followed by an increase to 850 °C at a rate of 25°C/min to incinerate all the residual carbon.

3.4 Fourier Transform Infrared Spectroscopy

FTIR spectroscopic analyses from 4000 to 500 cm^{-1} were performed in transmission mode using a Nicolet MAGNA 550 model operated by OMNIC® software. The KBr beam splitter guided the two respective beams to the sample through reflection at the parabolic mirror. Spectra were obtained for a defined area by co-adding up to 512 scans with a spectral resolution of 4 cm^{-1}. The standard setting of the recorded spectra was 512 scans/4 cm^{-1}. Background spectra were collected after each sample analysis. Peak assignments were based on the published literature [30, 31].

3.5 Biomarker Analyses

A suite of lignite, claystone, calcareous clay and shale samples was pulverized using a ball mill (Retsch PM100). A solution of dichloromethane (DCM) and methanol (9:1) was then added to each powdered sample followed by ultrasonication for 30 min. The extracted solution was decanted after allowing precipitation of the powdered sample. This process was repeated twice to pool the extractable organic matter from the rock to a feasible extent. The precipitation of asphaltenes was performed using *n*-hexane, and the extracts decanted were separated into saturate, aromatic fractions by liquid column chromatography using glass pipette ~13 cm column packed with activated silica gel (120 °C > 8 h). Aliphatic fraction was eluted with *n*-hexane (3/8 of column volume), and aromatic fraction was eluted with a solution of *n*-hexane and DCM (4:1; 4 column volume).

3.6 Gas Chromatography-Mass Spectrometry

The aliphatic and aromatic hydrocarbon fractions were analysed by gas chromatography-mass spectrometry (GC–MS) in full scan mode. The instrument comprised an Agilent 5975C MSD attached with an Agilent 7890A gas chromatograph. The column in the GC oven used for the separation of components was an HP–5MS fused silica capillary column (30 m × 0.25 mm i.d., 0.25-μm film thickness; J&W Scientific). Helium was used as the carrier gas with flow rate 1 ml/min. During each analysis, the initial temperature was held at 40 °C for 5 min and then ramped from 40 °C to 310 °C at 4 °C/min, which was held for 5.5 min.

The detector source was operated in EI-mode at 300 °C, ionization energy being 70 eV. Full scan analyses were conducted over a mass range of 50–550. Data was acquired and processed using ChemStation software. Peak assignments were based on GC retention time and mass spectral data including comparison with MS libraries and published literatures.

4 Results

4.1 Petrographical Studies

The petrographical characteristics of Vastan lignites are determined by frequency distribution of organic microconstituents or macerals (organo-petrographic entities), their composition and associated mineral matter contents together with maturation of lignite. The macerals are categorized into three main groups, namely huminite, liptinite and inertinite. The frequency distributions of macerals along with rank are shown in Tables 1 (upper seam) and 2 (lower seam). The representative photomicrographs are illustrated in Figs. 2 and 3.

The huminite group macerals appear in different shades of grey under normal reflectance mode. Certain fraction of huminite (low-reflecting granular/spongy textured) fluoresces in dark reddish-brown or brown colour with low intensity. The studied lignites from both the seams are rich in huminite (upper: avg. 48%; lower: avg. 62%), consisting of almost all the macerals of this group. The structured telohuminite (avg. 21%) and the detrital/unstructured detrohuminite (avg. 18%) are the chief constituents of huminite in the upper seam.

In the lower seam, on the other hand, only detrohuminite (avg. 45%) forms the main constituent, followed by telohuminite (avg. 13%). Telohuminite in both the seams is represented by ulminite (upper: 21%; lower: 12%) with extraordinary high proportion in sample nos. V/Z1/L/2 in the upper seam (75%) and V/Z2/L/09/7 in the lower seam (28%). Detrohuminite is chiefly constituted by attrinite (upper: avg. 13%; lower: avg. 31%) with lower frequencies in the middle parts of both the seams.

Gelohuminite, the gelified huminite, has been recorded in fair amount in both the seams (upper: avg. 8%; lower: avg. 4%) and mainly represented by corpohuminite (upper: 7%; lower: 4%) which is dominant in the upper part of the upper seam section.

The fluorescing fractions of the huminite group in studied lignites fluoresce (with weak intensity) either due to relics of cellulose and/or by resinous impregnations.

In the lower seam, the amount of perhydrous (or fluorescing) huminite is higher (11–20%; avg. 14%) than its concentration in the upper seam (2–20%; avg. 8%). However, no definite trend of distribution is observed for perhydrous huminite in the studied lignite seams.

The liptinite macerals are relatively darker in colour than huminite and are easily recognized by their morphographic features. They appear black in normal incident

Table 1 Maceral contents (vol. %), huminite reflectance ($R_{r\ mean}$ %), GI and TPI values of the upper seam, Vastan lignite mine

Sample no. / Macerals	V/Z1/ L/2	V/Z1/ L/9	V/Z1/ L/12	V/Z1/ L/18	V/Z1/ L/21	V/Z1/ L/27	Mean	Range
Huminite (H)	85 (88)	41 (75)	32 (67)	55 (57)	60 (62)	18 (19)	48.5 (61.3)	18–85 (19–88)
Telohuminite	75	13	5	14	19	4	21.6	4–75
Textinite	0	0	0	2	1	0	0.5	0–2
Ulminite	75	13	5	12	18	4	21.1	4–75
Detrohuminite	5	24	23	26	26	8	18.6	5–26
Attrinite	3	20	21	13	16	5	13	3–20
Densinite	2	4	2	13	10	3	5.6	2–13
Gelohuminite	5	4	4	15	15	6	8.1	4–15
Gelinite	3	0	0	0	0	0	0.5	0–3
Corpohuminite	2	4	4	15	15	6	7.6	2–15
Liptinite (L)	10 (10)	11 (20)	14 (29)	26 (27)	31 (32)	68 (71)	26.6 (31.5)	10–68 (10–71)
Sporinite	1	2	2	2	1	9	2.83	1–9
Cutinite	0	0	2	1	12	3	3	0–12
Suberinite	0	0	0	2	2	4	1.3	0–4
Resinite	1	1	3	11	12	17	7.5	1–17
Alginite	0	0	0	0	0	0	0	0
Liptodetrinite	8	8	7	10	4	35	12	4–35
Inertinite (I)	2 (2)	3 (5)	2 (4)	16 (16)	6 (6)	10 (10)	6.5 (7.1)	2–16 (2–16)
Semifusinite	0	0	0	10	4	3	2.8	0–10
Fusinite	0	0	0	0	1	0	0.1	0–1
Funginite	1	2	1	2	0	4	1.6	0–4
Micrinite	0	0	0	3	1	3	1.1	0–3
Inertodetrinite	1	1	1	1	0	0	0.6	0–1
Mineral matter (M)	3	45	52	3	3	4	18.3	2–52
Pyrite	1	6	7	1	0	1	2.6	0–7
Fluorescing H	7	6	2	9	20	6	8.3	2–20
Total fluorescing (H + L)	17	17	16	35	51	74	35	16–74
Non-fluorescing (H + I + M)	83	83	84	65	49	26	65	26–84
$R_{r\ mean}$ (%)	0.35	0.28	0.24	0.30	0.29	0.30	0.293	0.24–0.35
GI	16.2	0.954	0.50	1.60	1.913	1.625	3.798	0.50–16.2
TPI	0.025	0.108	0.143	0.71	0.523	0.75	0.376	0.025–0.75

Table 2 Maceral contents (vol. %), huminite reflectance ($R_{r\ mean}$, %), GI and TPI values of the lower seam, Vastan lignite mine

Sample no. / Macerals	V/Z2/ L/7	V/Z2/L/ 09/3	V/Z2/L/ 09/7	V/Z2/L/ 09/9	V/Z2/L/ 09/13	Mean	Range
Huminite (H)	52 (53)	57 (61)	64 (67)	67 (70)	74 (78)	62.8 (65.8)	52–74 (61–78)
Telohuminite	12	8	31	3	14	13.6	3–31
Textinite	2	1	3	0	0	1.2	0–3
Ulminite	10	7	28	3	14	12.4	3–28
Detrohuminite	34	46	29	62	55	45.2	29–62
Attrinite	24	24	22	47	40	31.4	22–47
Densinite	10	22	7	15	15	13.8	7–22
Gelohuminite	6	3	4	2	6	4.2	2–6
Gelinite	0	0	0	0	0	0	0
Corpohuminite	6	3	4	2	6	4.2	2–6
Liptinite (L)	41 (42)	29 (31)	29 (31)	22 (23)	16 (17)	27.4 (28.8)	16–41 (17–42)
Sporinite	6	3	4	3	2	3.6	2–6
Cutinite	2	0	1	1	0	0.8	0–2
Suberinite	3	4	3	2	0	2.4	0–4
Resinite	15	9	8	8	5	9	5–15
Alginite	0	0	0	0	P	0	0
Liptodetrinite	15	13	13	8	9	11.6	8–15
Inertinite (I)	5 (5)	7 (7)	2 (2)	7 (7)	5 (5)	5.2 (5.2)	2–7 (2–7)
Semifusinite	1	2	1	3	2	1.8	1–3
Fusinite	0	0	0	0	0	0	0
Funginite	4	5	1	4	2	3.2	1–5
Micrinite	0	0	0	0	0	0	0
Inertodetrinite	0	0	0	0	1	0.2	0–1
Mineral matter (M)	2	7	5	4	5	4.6	4–7
Pyrite	1	2	1	0	2	1.2	0–2
Fluorescing H	20	18	11	12	12	14.6	11–20
Total fluorescing (H + L)	61	47	40	34	28	42	28–61
Non-fluorescing (H + I + M)	39	53	60	66	72	58	39–72
$R_{r\ mean}$ (%)	0.31	0.30	0.30	0.30	0.30	0.30	0.30–0.31
GI	0.963	1.185	1.50	0.40	0.814	0.972	0.40–1.50
TPI	0.204	0.113	0.14	0.092	0.116	0.133	0.092–0.204

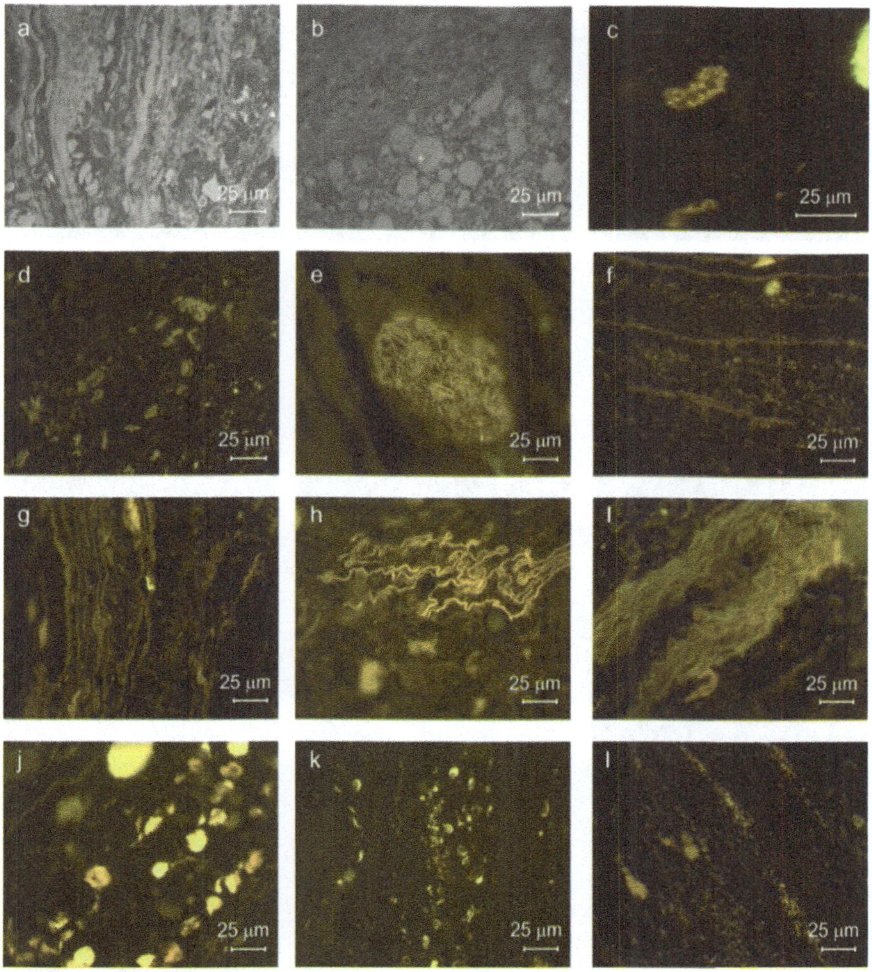

Fig. 2 Representative photomicrographs of macerals from Vastan lignites: **a** textinite, **b** attrinite associated with corpohuminite (lower part), **c** and **d** sporinite (individual pollen), **e** sporinite (pollen mass), **f** and **g** cutinite, **h** and **i** suberinite, **j–l** resinite (cell fillings) (photographs **a** and **b** are taken in normal incident light; **c–l** are taken in blue light excitation)

light and fluoresce in different colours and intensities on excitation with blue light. The total liptinite content in the studied lignites varies between 10 and 31% (with two higher values of 41 and 68%) with higher concentration in the upper seam (avg. 29%) than in the lower seam (avg. 24%). In both the seams, liptinite is constituted chiefly by liptodetrinite (upper: avg. 12%; lower: avg. 11%) and resinite (upper: avg. 7%; lower: avg. 9%), followed by sporinite (upper: 2%; lower: 3%), suberinite (upper: 1%; lower: 2%) and cutinite (upper: 3%; lower: 0.8%). Earlier, very low content of liptinite was reported from the Vastan lignite [32]. However, the macerals were not investigated under fluorescence light.

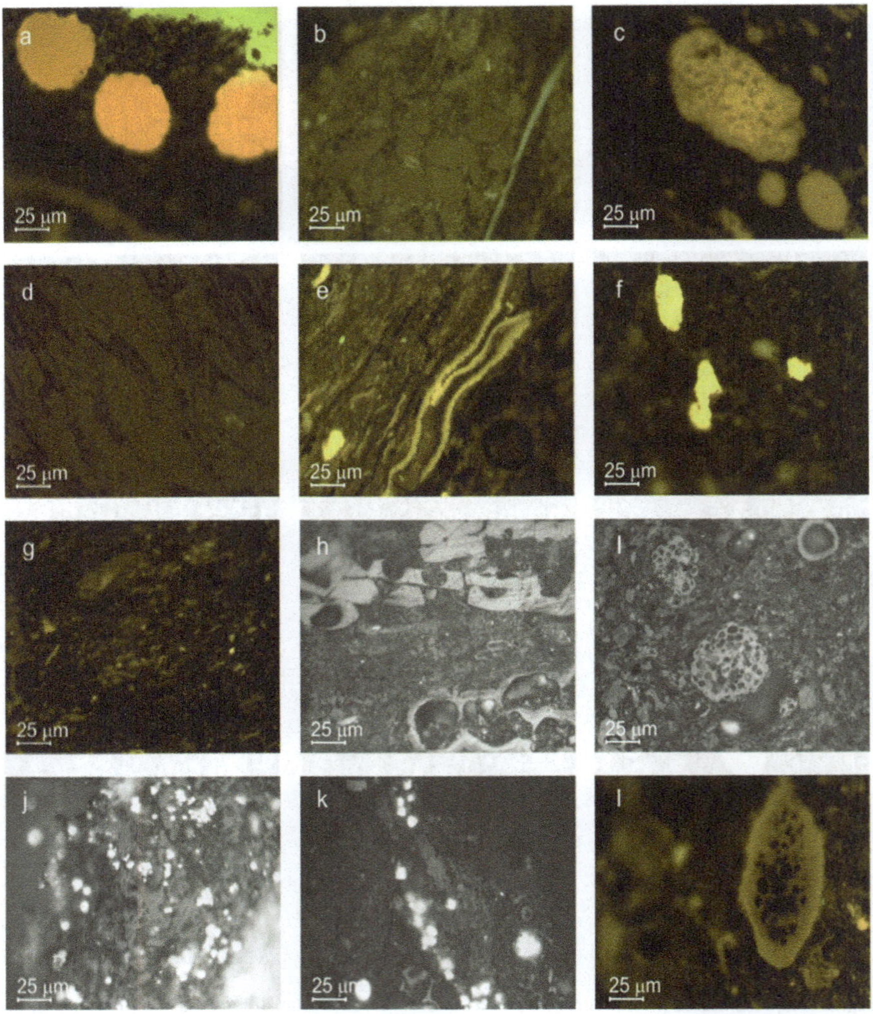

Fig. 3 Representative photomicrographs of macerals and minerals from Vastan lignites: **a–c** resinite (bodies of different colours), **d** perhydrous huminite (densinite), **e** perhydrous huminite along with sporinite and resinite, **f** alginate (*Botryococcus*), **g** liptodetrinite, **h** fusinite (brighter), **i** funginite associated with attrinite, **j** and **k** pyrite (framboids and crystals), **l** resinite filled with pyrite (photographs a–g and l are taken in blue light excitation; h–k are taken in normal incident light)

Resinites occurring as discrete bodies and as cell fillings are derived from oil secretions and resins (the metabolic products of plants). They differ widely in their physical and chemical properties and fluoresce in yellow, greenish-yellow and brown to reddish-brown colours. The chemical composition of the resinites found in these lignites suggested that they were derived from angiosperm tree family

Dipterocarpaceae [33–35]. Liptodetrinite includes the relics and fragmented parts of liptinite macerals which had lost their identity. It fluoresces in different colours depending on the nature of liptinites (resinite, sporinite, cutinite, etc.). Bituminite and alginite are also recorded along with sporadic fluorinite (associated with cutinite) and rarely exsudatinite. Bituminite is mostly accounted for by the perhydrous huminite. Alginite is observed (non-recordable) in one sample only in the lower seam. All the macerals of liptinite group show dominance towards the upper part of the upper seam and upper-middle portion of the lower seam.

Macerals of inertinite group, though not recorded in all the samples individually, appear greyish-white to white in normal incident light and do not fluoresce (appear black) in fluorescence light. In the upper seam, inertinite (avg. 6%) is represented by semifusinite, funginite, micrinite, inertodetrinite and fusinite (in decreasing order of dominance). In the lower seam (avg. 5%), it is represented mainly by funginite and semifusinite. Most common funginite (fungal sclerotia) occurs as oval and elliptical bodies. Single- to multi-celled fungal spores (teleutospores) are common. Inertinites do not show any specific trend of distribution in both the seams.

The studied lignites contain appreciable amount of mineral matter (upper seam: avg. 18%; lower seam: avg. 4%) represented chiefly by carbargilite (clay minerals; upper: avg. 16%; lower: avg. 3%) and carbopyrite (pyrite; upper: avg. 2%; lower: avg. 1%) along with sporadic carbankerite (calcite and siderite). Pyrite occurs in various forms—massive, framboidal and disseminated being the most common forms. Mineral matter is well recorded from middle part of the upper seam and does not show much variation in frequency in lower seam, except for the absence of pyrite in one sample.

Total fluorescing (perhydrous huminites + liptinites) and non-fluorescing (huminites + inertinites + mineral matters) contents of the lignites in upper seam vary from 16 to 74% (avg. 35%), and from 26 to 84% (avg. 65%), respectively. The lower seam has slightly lower proportion (avg. 58%) of non-fluorescing and higher amount (avg. 42%) of fluorescing macerals.

Rank of the lignites is determined by reflectance measurement on maceral huminite. The calculated mean ($R_{r\ mean}$) values range from 0.24 to 0.35% (avg. 0.29%; Table 1) for the upper seam, whereas the lower seam shows uniform values ($R_{r\ mean}$ 0.30%; Table 2). The reflectance values suggest that the studied lignites have attained 'brown coal' (German Standard) or 'lignitic' stage/rank (ASTM) and occur in early diagenetic zone of methane generation [36].

4.2 *Rock-Eval Pyrolysis Data*

The TOC contents of the lignite samples vary from ~3 to 54 wt%, whereas in the shale or claystone they are in the range ~0.5–4 wt% (Table 3; Fig. 4a). HI values are in the range 90–385 mg HC/g rock for lignites and 10–80 mg HC/g rock for shale or claystones (Table 3; Fig. 4b). It should be noted that the S_2 and TOC values are higher in lignites compared to claystone or shale samples and so are HI values. Therefore, it is evident that the source rock potential of the samples is variable. The T_{max} values of all the samples are low, ranging from 400 to 434 °C, except for one sample (V/Z1/14) showing very high T_{max} value ~514 °C (Table 3).

4.3 Fourier Transform Infrared Spectroscopy

The representative Fourier transform infrared spectroscopy (FTIR) spectra of the demineralized lignite samples from Vastan lignite mine section are presented in Fig. 5.

The spectra of the lignite samples are characterized by prominent absorption peaks at 3400–3320 cm^{-1}, 3000–2800 cm^{-1} and 1650–1560 cm^{-1}.

The broad band in the region 3400–3320 cm^{-1} represents hydroxyl stretching vibration, and relatively less intense peak representing aliphatic CH_x stretching vibration occurs in the range 3000–2800 cm^{-1}. The peaks at 2925 cm^{-1} and 2849 cm^{-1} correspond to asymmetric and symmetric stretching vibration of CH_2, respectively [37]. The absorption band at 1700–1740 cm^{-1} is attributed to aromatic C=O (carbonyl or carboxyl group, namely ketones) and mainly a combination of the aliphatic ester C=O groups at about 1740–1730 cm^{-1} and aromatic carbonyl/carboxyl C=O groups at about 1710–1700 cm^{-1} [38]. The intense peak in the region 1650–1560 cm^{-1} is attributed to aromatic C=C ring stretching vibration. The absorption band at 1465–1450 cm^{-1} is due to aliphatic CH_x asymmetric deformation. The peak at 1390–1370 cm^{-1} is due to absorption of symmetric deformational vibration of CH_3. The absorptions of asymmetric aromatic C–O stretching and –OH deformation in a carboxylic acid ester or ether occur at 1270–1240 cm^{-1} [37].

Although all the spectra show similar pattern, a few differences have been noticed. For instance, the absorption band at 1800–1650 cm^{-1} for aromatic carbonyl/carboxyl C=O or ketone is more prominent in the lower seam samples, so is the absorption peak of asymmetric aromatic C–O stretching and –OH deformation in a carboxylic acid ester or an ether. The cross-plot between 'A' factor (3000–2800 cm^{-1}/3000–2800 cm^{-1} + 1650–1520 cm^{-1}) and 'C' factor (1800–1650 cm^{-1}/1800–1650 cm^{-1} + 1650–1520 cm^{-1}) obtained from IR spectra is represented in Fig. 6.

4.4 Biomarker Analyses

The major biomarkers detected in aliphatic fraction of soluble organic matter of lignite, calcareous clay, claystone and shale samples from Vastan lignite mine include *n*-alkanes, acyclic isoprenoids and triterpenoids.

Table 3 Summarized results obtained from Rock-Eval pyrolysis of samples from Vastan lignite mine

Samples	TOC (wt%)	S_1 (mg/g)	S_2 (mg/g)	S_3 (mg/g)	T_{max} (°C)	HI (mg/g)	OI (mg/g)
V/Z1/32	3.81	0	1.33	4.42	431	35	116
V/Z1/L/27	49.68	5.54	146.14	22.61	423	294	46
V/Z1/25	3.99	0.03	2.48	4.16	428	62	104
V/Z1/L/23	48.5	11.35	165.49	21.75	400	341	45
V/Z1/L/21	54.09	5.6	103.36	28.12	421	191	52
V/Z1/19	7.7	3.13	10.23	5.43	417	133	71
V/Z1/L/18	51.19	0.49	73.04	32.42	428	143	63
V/Z1/16	1.87	0.05	0.37	2.45	413	20	131
V/Z1/15	0.87	0	0.18	1.00	413	21	115
V/Z1/14	0.32	0.01	0.11	0.7	514	34	219
V/Z1/13	0.58	0.01	0.16	1.05	428	28	181
V/Z1/L/12	3.53	0	0.4	4.31	421	11	122
V/Z1/11	2.96	0	1.17	2.86	429	40	97
V/Z1/L/9	37.44	0.94	33.81	33.39	419	90	89
V/Z1/7	0.98	0	0.14	1.07	424	14	109
V/Z1/6	0.78	0.01	0.35	1.64	439	45	210
V/Z1/5	0.91	0.01	0.09	0.78	414	10	86
V/Z1/4	0.58	0.01	0.08	1.04	423	14	179
V/Z1/L/2	54.28	3.92	103.2	23.36	425	190	49
V/Z2/L/7	52.23	9.44	186.78	25.83	419	358	49
V/Z2/L/09/15	40.51	1.36	82.17	13.86	423	203	34
V/Z2/L/09/13	59.17	2.16	121.42	19.78	422	205	33
V/Z2/L/09/11	50.14	1.21	100.16	18.62	422	200	37
V/Z2/L/09/9	49.37	3.74	190.19	12.18	428	385	25
V/Z2/L/09/7	44.37	1.26	68.45	24.32	422	154	55
V/Z2/Sh/09/5	7.36	0.17	5.89	5.53	415	80	75
V/Z2/Sh/09/4	1.1	0.01	0.36	1.76	434	33	160
V/Z2/L/09/3	48.72	2.88	149.05	15.05	423	306	31
V/Z2/Sh/09/1	3.43	0.05	1.08	2.36	412	31	69

The *n*-alkane distribution of the samples confirmed by *m/z* 57 is presented in Fig. 7 which shows distribution in the range of C_{10}–C_{35}. However, C_{10}–C_{13} are present in very low abundance and on certain occasions beyond the level of detection. All the samples exhibit similar bimodal *n*-alkane distribution excluding calcareous clay sample.

The first mode which is dominant in all the samples ranges from C_{13} to C_{18} (short-chain alkane region), C_{16} being the most abundant component. This mode predominates over mid- and long-chain *n*-alkanes (C_{19}–C_{24} and C_{25}–C_{35}, respectively) in most of the samples as suggested by the relative proportion of short-chain

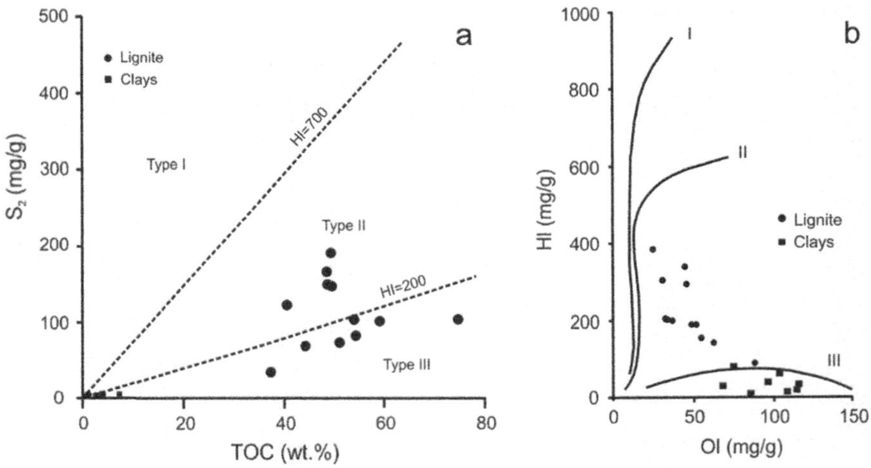

Fig. 4 **a** S_2 versus TOC plot and **b** classification of kerogen types on a van Krevelen plot of HI versus OI for lignites and clays (claystone, calcareous clay, shale) from Vastan lignite mine

n-alkanes to the sum of *n*-alkanes (Table 4) and shows even-carbon predominance (Table 4). The relative proportion of mid-chain *n*-alkanes (C_{19}–C_{24}) to the sum of *n*-alkanes is quite high in the samples; however, this range of *n*-alkanes does not exhibit any carbon-chain predominance (Table 4). The second mode comprised of long-chain *n*-alkanes in the range C_{25}–C_{35} shows odd-carbon predominance maximizing either at C_{29} or C_{31} (Fig. 7). This phenomenon is not observed in calcareous clay sample which had experienced marine influence during deposition [3, 24]. It lacks a second envelope with odd-carbon predominance in the long-chain region; however, even carbon-chain *n*-alkane predominance is observed that maximizes at C_{16} (Fig. 7). The carbon preference index (CPI) values have been determined following published literature [39] and range from ~0.55 to 2.73. The samples with CPI values > 1 indicate odd-carbon predominance, whereas samples with CPI value ~0.55–0.98 denote even-carbon predominance. Pristane/phytane (Pr/Ph) ratios of all the samples vary widely, with the lowest value of 0.84 to as high as 5.52. Phytyl side-chain of chlorophyll a or bacteriochlorophyll a/b converts to pristane and phytane in oxic and reducing condition, respectively [40]. Therefore, Pr/Ph > 1.0 indicates oxic depositional condition while Pr/Ph < 1 denotes anoxic environment. Furthermore, Pr/Ph > 3.0 denotes terrigenous source input in oxic condition.

Pentacyclic triterpenoids with five- or six-membered E-ring are detected in the aliphatic hydrocarbon fraction of all the samples in the *m/z* 191 partial mass chromatograms (Fig. 8). The identified compounds are listed in Table 5. The extracts of all the samples show similar type of distribution pattern of compounds except for the claystone sample which contains a few more compounds.

The hopane distribution in all the samples is characterized by 17β(H),21β(H)–hopane series. The hopane series is ranging from C_{27} to C_{32} except for the C_{28}

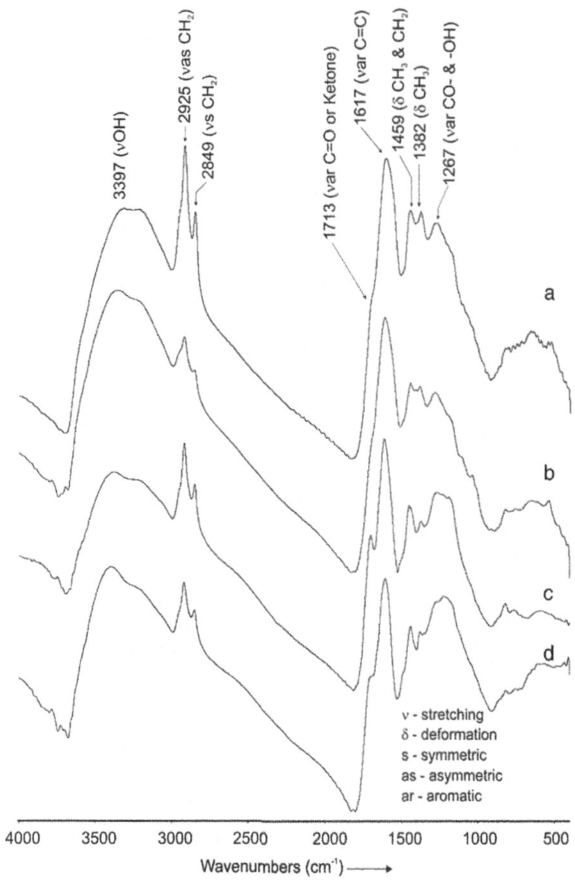

Fig. 5 FTIR spectra of lignite samples from the upper seam **a**—V/Z1/L/27; **b**—V/Z/L/21 and lower seam **c**—V/Z2/L/09/13; **d**—V/Z2/L/09/7 of Vastan lignite mine

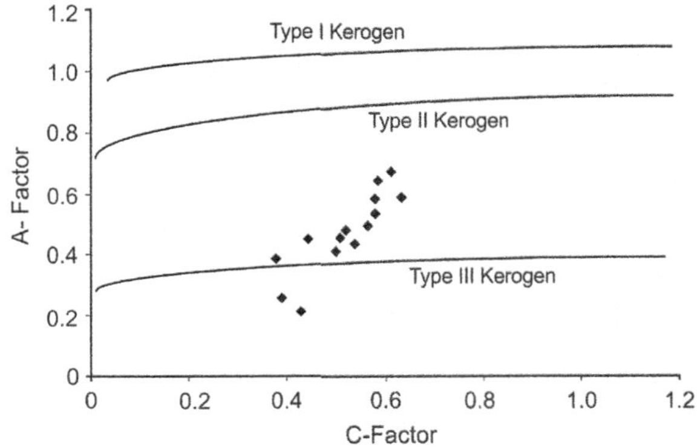

Fig. 6 Pseudo van Krevelen plot of A-factor versus C-factor obtained from IR spectra of lignites from Vastan lignite mine

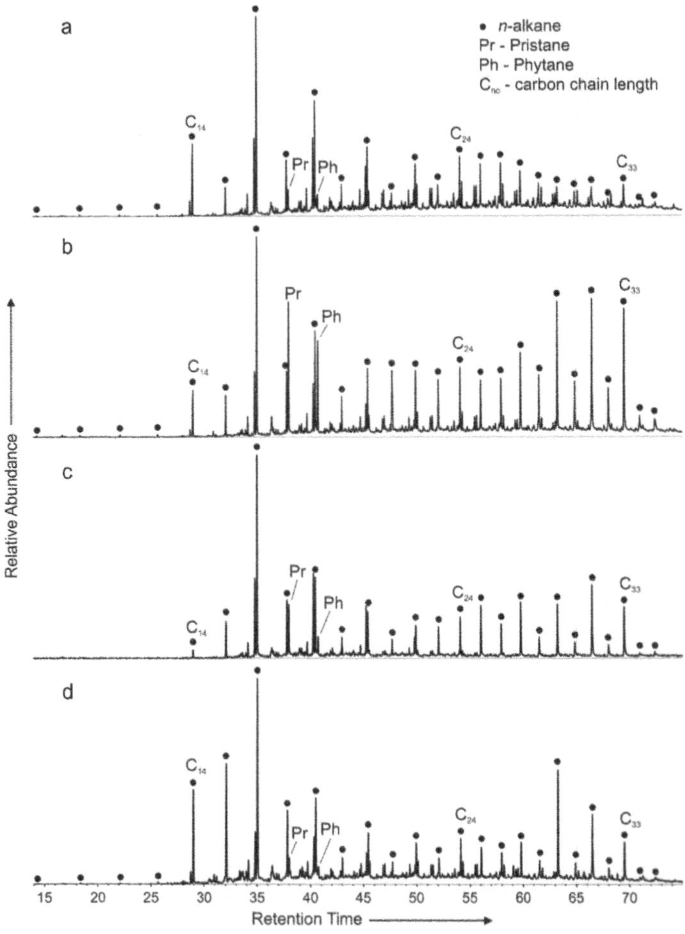

Fig. 7 Partial mass chromatograms (*m/z* 57) showing distribution of *n*-alkanes of **a** calcareous clay (V/Z1/15), **b** shale (V/Z2/Sh/09/5), **c** claystone (V/Z1/11) and **d** lignite (V/Z1/L/21) samples from Vastan lignite mine

homolog, while only C_{30} 17α(H), 21β(H)–hopane and C_{31} 17α(H),21β(H)–homohopane (both 22S and 22R epimers) have been detected in relatively low amount. Only one moretane, 17β(H),21α(H)–30–normortane, has been detected in the samples which co-elutes with urs–12–ene.

8,14–Secohopane co-eluting with olean–18–ene has been detected in the calcareous clay. 8,14–Secohopane has been reported from Nigerian crude oils [41], and it was presumed that hopanes cleave at the C-ring to form secohopane.

Some monounsaturated hopenes are also abundant in the samples and comprise 22,29,30–trisnorhop–13(18)–ene, 28,30–bisnorhop–13(18)–ene, 30–norneohop–13 (18)–ene, hop–17(21)–ene and neohop–13(18)–ene. 30–Norneohop–13(18)–ene co-elutes with olean–13(18)–ene.

Table 4 Geochemical parameters of *n*-alkanes, isoprenoid alkanes and triterpenoids of samples from Vastan lignite mine

Sample no.	A	B	C	D	E	F	G	H
V/Z1/32	0.96	51.94	18.63	29.42	0.32	0.55	1.81	2.42
V/Z1/L/27	2.47	43.95	17.20	38.84	0.28	0.35	4.40	2.10
V/Z1/25	2.43	21.80	10.28	67.91	0.35	0.67	3.36	2.47
V/Z1/L/21	2.03	51.71	13.58	34.69	0.49	0.49	3.11	1.62
V/Z1/19	0.75	58.85	25.57	15.58	0.32	0.30	1.68	1.65
V/Z1/L/18	2.09	32.70	17.18	50.11	0.30	0.37	3.52	2.25
V/Z1/16	0.96	57.71	23.34	18.94	0.27	0.34	2.34	1.58
V/Z1/15	0.92	43.49	21.50	35.00	0.22	0.42	1.40	1.32
V/Z1/14	0.58	41.35	37.91	20.73	0.25	0.34	1.30	0.84
V/Z1/13	0.87	45.53	20.34	34.13	0.33	0.34	1.06	1.31
V/Z1/L/12	0.98	48.61	29.09	22.29	0.26	0.40	1.68	1.33
V/Z1/11	2.73	37.38	17.07	45.53	0.33	0.58	3.14	2.73
V/Z1/L/9	1.28	14.35	10.98	74.66	0.34	0.79	3.00	5.44
V/Z1/7	0.82	68.99	21.03	9.96	0.11	0.32	1.94	1.34
V/Z1/6	0.55	69.77	18.68	11.54	0.48	0.34	0.99	1.66
V/Z1/5	0.72	53.03	24.52	22.43	0.20	0.31	1.12	1.34
V/Z1/4	0.68	71.97	22.81	5.21	0.19	0.24	1.92	1.53
V/Z1/L/2	1.61	49.17	17.41	33.40	0.30	0.60	2.76	4.84
V/Z2/L/7	1.78	44.20	27.41	28.38	0.20	0.26	3.44	1.17
V/Z2/L/09/13	0.83	76.72	13.75	9.51	0.05	0.25	3.07	1.30
V/Z2/L/09/9	0.75	57.69	23.86	18.44	0.17	0.37	1.34	0.97
V/Z2/L/09/7	0.62	87.88	8.90	3.22	0.06	0.32	1.46	1.42
V/Z2/Sh/09/5	2.03	24.51	20.50	54.98	0.30	0.96	2.20	1.42
V/Z2/Sh/09/4	0.60	72.31	19.66	8.02	0.32	0.29	1.34	1.30
V/Z2/L/09/3	1.63	34.03	22.91	43.05	0.45	0.60	3.06	5.52
V/Z2/Sh/09/1	1.13	37.59	18.37	44.03	0.25	0.49	2.32	2.01

Sample no.	*I*	*J*	*K*	*L*	*M*	*N*	*O*
V/Z1/32	5.10	0.16	0.34	0.63	3.61	0.61	0.58
V/Z1/L/27	0.76	0.12	0.17	0.90	1.11	1.85	0.81
V/Z1/25	0.54	0.18	0.26	0.80	0.54	0.87	0.93
V/Z1/L/21	0.31	0.19	0.11	0.90	1.87	1.48	0.83
V/Z1/19	0.46	0.13	0.08	0.85	6.03	1.28	1.2
V/Z1/L/18	0.45	0.10	0.07	0.92	0.56	1.54	0.51
V/Z1/16	0.73	0.18	0.17	0.79	2.91	0.76	1.66
V/Z1/15	0.52	0.20	0.28	0.75	1.20	0.63	1.48
V/Z1/14	0.52	0.21	0.25	0.68	3.22	1.66	2.10
V/Z1/13	0.54	0.22	0.25	0.61	1.06	1.25	1.62
V/Z1/L/12	0.74	0.21	0.35	0.63	2.22	0.99	1.71
V/Z1/11	0.98	0.27	0.13	0.79	0.93	0.71	0.85

(continued)

Table 4 (continued)

Sample no.	A	B	C	D	E	F	G	H
V/Z1/L/9	5.79	0.67	0.16	0.88	0.98	0.29	1.32	
V/Z1/7	0.82	0.14	0.36	0.70	2.93	0.80	N.D.	
V/Z1/6	0.29	0.16	0.35	0.53	17.24	1.07	1.36	
V/Z1/5	0.50	0.16	0.24	0.62	2.41	1.89	0.49	
V/Z1/4	0.59	0.10	0.22	0.68	4.18	1.20	0.52	
V/Z1/L/2	2.35	0.31	0.05	0.89	0.61	1.08	0.39	
V/Z2/L/7	0.61	0.13	0.17	0.87	1.41	5.15	0.39	
V/Z2/L/09/13	0.80	0.09	N.D.	N.D.	2.37	0.73	N.D.	
V/Z2/L/09/9	0.50	0.23	0.09	0.83	1.92	1.67	0.31	
V/Z2/L/09/7	0.79	0.12	0.35	0.79	4.86	1.29	N.D.	
V/Z2/Sh/09/5	2.34	1.07	0.21	0.79	0.68	0.85	1.44	
V/Z2/Sh/09/4	0.59	0.20	0.22	0.75	8.49	0.97	1.39	
V/Z2/L/09/3	3.50	0.34	0.12	0.87	0.93	1.56	1.41	
V/Z2/Sh/09/1	1.13	0.30	0.24	0.79	1.06	0.64	0.17	

A—CPI = $2(C_{23} + C_{25} + C_{27} + C_{29})/[C_{22} + 2(C_{24} + C_{26} + C_{28}) + C_{30}]$
B—$\sum nC_{13-18}/\sum nC_{10-35}$
C—$\sum nC_{19-24}/\sum nC_{10-35}$
D—$\sum nC_{25-35}/\sum nC_{10-35}$
E—nC_{13-18} odd/even
F—nC_{19}–C_{24} odd/even
G—nC_{25}–C_{35} odd/even
H—Pr/Ph
I—Pr/nC_{17}
J—Ph/nC_{18}
K—C_{31} 17α(H),21β(H)–hopane ratio [2S/(22S + 22R)]
L—C_{30}–hopane ratio [ββ/(αβ + ββ)]
M—nC_{17}/nC_{27}
N—nC_{29}/nC_{31}
O—Hopene/oleanene
N.D.—Not detected

30–Norneohop–13(18)–ene present in these samples is originally derived by demethylation of neohop–13(18)–ene. Neohop–13(18)–ene and hop–17(21)–ene are diagenetic products of diplotene (hop–22(29)–ene) and are derived due to acid (clay)-catalysed rearrangement of the same [42, 43].

The six-membered E-ring pentacyclic triterpenoids detected in the present analysis comprise oleanane- and ursane-type compounds which include olean–13 (18)–ene, olean–12–ene, olean–18–ene and urs–12–ene.

It has been suggested that during early diagenesis of sediments, olean–12–ene is originated by isomerization of β–amyrin (olean–12–en–3β–ol) [44]. With progressive diagenesis, further isomerization of this olean–12–ene leads to the formation of olean–13(18)–ene and olean–18–ene. Likewise, isomerization of α–amyrin (urs–12–en–3β–ol) yields urs–12–ene. Although only monounsaturated compounds are predominant in the extracts, two di-unsaturated compounds,

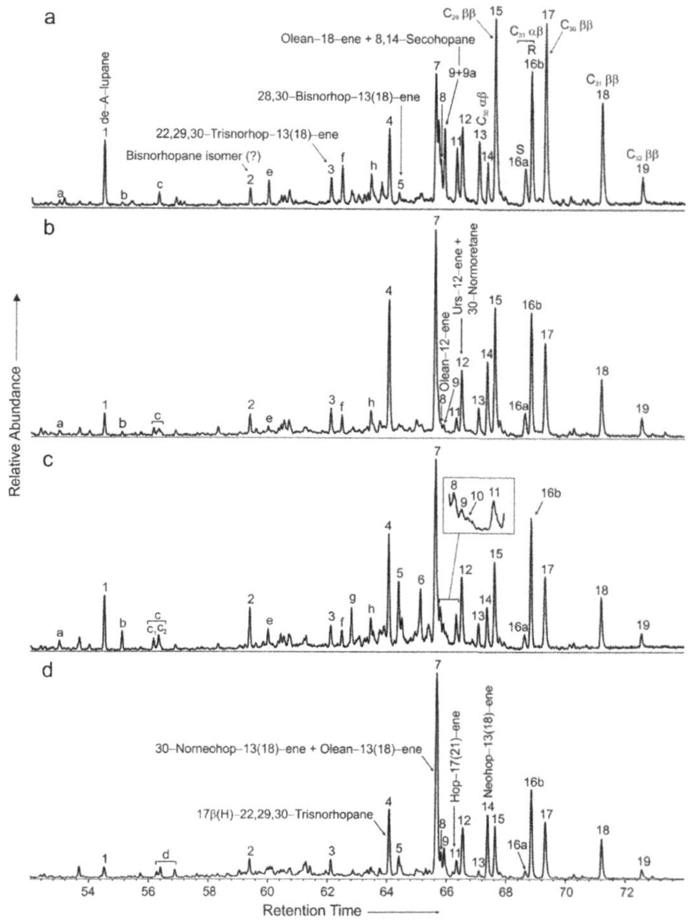

Fig. 8 Partial mass chromatograms (*m/z* 191) showing distribution of triterpenoids of **a** calcareous clay (V/Z1/15), **b** shale (V/Z2/Sh/09/5), **c** claystone (V/Z1/11) and **d** lignite (V/Z1/L/21) samples from Vastan lignite mine

olean–2,13(18)–diene and urs–2,12–diene, have been identified in the claystone sample. Olean–2,12–diene and urs–2,12–diene are derived from the precursors β– and α–amyrin, respectively, by dehydration of alcohol group [45]. These dienes transformed into olean–12–ene and urs–12–ene with progressive diagenesis. Acid-catalysed double bond rearrangement of olean–2,12–diene can produce olean–2,13(18)–diene.

Some tetracyclic C_{24} terpenoids (MW 330) are detected in low abundance in the *m/z* 191 partial mass chromatograms and identified as de–A–lupane following published work [46]. De–A–lupane is predominant in claystone, shale and calcareous clay samples. The mass spectra of other two C_{24} terpenoids resemble with the mass spectra of de–A–oleanane [47]. Minor amount of few other compounds (MW 328) is

Table 5 Major biomarkers present in the samples from Vastan lignite mine

Peak no.	Compound name	Molecular weight	Base peak
1.	de-A-Lupane	330	123
2.	Bisnorhopane isomer (?)	384	191
3.	22,29,30-Trisnorhop-13(18)-ene	368	191
4.	17β(H)-22,29,30-Trisnorhopane	370	149
5.	28,30-Bisnorhop-13(18)-ene	382	191
6.	Olean-2,13(18)-diene	408	189
7.	30-Norneohop-13(18)-ene & olean-13(18)-ene	396 and 410	191 and 205
8.	Olean-12-ene	410	218
9.	Olean-18-ene	410	204
9a.	8,14-Secohopane	414	123
10.	Ursa-2,12-diene	408	218
11.	Hop-17(21)-ene	410	191
12.	Urs-12-ene &17β(H),21α(H)-30-normoretane	410 and 398	218 and 177
13.	17α(H),21β(H)-Hopane	412	191
14.	Neohop-13(18)-ene	410	191
15.	17β(H),21β(H)-30-Norhopane	398	177
16a.	17α(H),21β(H)-Homohopane (S)	426	191
16b.	17α(H),21β(H)-Homohopane (R)	426	191
17.	17β(H),21β(H)-Hopane	412	191
18.	17β(H),21β(H)-Homohopane	426	205
19.	17β(H),21β(H)-Bishomohopane	440	219
19a.	de-A-Olean-13(18)-ene (?)	328	313
$19a_1$.	Isomer of a	328	313
$19a_2$.	Isomer of a	328	109
$19a_3$.	Isomer of a	328	313
19b.	de-A-Oleanene isomer (?)	328	328
$19c_1$.	de-A-Oleanane isomer (?)	330	109
$19c_2$.	de-A-Oleanane isomer (?)	330	136
19d.	Tetrakisnorhopane (?)	356	191
19e.	Unknown compound	382	229
19f.	Unknown tricyclic terpenoid	382	191
19g.	Unknown compound	408	190
19h.	Unknown tricyclic terpenoid	382	109

present in the samples which are possibly de–A–olean–13(18)–ene and its isomers [48]. De–A–oleananes and de–A–lupanes are derived from pentacyclic oleanane- and lupane-type terpenoids. De–A–lupane probably originates from 3–oxygenated precursor, lupan–3–one by photochemical degradation [49].

5 Discussions

5.1 Type of Kerogen and Source Rock Potential

Rock-Eval parameters together with FTIR results aid in determining type of kerogen and source rock potential. The FTIR spectra of all lignite samples indicate abundance of aromatic and phenolic components. Kerogen classification has been done following the technique of Ganz and Kalkreuth [50] and by calculating A-factor and C-factor. However, in the present study, instead of peak intensity, band area has been used following the work of Guo and Bustin [37]. A-factor represents the relative abundance of aliphatic components, while C-factor denotes changes in the C=O groups. 'A' and 'C' factors of the samples have been plotted in a diagram (Fig. 6) similar to that of a traditional van Krevelen H/C–O/C plot. The lignite samples of Vastan represent a mixed type II and III kerogen (Fig. 6). This proposition is also in accordance with the results obtained from cross-plots between hydrogen index (HI) versus oxygen index (OI) (Fig. 4b) and S_2 versus TOC plot (Fig. 4a). The S_2 versus TOC plot describes more clearly that some of the samples are classified as type II kerogen, others as type III kerogen. This would indicate relative abundance of aliphatic components which could be better explained by the maceral composition of these lignites obtained from petrographic studies. Huminite (avg. ~57%) is the prevailing maceral group though substantial proportion of hydrogen-rich liptinite group macerals (avg. ~26%), and especially alginate, sporinite, cutinite, suberinite, resinite, liptodetrinite are also present in the studied lignites. The overall maceral composition of these lignites, therefore, corroborates the geochemical data representing type III kerogen with subordinate amount of type II kerogen. Pr/Ph ratio of the samples indicates oxic depositional environment, and the Pr/nC_{17} versus Ph/nC_{18} cross-plot (Fig. 9) of the samples denotes early mature stage of the samples (Pr/nC_{17} and Ph/nC_{18} decreases with increasing maturity) with terrigenous type III to mixed type II/III organic matter.

The hydrocarbon-generating potential of coal-bearing sequences depends on the maceral composition. Liptinite macerals with a volume percentage of 15–20% are considered to have potential for generating oil [51, 52]. However, liptinite is not the sole producer of oil. Some researchers have shown that perhydrous vitrinite (hydrogen-rich) can also produce oil [9, 53, 54]. Vastan lignites show liptinite content varying from 10 to 31%. In fact, the lower seam samples are quite rich in perhydrous huminite (avg. ~13%). Besides, geochemical studies reveal a mixed type II/III kerogen. While type III kerogen is gas-prone, type II kerogen produces both oil and gas. This in turn suggests potential for generating both oil and gas of these lignites. T_{max} values of the samples are low suggesting immature source rock. Furthermore, the immature nature of source rock has been affirmed by the presence of hopenes, ββ–hopanes and predominance of ββ–isomers for C_{30}–C_{31} hopanes over αβ-hopanes in the solvent extracts (Fig. 8). Vitrinite reflectance (R_r) values of

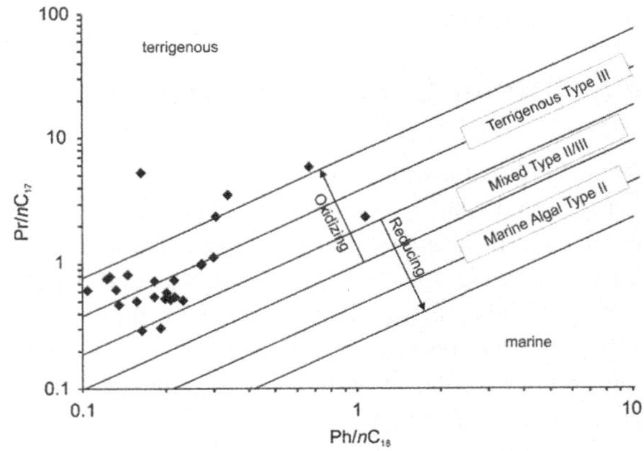

Fig. 9 Pr/nC_{17} versus Ph/nC_{18} plot of samples from Vastan lignite mine

the lignite samples (upper seam: 0.24–0.35%; lower seam: 0.30%; Tables 1 and 2) as well as 22S/[22S + 22R]–homohopane ratio (0.05–0.35; Table 4; ratio increases with progressive diagenesis, and equilibrium value 0.6 occurs at vitrinite reflectance of 0.5–0.6%) also point towards the immature nature of Vastan lignites.

5.2 Source Vegetation and Depositional Environment

Biomarkers along with maceral composition can be an excellent tool for determining source vegetation and depositional environment. Predominance of huminite in general indicates wood-dominated ancient vegetation as the main source for the formation of Vastan lignites. Although *n*-alkanes are widespread in biosphere, the source input of a particular organic matter can be assessed based on distribution pattern, carbon preference index (CPI), carbon number maximum (C_{max}) and odd–even predominance of *n*-alkane series. Representative lignite, shale and claystone samples from Vastan lignite mine show a bimodal *n*-alkane distribution (see Sect. 4.4). In majority of the samples, the first mode predominates as indicated by the high relative proportion, almost 14–87% (Table 4), of short-chain *n*-alkanes (C_{13}–C_{18}) to sum of *n*-alkanes (C_{10}–C_{35}). Short-chain *n*-alkanes in this range with C_{max} at C_{16} or C_{18} and even-chain predominance are usually attributed to algae and micro-organisms [55, 56]. However, even-over-odd short-chain *n*-alkanes with C_{max} at C_{16} or C_{18} have also been reported from woody vegetation and grasses [57] as well as produced by the bacterial reduction of fatty acids and alcohols (C_{14}–C_{18}) from waxes in anoxic depositional condition [58]. Algae can be discarded as a possible source here as very few alginite macerals have been found in the samples (Tables 1 and 2). Besides, algal sourced short-chain *n*-alkanes are characterized by high CPI values. Relatively low CPI values of the samples (0.5–2.7) argue against this proposition. Formation of short-chain *n*-alkanes due to reduction of fatty acid and alcohols from waxes in reducing condition is often associated with low Pr/Ph values. Pr/Ph values of the samples ranged from 0.8 to 2.7; only 3 lignite samples show high

Pr/Ph ratio, in the range 4.8–5.5. Therefore, it is very likely that these short-chain *n*-alkanes are derived from bacteria and micro-organisms. Microbial input in the samples can also be corroborated from the presence of high amount of hopanoids in the samples. Hopenes detected in the samples are also indicative of microbial activity as they are the diagenetic products of bacterially derived diplotene (hop–22 (29)–ene). Although the presence of diplotene has been reported in living organisms as well as diagenetic product in sediments, the main source of hopanes is however bacterial cell membrane [59]. Likewise, ββ–hopane series present in the samples has biogenic configuration of their supposed precursor, bacteriohopanetetrol, synthesized only by bacteria [60, 61]. However, hopenes are present here along with the ββ–hopanes in high concentration. This roughly indicates that these hopanes might have been derived from an additional microbial precursor such as bacteriohopane polyol rather than a C_{30} precursor like diplotene. *n*–Alkanes of intermediate molecular weight (C_{19}–C_{24}) are found to be present in relative proportion of 8–37% (Table 4) with a slight even-carbon predominance most likely derived from their reported precursor *Sphagnum* or aquatic macrophytes [62, 63].

Conversely, the land plants which synthesize solid waxes as external lipids to prevent transpiration, minimize mechanical damage and inhibit fungal or insect attack show a strong predominance of odd-carbon chain in the range of C_{27}–C_{33} which usually maximize at C_{29} or C_{31} [64]. Plants synthesize these leaf epicuticular waxes directly, or they are derived from even-carbon fatty acids from cuticular waxes which during diagenesis produce odd-carbon *n*–alkanes by defunctionalization. C_{max} at C_{31} is usually attributed to C_4 grasses, whereas C_{max} at C_{27} or C_{29} indicates dominance of woody species of angiosperms [57, 65]. The second mode in the range C_{25}–C_{35} with odd-carbon predominance (Table 4) and C_{max} at C_{29} or C_{31} of the samples therefore is attributed to higher plant input, both C_4 grass and C_3 woody angiosperm. Most of the samples are characterized by either near-equal intensities of C_{29} and C_{31} or a predominance of C_{31} (Table 4) representing significant C_3 plant (woody angiosperms) input additional to C_4 grasses. Hence, predominance of short-chain *n*–alkanes over long-chain *n*–alkane in most of the samples (nC_{17}/nC_{27} ratio; Table 4), especially claystone samples, infers higher microbial source input compared to higher plant wax input. Higher plant contribution can further be confirmed by the detection of oleanenes and ursenes in the soluble organic matter. Basically, the β– and α–amyrins present in the angiosperms in turn give rise to oleanenes and ursenes with progressive diagenesis. De–A–oleananes and de–A–lupanes are also derived from their corresponding terpenoids present in angiosperms. Furthermore, higher amount of detrohuminite in lignites suggests the contribution of herbaceous and pteridophytic plants growing profusely in the dense angiospermic forests (corroborated by palynological studies [21]). A few compounds with MW 382, base peak 243 and fragmentation pattern quite similar to fernenes are present in the samples in minor amount. These compounds may be degraded C_{28} fernenes [66]. Fernenes are mainly produced from ferns; however, they have been reported from some other organisms such as lichens and some anaerobic photosynthetic bacteria (cited by [66, 67]). Besides, neohop–13 (18)–ene and neohop–12–ene detected in the samples have been reported from ferns

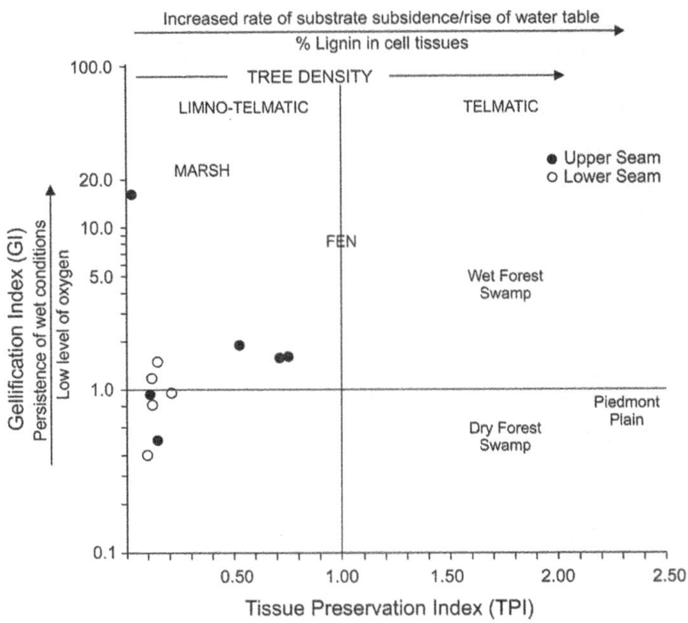

Fig. 10 GI versus TPI plot of studied lignites from Vastan lignite mine (arrows indicate the change of the depositional environment through time)

[68]. Neohop–13(18)–ene may also be produced by acid-catalysed rearrangement of fernene [42]. Corpohuminite, the oxidation and condensation products of secretions from wood parenchyma, medullary rays and cork cells mainly from mangrove and some angiospermic plants also point towards the existence of mangrove mixed angiospermic forest during Vastan lignite formation.

The presence of 17β(H), 21β(H)–hopane series in the samples indicates microbial origin and immature nature of organic matter or mild thermal history during burial and deposition of sediments [69, 70]. 17α(H),21β(H)–hopane stereo isomers might be formed due to isomerization of ββ–hopane configuration produced by the rearrangement in the presence of acid-activated clay minerals which act as catalysts. Likewise, high detrohuminite and liptodetrinite fractions as well as low tissue preservation index (0.025–0.75: Tables 1 and 2; Fig. 10) suggest high biological and bacterial degradation of ancient peat and elevated pH conditions of swamp.

Moderate gelification index (0.4–1.91), with marked increase at the top (16.2) and bottom (1.9) parts of upper seam and slight increase at the top of lower seam (1.5), indicates increased wetness in the basin swamp. High amount of ulminite in the lignites well corroborates with the high gelification index values. Low amount of inertinite and its close association with detrohuminite point towards the formation of inertinite at early diagenetic stage during the intermittent aerial exposure of peat surface [36, 71]. This is also evident from Pr/Ph ratios of the samples which are always >1.0 indicating oxic depositional environment. The Pr/nC_{17} ratios < 0.5 and >1 are usually attributed to open-marine environment and inland peat swamps, respectively. Here, the Pr/nC_{17} ratios of the samples vary from 0.3 to 2.34 indicating deposition in inland peat swamps except the one with Pr/nC_{17} ratio 0.3 indicates marine influence. The ratio of 17β(H)21β(H) to (17β(H)21β(H) + 17α(H) 21β(H) C_{30}–hopane (ββ/ββ + αβ C_{30}–hopane) is slightly higher (0.63–0.90; Table 4) than the range generally measured in lignites (0.5–0.7; [72]), consistent with deposition in a suboxic to slightly oxic environment.

6 Conclusions

The Eocene lignite-bearing sequence in Cambay Basin is analysed for comprehending the source vegetation, source rock potential as well as the palaeoenvironment of deposition. The maceral compositions of the lignites indicate that huminite is the dominant maceral present along with liptinite and inertinite. Huminite is indicative of wood-dominated vegetation. Reflectance values of huminites as well as Rock-Eval T_{max} suggest that the lignites occur in the early diagenetic zone of methane generation. The kerogen is classified as mixed type II/ type III as denoted by Rock-Eval parameters. The Rock-Eval pyrolysis results clearly suggest that the lignite-bearing sequence has mixed oil and gas-generating potential upon maturation. FTIR analysis demonstrates the functional groups present and also corroborates the kerogen type. The homologous series of normal alkanes display a bimodal distribution reflecting microbial as well as higher plant sources. Several parameters (tissue preservation index and gelification index) determined from maceral data reflect wet and elevated acidic condition of the depositional environment. The biomarker distribution is characterized by hopanes, monounsaturated hopenes and six-membered E-ring pentacyclic triterpenoids bearing mono- and di-unsaturations. The hopanes and hopenes are primarily derived from microbial components like bacteria. The presence of hopenes suggests immature source rock. However, hopanes are products generated at matured stages of diagenesis. The concurrent presence of hopanes and hopenes may suggest that they share separate precursor and genesis. The six-membered E-ring compounds present, viz. olean–13(18)–ene, olean–12–ene, olean–18–ene, urs–12–ene, olean–2,13(18)–diene, urs–2,12–diene, are intermediate diagenetic products derived from angiosperm precursors like α-amyrin and β-amyrin. Compounds with cleaved A-ring such as de–A–oleanane and de–A–lupane have been detected which also share angiosperm precursor. Fernenes supposedly derived from ferns are recognized in the samples, thus denoting pteridophytic input as well into the organic matter. Finally, it is concluded that angiosperm dominated vegetation served as the major source for the lignites.

Acknowledgements CSIR-NGRI, Dr. V.M. Tiwari, Director of CSIR-NGRI, and Dr. E.V.S.S.K. Babu are acknowledged for providing support to M. Mallick. The authors are thankful to IIT Bombay, India, for providing infrastructure.

References

1. Prasad V, Farooqui A, Tripathi SKM, Garg R (2009) Evidence of Late Palaeocene-Early Eocene equatorial rain forest refugia in south-western Ghats, India. J Biosci 34:1–21
2. McGowran B (1989) Silica burp in the Eocene ocean. Geology 17:857–860
3. Sahni A, Saraswati PK, Rana RS, Kumar K, Singh H, Alimohammadian H, Sahni N, Rose KD, Singh L, Smith T (2006) Temporal constraints and depositional palaeoenvironments of the Vastan Lignite Sequence, Gujarat: Analogy for the Cambay Shale Hydrocarbon Source Rock. Ind J Pet Geol 15:1–20
4. Ahmed M, Volk H, George SC, Faiz M, Stalker L (2009) Generation and expulsion of oils from Permian coals of the Sydney Basin, Australia. Org Geochem 40:810–831
5. Brooks JD, Smith JW (1969) The diagenesis of plant lipids during the formation of coal, petroleum and natural gas: II. Coalification and the formation of oil and gas in the Gippsland Basin. Geochim Cosmochim Acta 33:1183–1194
6. Shanmugam G (1985) Significance of coniferous rain forests and related organic matter in generating commercial quantities of oil, Gippsland Basin, Australia. AAPG Bull 69:1241–1254
7. Hirner AV, Lyon GL (1989) Stable isotope geochemistry of crude oils and of possible source rocks from New Zealand: 1. Carbon. Appl Geochem 4:109–120
8. Hirner AV, Robinson BW (1989) Stable isotope geochemistry of crude oils and of possible source rocks from New Zealand: 2. Sulfur. Appl Geochem 4:121–130
9. Killops SD, Woolhouse AD, Weston RJ, Cook RA (1994) A geochemical appraisal of oil generation in the Taranaki Basin, New Zealand. AAPG Bull 78:1560–1585
10. Hoffmann CF, Mackenzie AS, Lewis CA, Maxwell JR, Oudin JL, Durand B, Vandenbroucke M (1984) A biological marker study of coals, shales and oils from the Mahakam Delta, Kalimantan, Indonesia. Chem Geol 42:1–23
11. Horsfield B, Yordy KL, Crelling JC (1988) Determining the petroleum-generating potential of coal using organic geochemistry and organic petrology. Org Geochem 13:121–129
12. Peters KE, Snedden JW, Sulaeman A, Sarg JE, Enrico RJ (2000) A new geochemical-sequence stratigraphic model for the Mahakam Delta and Makassar Slope, Kalimantan, Indonesia. AAPG Bull 84:12–44
13. Odden W, Patience RL, van Graas GW (1998) Application of light hydrocarbons (C_4-C_{13}) to oil/source rock correlations: a study of the light hydrocarbon compositions of source rocks and test fluids from offshore Mid-Norway. Org Geochem 28:823–847
14. Petersen HI, Rosenberg P, Andsbjerg J (1996) Organic geochemistry in relation to the depositional environments of Middle Jurassic coal seams, Danish Central Graben, and implications for hydrocarbon generative potential. AAPG Bull 80:47–62
15. Petersen HI, Andsbjerg J, Bojesen-Koefoed JA, Nytoft HP (2000) Coal-generated oil: source rock evaluation and petroleum geochemistry of the Lulita oilfield, Danish North Sea. J Pet Geol 23:55–90
16. Petersen HI, Brekke T (2001) Source rock analysis and petroleum geochemistry of the Trym discovery, Norwegian North Sea: A Middle Jurassic coal-sourced petroleum system. Mar Petrol Geol 18:889–908
17. Hendrix MS, Brassell SC, Carroll AR, Graham SA (1995) Sedimentology, organic geochemistry, and petroleum potential of Jurassic coal measures: Tarim, Junggar, and Turpan Basins, Northwest China. AAPG Bull 79:929–959

18. Ramaswamy G (2005) Some fields in India's North Cambay Basin have oil derived from nearby lignite seams. Oil Gas J 103:37–42
19. Bajpai S, Kay RF, Williams BA, Das DP, Kapur VV, Tiwari BN (2008) The oldest Asian record of Anthropoidea. PNAS 105:11093–11098
20. Rose KD, DeLeon VB, Missiaen P, Rana RS, Sahni A, Singh L, Smith T (2008) Early Eocene lagomorph (Mammalia) from western India and the early diversification of Lagomorpha. Proc R Soc Lond B 275:1203–1208
21. Mandal J, Guleria JS (2006) Palynology of Vastan lignite (Surat District), Gujarat: its age, palaeoecology and depositional environment. Palaeobotanist 55:51–66
22. Garg R, Ateequzzaman K, Singh V, Tripathi SKM, Singh IB, Jauhri AK, Bajpai S (2008) Age-diagnostic dinoflagellate cysts from the lignite-bearing sediments of the Vastan Lignite Mine, Surat District, Gujarat, Western India. J Palaeontol Soc Ind 53:99–105
23. Clementz M, Bajpai S, Ravikant V, Thewissen JGM, Saravanan N, Singh IB, Prasad V (2010) Early Eocene warming events and the timing of terrestrial faunal exchange between India and Asia. Geology 39:15–18
24. Punekar J, Saraswati PK (2010) Age of the Vastan lignite in context of some oldest Cenozoic fossil mammals from India. J Geol Soc India 76:63–68
25. Samanta A, Bera MK, Ghosh R, Bera S, Filley T, Pande K, Rathore SS, Rai J, Sarkar A (2013). Do the large carbon isotopic excursions in terrestrial organic matter across Palaeocene-Eocene boundary in India indicate intensification of tropical precipitation? Palaeogeogr Palaeoclimatol Palaeoecol 387:91–103
26. ICCP (International Committee for Coal and Organic Petrology). International Handbook of Coal Petrography (1975) Centre National de la Recherche Scientifique (CNRS), Paris, France, 1st suppl. to 2nd edition (1971), 197 pp, 2nd suppl. to 2nd edition, 60 pp
27. Stach E, Mackowsk M-Th, Teichmüller M, Taylor GH, Chandra D, Teichmüller R (1982) Stach's Textbook of Coal Petrology, 3rd edn. GebrüderBorntraeger, Berlin, p 535
28. Sýkorová I, Pickel W, Christanis K, Wolf M, Taylor GH, Flores D (2005) Classification of huminite—ICCP System. Int J Coal Geol 62:85–106
29. Lafargue E, Marquis F, Pillot D (1998) Rock-Eval 6 applications in hydrocarbon exploration, production and soil contamination studies. Oil Gas Sci Technol 53:421–437
30. Painter P, Starsinic M, Coleman M (1985) Determination of functional groups in coal by Fourier Transform Interferometry. In: Ferraro JR, Basile LJ (eds) Fourier transform infrared spectroscopy, vol. 4. applications to chemical systems, pp 169–241
31. Dutta S, Hartkopf-Fröder C, Witte K, Brocke R, Mann U (2013) Molecular characterization of fossil palynomorphs by transmission micro-FTIR spectroscopy: Implications for hydrocarbon source evaluation. Int J Coal Geol 115:13–23
32. Singh PK, Singh MP, Singh AK (2010) Petro-chemical characterization and evolution of Vastan lignite, Gujarat, India. Int J Coal Geol 82:1–16
33. Dutta S, Mallick M, Bertram N, Greenwood PF, Mathews RP (2009) Terpenoid composition and class of Tertiary resins from India. Int J Coal Geol 80:44–50
34. Dutta S, Tripathi SM, Mallick M, Mathews RP, Greenwood PF, Rao MR, Summons R (2011) Eocene out-of-India dispersal of Asian dipterocarps. Rev PalaeobotPalynol 166:63–68
35. Mallick M, Dutta S, Greenwood PF, Bertram N (2009) Pyrolytic and spectroscopic studies of Eocene resin from Vastan lignite mine, Cambay Basin, western India. J Geol Soc India 74:16–22
36. Taylor GH, Teichmüller M, Davis A, Diessel CFK, Littke R, Robert P (1998) Organic Petrology. GebrüderBorntraeger, Berlin and Stuttgart
37. Guo Y, Bustin RM (1998) Micro-FTIR spectroscopy of liptinite macerals in coal. Int J Coal Geol 36:259–275
38. Guo Y, Renton JJ, Penn JH (1996) FTIR microspectroscopy of particular liptinite-(lopinite) rich, Late Permian coals from southern China. Int J Coal Geol 29:187–197
39. Peters KE, Walters CC, Moldowan JM (2005) The biomarker guide. Volume 2: biomarkers and isotopes in the petroleum exploration and earth history, 2nd edn. Cambridge University Press, Cambridge, UK

40. Powell TG, McKirdy DM (1973) Relationship between ratio of pristane to phytane, crude oil composition and geological environment in Australia. Nature 243:37–39
41. Schmitter JM, Sucrow W, Arpino PJ (1982) Occurrence of novel tetracyclic geochemical markers: 8, 14-secohopanes in a Nigerian crude oil. Geochim Cosmochim Acta 46:2345–2350
42. Ageta H, Shiojima K, Arai Y (1987) Acid-induced rearrangement of triterpenoid hydrocarbons belonging to the hopane and migrated hopane series. Chem Pharm Bull 35:2705–2716
43. Moldowan JM, Fago FJ, Carlson RMK, Young DC, van Duyne G, Clardy J, Schoell M, Pillinger CT, Watt DS (1991) Rearranged hopanes in sediment sand petroleum. Geochim Cosmochim Acta 55:3333–3353
44. Rullkötter J, Peakman TM, ten Haven HL (1994) Early diagenesis of terrigenous terpenoids and its implications for petroleum geochemistry. Org Geochem 21:215–223
45. ten Haven HL, Peakman TM, Rullkötter J (1992) Δ2-Triterpenes: Early intermediates in the diagenesis of terrigenous triterpenoids. Geochim Cosmochim Acta 56:1993–2000
46. Schmitter JM, Arpino PJ, Guiochon G (1981) Isolation of degraded pentacyclic triterpenoid acids in a Nigerian crude oil and their identification as tetracyclic carboxylic acids resulting from ring A cleavage. Geochim Cosmochim Acta 45:1951–1955
47. Pearson MJ, Alam M (1993) Bicadinanes and other terrestrial terpenoids in immature Oligocene sedimentary rocks and a related oil from the Surma Basin, N.E. Bangladesh. Org Geochem 20:539–554
48. Stout SA (1992) Aliphatic and aromatic triterpenoid hydrocarbons in Tertiary angiospermous lignite. Org Geochem 18:51–66
49. Corbet B, Albrecht P, Ourisson G (1980) Photochemical or photometric fossil triterpenoids in sediments and petroleum. J Am Chem Soc 102:1171–1173
50. Ganz H, Kalkreuth W (1987) Application of infrared spectroscopy to the classification of kerogen-types and the evolution of source rock and oil-shale potentials. Fuel 66:708–711
51. Hunt JH (1991) Generation of gas and oil from coal and other terrestrial organic matter. Org Geochem 17:673–680
52. Mukhopadhyay PK, Hatcher PG (1993) Composition of coal. In: Law BE, Rice DD (eds) Hydrocarbons from Coal, vol 38. American Association of Petroleum Geologists, pp 79–118
53. Killops SD, Funnell RH, Suggate RP, Sykes R, Peters KE, Walters C, Woolhouse AD, Weston RJ, Boudou J-P (1998) Predicting generation and expulsion of paraffinic oil from vitrinite-rich coals. Org Geochem 29:1–21
54. Newman J, Price LC, Johnston JH (1997) Hydrocarbon source potential and maturation in Eocene New Zealand vitrinite-rich coals. J Pet Geol 20:137–163
55. Cranwell PA (1977) Organic geochemistry of CamLoch (Sutherland) sediments. Chem Geol 20:205–221
56. Cranwell PA (1984) Lipid geochemistry of sediments from Upton Broad, a small productive lake. Org Geochem 7(1):25–37
57. Kuhn TK, Krull ES, Bowater A, Grice K, Gleixner G (2010) The occurrence of short chain n-alkanes with an even over odd predominance in higher plants and soils. Org Geochem 41:88–95
58. Bechtel A, Sachsenhofer RF, Kolcon I, Gratzer R, Otto A, Püttmann W (2002) Organic geochemistry of the Lower Miocene Oberdorf lignite (Styrian Basin, Austria): its relation to petrography, palynology and the paleoenvironment. Int J Coal Geol 51:31–57
59. del Rio JC, Gonzalez-Vila FJ, Martin F (1992) Variation in the content and distribution of biomarkers in two closely situated peat and lignite deposits. Org Geochem 18:67–78
60. Seifert WK, Moldowan JM (1980) The effect of thermal stress on source-rock quality as measured by hopane stereochemistry. Phys Chem Earth 12:229–237
61. Seifert WK, Moldowan JM (1986) Use of biological markers in petroleum exploration. In: Johns RB (ed) Methods in Geochemistry and Geophysics, vol 24, pp 261–290

62. Nott CJ, Xie S, Avsejs LA, Maddy D, Chambers FM, Evershed RP (2000) n-Alkane distributions in ombrotrophic mires as indicators of vegetation change related to climatic variation. Org Geochem 31:231–235
63. Ficken KJ, Li B, Swain DL, Eglinton G (2000) An n-alkane proxy for the sedimentary input of submerged/floating freshwater aquatic macrophytes. Org Geochem 31:745–749
64. Hunt JM (1995) Petroleum geochemistry and geology. W.H. Freeman, New York
65. Cranwell PA (1973) Chain-length distribution of n-alkanes from lake sediments in relation to post-glacial environmental change. Freshwater Biol 3:259–265
66. Raymond A, Phillips MK, Gennett JA, Comet PA (1997) Palynology and paleoecology of lignites from the Manning Formation (Jackson Group) outcrop in the Lake Somerville spillway of east-central Texas. Int J Coal Geol 34:195–223
67. Hauke V, Graff R, Wehrung P, Trendel JM, Albrecht P, Riva A, Hopfgartner G, Gülacar FO, Buchs A, Eakin PA (1992) Novel triterpene-derived hydrocarbons of the arborane/ fernane series in sediments: Part II. Geochim Cosmochim Acta 56:3595–3602
68. Ageta H, Shiojima K, Arai Y (1968) Fern constituents: neohopene, hopene-II, neohopadiene and fernadiene isolated from Adiantum species. J Chem Soc, Chem Commun 1968:1105–1107. https://doi.org/10.1039/C19680001105
69. Ensminger A, van Dorsselaer A, Spyckerelle C, Albrecht P, Ourisson G (1973) Pentacyclic triterpenes of the hopanetypeas ubiquitous geochemicalmarkers: origin and significance. In: Tissot B, Bienner F (eds) Advances in Organic Geochemistry. Editions Technip, Paris, pp 245–260
70. Rohmer M, Dastillung M, Ourisson G (1980) Hopanoids from C30 to C35 in recent muds, chemical markers and bacterial activity. Naturewissenschaften 67:456–458
71. Teichmüller M (1989) The genesis of coal from the view point of coal petrology. Int J Coal Geol 12:1–87
72. Mackenzie AS, Patience RL, Maxwell JR (1981) Molecular changes and the maturation of sedimentary organic matter. In: Atkinson G, Zuckermann JJ (eds) Origin and chemistry of petroleum. Proc. 3rd Annual Karcher Symposium. Pergamon Press, Oxford, pp 1–31

Molecular Insight of Coal: A Spectroscopic Approach for Evaluating Maturity Parameters

Archchi Sarkar, Ushma Patel, Suryendu Dutta, B. D. Singh, Samir Kumar Pal, and Uttam K. Bhui

1 Introduction

The global demand for energy continues to increase, primarily due to the growing level of industrialization of the developing nations. With the oil reserves dwindling all over the world following its high demand, it is essential to find out new reserves and resources. Coal has always been one of the cheapest and most efficient energy resources in the world, and it already accounts for most of this demand, especially in countries like India and China. The widely distributed and abundant coal reserves, relatively low price, versatility as a fuel and scarcity of petroleum products are possibly the most recurrent advantages for continuing the use of coal. Use of

A. Sarkar · U. Patel · U. K. Bhui (✉)
School of Petroleum Technology, Pandit Deendayal Energy University (PDEU, Formerly Pandit Deendayal Petroleum University-PDPU), Raisan, Gandhinagar 382007, Gujarat, India
e-mail: Uttam.bhui@spt.pdpu.ac.in

A. Sarkar
e-mail: archchi.sphd18@spt.pdpu.ac.in

U. Patel
e-mail: Ushma.pmtpe15@spt.pdpu.ac.in

S. Dutta
Department of Earth Sciences, IIT Bombay, Powai, Mumbai 400076, India
e-mail: s.dutta@iitb.ac.in

B. D. Singh
Birbal Sahni Institute of Palaeosciences, 53 University Road, Lucknow 226007, India
e-mail: bdsinghbsip@yahoo.co.in

S. K. Pal
Department of Chemical Biological and Macromolecular Sciences, Technical Research Centre (TRC), S. N. Bose National Centre for Basic Sciences, JD Block, Sector III, Salt Lake, Kolkata (Calcutta) 700106, India
e-mail: skpal@bose.res.in

U. K. Bhui (ed.), *Macromolecular Characterization of Hydrocarbons for Sustainable Future*, Green Energy and Technology,
https://doi.org/10.1007/978-981-33-6133-1_12

coal as a major chemical feedstock is gaining popularity, and the economic giant China is already on the run to lead this industry [1]. In this competitive scenario, 'clean' coal and its characterization are of prime importance. Although some of the methods applied to coal have changed in response to new technological developments and improvements in analytical techniques, coal science still has a strongly traditional basis. Maturity parameters like vitrinite reflectance (VR_o), lipid biomarker, thermal alteration index (TAI) already have been established firmly in this domain and have been used for characterization purpose since long. With the advancement of technology, it is now possible to thoroughly analyse and more precisely determine the structural parameters to estimate thermal maturity, along with the extraction of information hidden in the innate molecular structure [2].

The macromolecular network of coal is typically made up of aromatic–hydroaromatic clusters containing heteroatoms and side groups, linked by flexible chemical bridges [3]. The relation of kerogen to this structure of coal is important as it is defined as the organic matter present in coal and the definitive structural components are part of the type of kerogen present in the coal [4]. In previous studies, the importance of heteroatomic functionalities in modelling and understanding the kerogen structure was tested greatly. The kinetics of thermal maturation is similar to that of cracking, and larger kerogen molecules are broken into smaller molecules [5]. This decomposition occurs at various places inside the molecular network. Carbonyl and carboxyl groups are eliminated with progressive burial; isoprenoid and other polycyclic aromatic hydrocarbon (PAH) markers are generated during maturation indicating a possible link with the kerogen backbone via ester functionalities. PAHs are molecules containing two or more simple aromatic rings fused together by sharing two neighbouring carbon atoms. In fused aromatics, all carbon–carbon bonds are not necessarily equivalent, as the electrons are not delocalized over the entire molecule. Aromaticity of these molecules can be explained using their orbitals. Like benzene and other monocyclic aromatic molecules, PAHs have a cyclic conjugated π-system with p-orbital overlap above and below the plane of the ring. Moreover, with increasing aromaticity, carboxyl content decreases and ether content increases; a negligible amount of total carbonyl content remains at 80% aromaticity, along with traces of aliphatic sulphur and an increased volume of pyrrolic nitrogen [3, 6, 7].

Spectroscopic investigations are applied in the study of molecular structures for a very long time, especially for the clay minerals [8, 9]. The application of reflectance spectrometry to the mid-IR spectra of coals and other carbonaceous material has been able to produce reliable results [10]. Fluorescence spectroscopy has been commonly used as a selective diagnostic tool for the analysis of PAH molecules in a complex aromatic system. Synchronous fluorescence scan (SFS) spectra allow a general comparison between the different emitting species (molecular components) present. Thus, the aim of this study was to chemically observe the macromolecular network and distinguish the moieties with the help of spectroscopic studies, in order to provide a holistic idea regarding the structural information of the components present in coal with their link to rank or maturity evolution.

2 Materials, Methods and Instrumentation

2.1 Materials

Two coal samples (P1 and P2) were collected from Raniganj which is part of the Gondwana coalfields of Permian age. One coal sample (T1) was collected from the Tertiary coalfields of Shillong, Meghalaya. The samples have different geological origins and different depositional environments. The Raniganj coalfields host one of the best coal measures in India and are characterized as non-coking and coking coals of medium to high rank bituminous. The Tertiary coals have a later origin and are less mature than the Raniganj coals [11–13].

2.2 Methods

For Rock-Eval pyrolysis and spectroscopic analyses, the coal samples were pulverized to 72 mesh, 212-micron size and 1 gm each of the powders was used for the experiments. For petrographic measurements, pellets of 0.85–1 mm size samples were used. FTIR spectroscopy was carried out using bulk samples (on dry basis). For UV–visible and fluorescence spectroscopy, a soluble organic fraction and the residual demineralized fraction were separated. 30 ml of chloroform ($CHCl_3$) was used to extract a soluble fraction from the samples. Individual extract was filtered, and the residue was dried and demineralized using concentrated (69%) nitric acid. After demineralization, samples were washed with 1 N nitric acid and deionized water to remove clay and sulphide minerals. Residues were dried and subsequently dissolved in toluene. Each of the extracts and demineralized residues was analysed in the UV–visible and fluorescence spectrometers.

2.3 Instrumentation

FTIR Spectroscopy: The samples were analysed on dry basis in a Bruker Alpha-II Spectrometer using the attenuated total reflection (ATR) method.
UV–Vis *Spectroscopy*: Instrument used in UV–visible spectroscopy was a Perkin Elmer Lambda 35 UV–Visible Double Beam Spectrometer. Auto-correction of the samples was done prior to the analysis. Graphs were recorded along with baseline correction for all the samples.

Fluorescence Spectroscopy: The fluorescence spectroscopy was carried out using a Perkin Elmer LS 55 spectrophotometer. The emission spectra were acquired from 280 to 500 nm, and the excitation wavelength was constant with the slit width of 10 nm for both the solvent-extracted fraction and the demineralized residual fraction. Auto-correction was done prior to the analysis.

All spectroscopic examinations were undertaken at room temperature for both the solvent-extracted and demineralized residual fractions.

3 Results and Discussion

3.1 Rock-Eval Pyrolysis and VR_o

From the pyrolysis data, it was found that the samples are ranging from immature to mature, oil-prone, in terms of maturity, corresponding to the VR_o values indicating high-volatile bituminous A to high-volatile bituminous C type. Data from Rock-Eval pyrolysis is given in Table 1. Plots of HI versus Tmax to indicate maturity of the samples is shown in Fig. 1.

3.2 FTIR Spectroscopy

The IR spectra show distinguishing features for the Gondwana and tertiary samples. Sharp, small absorption peaks at 3623, 3610, 3631 cm^{-1} correspond to O–H stretching vibrations of inner structural hydrogen-bonded hydroxyl groups in polymeric association and isolated OH groups [14]. A moderate peak at 3439 and weak, broad peak at 3500 might correspond to the O–H stretching vibration of adsorbed water in interlayer positions. Peak at 3384 corresponds to O–H and N–H stretching vibrations [15]. In sample P1, a weak absorption band at 3056 corresponds to C–H stretching in aromatic rings and possibly N–H stretching in amines [14, 16]. This particular band is absent in both samples P2 and T1, indicating their condensed and highly substituted nature [17]. Sharp, moderate peaks at 2911 and 2849 can be assigned to aliphatic and alicyclic CH_3, CH_2 and CH groups although the major contribution is expected to be due to CH_2 group. Greater intensity of the 2911 peak indicates the presence of longer aliphatic chains [18]. The biggest peak observed at 2356 in case of samples P2 and T1 might indicate the presence of an admixture of carbonates [19] or C=C=C and C=C=O indicating the presence of carbonyl groups [20]. The absorption peaks at 1600 cm^{-1} are observed due to the aromatic C=C, vinyl C=C skeletal vibrations [17, 21] and possibly due to other

Table 1 Data from Rock-Eval pyrolysis and vitrinite reflectance

S. no.	Qty (mg)	S1 (mg/g)	S2 (mg/g)	S3 (mg/g)	T_{max} (°C)	TOC (%)	HI	OI	PI	VR_o
P1	6.5	0.62	176.03	4.19	448	69.77	252	6	0.00351	0.81
P2	6.4	0.63	172.89	3.06	445	64.15	270	5	0.003631	0.77
T1	6.3	1.39	234.42	3.43	427	63.45	369	5	0.005895	0.54

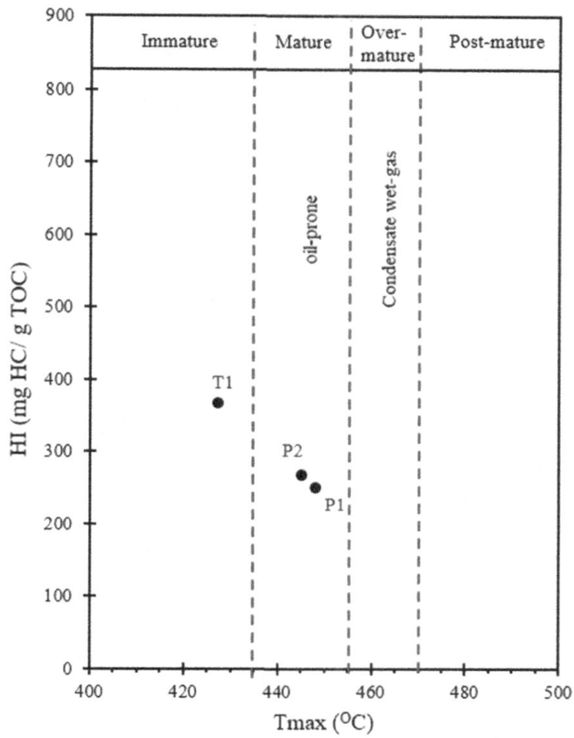

Fig. 1 HI versus T_{max} plot of the bulk coal samples

O-containing functional groups [22]. Strong absorption bands at 1437 (in samples P1 an P2) and 1451 cm^{-1} (in sample T1) were observed mainly due to asymmetric bending of CH3 and CH_2 in bridges but can also be partly due to aromatic C=C and strongly hydrogen-bonded O–H groups [17, 23]. Peak at 1027 cm^{-1} corresponds to Si–O stretching vibrations and presence of quartz and kaolinite; this also indicates the C–O–H deformation in matured cellulose or aromatic C–H in-plane deformation. The bands between 860 and 735 cm^{-1} are assigned to vibrations of aromatic out-of-plane rings with two neighbouring C–H groups. Most of the peaks in the spectra between 400 and 1100 cm^{-1} can be assigned to quartz and clay minerals such as kaolinite, illite and montmorillonite [24]. The small peak at 663 in sample T1 can be assigned to traces of magnetite [17]. Strong and sharp peak at 541 cm^{-1} is due to the Si–O–Al^{VI} bending vibration of clay minerals with Al in octahedral coordination (see Table 2, Fig. 2).

3.3 UV–Visible Spectroscopy

Solvent Extract: For sample P1, there is a cluster of peaks observed at the wavelength range of 247–290 nm with high absorbance value indicating the presence of

Table 2 Band assignments for the FTIR spectra of coal samples

Peak wavelength (nm)			Band assignment(s)
P1	P2	T1	
3610	3631	3623	O–H stretching vibrations of H-bonded hydroxyl groups (inner and isolated)
3439	–	3500	O–H stretching vibrations of H-bonded hydroxyl groups (outer)
–	–	3384	O–H, N–H stretching vibrations
3056	–	–	Symmetric aromatic stretching of C–H; N–H stretching in amine
2911	2911	2911	Symmetric aliphatic and alicyclic stretching of CH_3
2849	2849	2849	Symmetric aliphatic and alicyclic stretching of CH_2
2356	2356	2356	C=C, C=O, carbonyl groups
1600	1600	1600	Aromatic and vinyl skeletal vibrations
1437	1437	1451	CH_3 asymmetric deformation; CH_2 bending vibrations; aromatic C=C and strongly hydrogen-bonded O–H groups
1027	1027	1027	Si–O stretching vibration; C–O–H deformation in cellulose
860–735	860–735	860–735	Vibration in aromatic rings with two neighbouring C–H groups (out-of-plane)
746	–	–	Aromatic C–H bending vibration
541	541	541	Si–O–Al^{VI} vibrations of clay minerals (Al in octahedral coordination)

carbonyl compounds along with some aromatic traces. Another cluster identified in the range of 310–325 nm with high absorbance values shows the presence of alkenes, alkynes and carbonyl compounds. Other prominent peaks in the range of 340–383 nm and the single peak at 421 nm depict the presence of asphaltene porphyrins and heavy aromatic components. The peak corresponding to porphyrins at 421 nm is absent in the extract of sample P2. For sample P2, the peaks at wavelengths of 260, 275 and 290 nm indicate the possibility of presence of carbonyl compounds along with aromatic traces. Other peaks identified at 325 and 340 nm show the presence of alkenes, alkynes and carbonyl compounds, similar to sample P1. For sample T1, peaks in the 200–383 nm region become saturated when normalized, possibly indicating that the corresponding components are present in a higher concentration. The abrupt increase in the absorbance value in all the samples near the 260 nm region can be due to the absorption cut-off of $CHCl_3$ [25, 26] (see Fig. 3).

Demineralized Residue: All samples exhibit negligible absorbance except for the absorption peak of toluene at 285 nm. The absorbance value is higher in the 373–600 nm range in the residue of samples P1 and P2 compared to sample T1, possibly indicating the presence of bigger PAH ring structures in the extracted fractions of the said samples.

Fig. 2 Infrared spectra of coal samples (dry basis) with major peaks

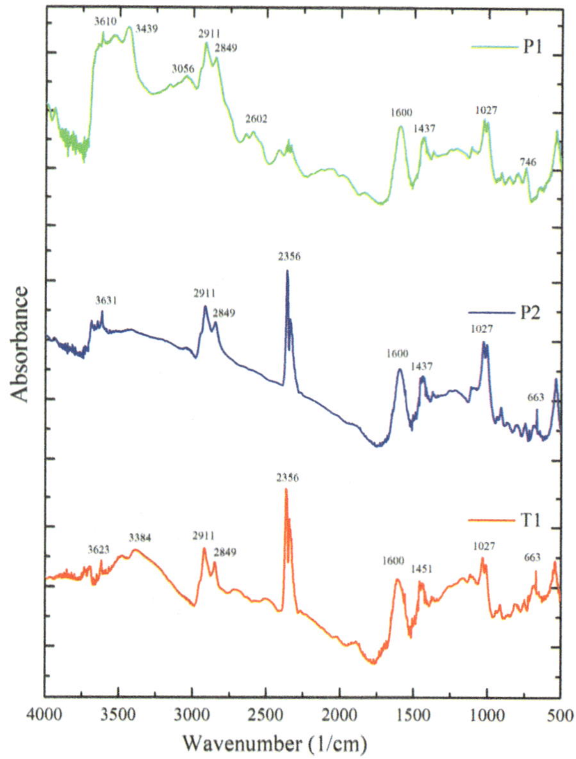

3.4 Fluorescence Spectroscopy

Solvent Extract: Fluorescence emission intensity (at 300 nm excitation wavelength; see Fig. 4) showing a broad wavelength range of 450–675 nm indicates the presence of different PAH structures. The most prominent and highest intensity region is depicted by a broad λ_{max} in the range between 525 and 575 nm, indicating the dominance of 4- and 5-ring PAH structures, i.e. naphthacene and perylene (see Table 3).

Demineralized Residue: Maximum emission intensity is observed in the wavelength range of 420–450 nm, indicating the presence of typical 3-ring PAH structures like anthracene in sample P1. The variation in the nature of the curves for sample P2 and T1 depicting variable intensity possibly corresponds to varying concentration of the PAHs in the samples. Prominent peaks of higher intensity compared to sample P1 are observed in samples P2 and P3 in the wavelength ranges of 420–450 nm and 450–460 nm, confirming the presence of anthracene-like PAH structures (see Table 3).

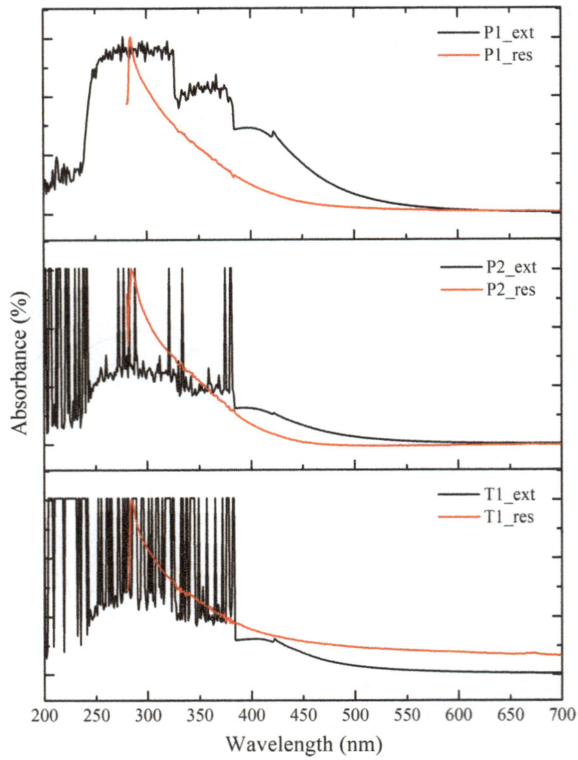

Fig. 3 Normalized UV–visible absorption spectra for solvent extracts (_ext) and demineralized residues (_res)

4 Summary

In the present study, an attempt was made to correlate the pyrolysis and VR_o data indicating the maturity of coal samples (dry basis) with spectroscopic information. T_{max} of the Gondwana samples implies a mature, oil-prone stage, while the Tertiary coal sample appears to be in the immature stage. From the VR_o data, the Gondwana samples were established to be of high-volatile bituminous A rank while the Tertiary sample belongs to high-volatile bituminous C rank. From the IR spectra of the samples, it was observed that the P1 sample lacks carbonyl content (depicted by sharp peak at 2356 cm^{-1}) compared to the other two samples, viz. P2 and T1. Supported by previous studies [3, 7, 27, 28], it indicates a higher degree of aromatization as the carbonyl content tends to decrease with increasing aromaticity. C–H stretching in the aromatic rings was observed in case of P1 sample, and this also successfully corresponds to the highly aromatic nature of the mature Gondwana coal. In terms of solvent extraction, it was observed that a higher degree of polyaromaticity leads to a lower degree of extraction. $CHCl_3$ being a highly polar solvent is capable of extracting more components from the less mature tertiary coal compared to more mature permian samples, as evident from the higher absorbance values of peaks in solvent extract of sample T1. Further study involving more

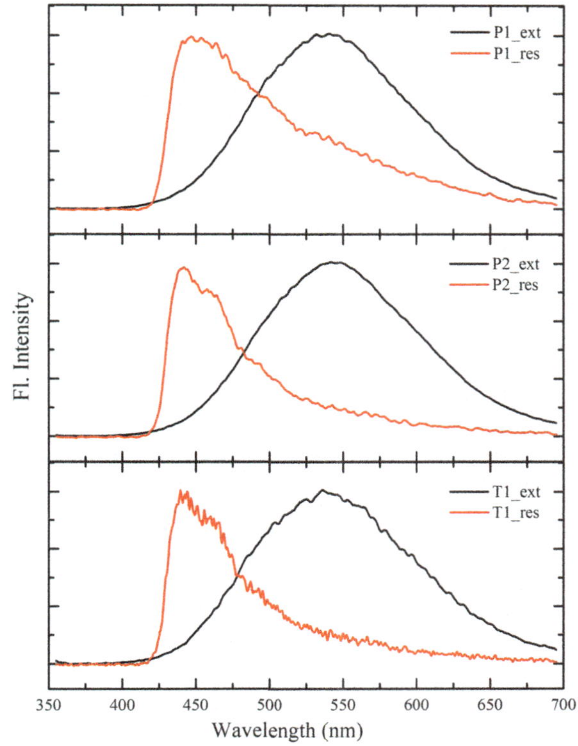

Fig. 4 Normalized fluorescence emission spectra for solvent extracts (_ext) and demineralized residues (_res) at 300 nm excitation wavelength

Table 3 Peak assignments from fluorescence emission spectra of coals

S. no.	λ_{max} (nm)	PAH ring structure	Possible component(s)
Solvent-extracted fraction			
P1	520–540	4-ring	Naphthacene
P2	520–540	4-ring	Naphthacene
T1	500–550	4- and 5-ring	Naphthacene and perylene
Demineralized residual fraction			
P1	425–500	3- to 4-ring	Anthracene and naphthacene
P2	425–475	3- to 4-ring	Anthracene and naphthacene
T1	425–475	3- to 4-ring	Anthracene and naphthacene

detailed and systematic observation of differently ranked coals is required to gain more precise insights on the rank-wise variation of coal with polyaromaticity.

Acknowledgements AS is thankfully acknowledging PDPU for providing fellowship for her doctoral research. The authors are indebted to their respective institutes for the support of instrumental facilities used in this work. Facilities of Petroleum Engineering Laboratory and Reservoir Characterization Laboratory (RCL) of SPT, PDPU, for UV–visible and fluorescence spectroscopy are especially acknowledged.

References

1. Levi PG, Cullen JM (2018) Mapping global flows of chemicals: from fossil fuel feedstocks to chemical products. Environ Sci Technol 52:1725–1734
2. Wei Z, Gao X, Zhang D, Da J (2005) Assessment of thermal evolution of kerogen geopolymers with their structural parameters measured by solid-state 13C NMR spectroscopy. Energy Fuels 19(1):240–250
3. Burnham AK (2017) Structures of coal, kerogen, and asphaltenes. In: Global chemical kinetics of fossil fuels. Springer, Heidelberg, pp 75–105
4. Robin PL, Rouxhet PG (1978) Characterization of kerogens and study of their evolution by infrared spectroscopy: carbonyl and carboxyl groups. Geochim Cosmochim Acta 42(9): 1341–1349
5. Burnham AK, Clarkson JE, Singleton MF, Wong CM, Crawford RW (1982) Biological markers from Green River kerogen decomposition. Geochim Cosmochim Acta 46:1243–1251
6. Burnham AK (1989) On the validity of the pristine formation index. Geochim Cosmochim Acta 53:1693–1697
7. Kelemen SR, Afeworki M, Gorbaty ML, Sansone M, Kwiatek PJ, Walters CC, Freund H, Siskin M, Bence AE, Curry DG, Solum M, Pugmire RJ, Vandenbroucke M, Leblond M, Behar F (2007) Direct characterization of kerogen by X-ray and solid-state 13C nuclear magnetic resonance methods. Energy Fuels 21:1548–1561
8. Lazarev AN (1972) Vibrational spectra and structure of silicates
9. Farmer VC (1974) The layer silicates. Infrared spectra miner, pp 331–363
10. Manoj B, Kunjomana AG (2011) Analytical study of two differently ranked coals using UV–Vis–NIR spectroscopy. J Min Mater Charact Eng 10(10):905–911
11. Coal resource position in India, Geological Survey of India. https://fossil.energy.gov/international/Publications/ucg_1106_gsi.pdf
12. Chattaraj S, Mohanty D, Kumar T, Halder G, Mishra K (2019) Comparative study on sorption characteristics of coal seams from Barakar and Raniganj formations of Damodar Valley Basin. India Int J Coal Geol 212:103202
13. Panwar DS, Saxena VK, Suman S, Kumar V, Singh AK (2017) Physicochemical study of coal for CBM extraction in Raniganj coal field, India. Energy Sources, Part A: Recover, Util, Environ Eff 39(11):1182–1189
14. Coates J (2006) Interpretation of infrared spectra, a practical approach. Encycl Anal Chem: Appl, Theory Instrum
15. Sanyal S, Bhui UK, Balaga D, Kumar SS (2019) Interaction study of montmorillonite-crude oil-brine: Molecular-level implications on enhanced oil recovery during low saline water flooding from hydrocarbon reservoirs. Fuel 254:115725
16. Okolo GN, Neomagus HW, Everson RC, Roberts MJ, Bunt JR, Sakurovs R, Mathews JP (2015) Chemical–structural properties of South African bituminous coals: Insights from wide angle XRD–carbon fraction analysis, ATR–FTIR, solid state 13C NMR, and HRTEM techniques. Fuel 158:779–792
17. Saikia BK, Boruah RK, Gogoi PK (2007) FT-IR and XRD analysis of coal from Makum coalfield of Assam. J Earth Syst Sci 116(6):575–579
18. Qiu Y, Zhang Q, Tian Y, Zhang J, Cao J, Xiao T (2011) Composition and structure of Luxing coal with different particle sizes. Pet & Coal 53(1): X–XX
19. Chukanov NV, Chervonnyi AD (2016) Infrared spectroscopy of minerals and related compounds. Springer, Cham
20. Li X, Ju Y, Hou Q, Li Z, Fan J (2012) FTIR and Raman spectral research on metamorphism and deformation of coal. J Geol Res 2012:8
21. Urbanski T, Kuczynski W, Hofman W, Urbanik H, Witanowski M (1959) The infrared absorption spectra of extracted coals. Bull Acad Pol Sci, Ser Sci, Chim, Geol Geogr 7: 207–214

22. Baruah MK, Kotoky P, Borah GC (2003) Distribution and nature of organic/mineral bound elements in Assam coals. India. Fuel 82(14):1783–1791
23. Yang F, Zhao M, Zheng B, Xiao D, Wu L, Guo Y (2012) Influence of pH on the fluorescence properties of graphene quantum dots using ozonation pre-oxide hydrothermal synthesis. J Mater Chem 22(48):25471–25479
24. Sonibare OO, Haeger T, Foley SF (2010) Structural characterization of Nigerian coals by X-ray diffraction. Raman FTIR Spectrosc Energy 35(12):5347–5353
25. Olajire AA, Ameen AB, Abdul-Hammed M, Adekola FA (2007) Occurrence and distribution of metals and porphyrins in Nigerian coal minerals. J Fuel Chem Technol 35(6):641–647
26. Ito O (1993) UV-visible and near-IR spectra of heat-treated pitches during mesophase formation. Carbon 31(3):401–406
27. Savest N, Hruljova J, Oja V (2009) Characterization of thermally pretreated kukersite oil shale using the solvent-swelling technique. Energy Fuels 23(12):5972–5977
28. Solomon PR, Serio MA, Despande GV, Kroo E (1990) Cross-linking reactions during coal conversion. Energy Fuels 4(1):42–54

Greenfield Energies from Underground Coal Gasification and Liquefaction of Solid Fossil Fuels—Basics and Future Potentiality in India

Sudip Bhattacharyya

1 Introduction

Coal has a complex macro-molecular network, consisting of polyaromatic clusters linked with side chains and functional groups on peripheral positions by covalent and non-covalent bonds [1]. The monomer organic compounds trapped in the porous structure of coals may be products and residues of the coalification process, which are thought to be related to the macromolecular precursor material. A spate of models has been proposed for macromolecular structure of coal based on experimental studies on structural and chemical analysis of coal [2]. Chemical identification of functional groups and general information on the carbon skeleton are established, but the structural examples given must be seen as attempts only to give the general framework of the coal structure. They are mainly intended to describe structures of hard brown coal to bituminous coal. The aromatic centres are generally linked by hydroaromatic and methylene bridges and fringed with methyl, hydroxyl, carboxyl, keto, amine and other functional groups.

Petrographic characteristics of coal play a key role in defining its suitability for its various industrial applications. Thus, importance of coal characterization is well established and is crucial for making metallurgical-quality coke for blast furnace applications, in thermal power plant electric power generation, coal beneficiation, hydrocarbon exploration and assessments including coal bed methane, underground coal gasification and coal liquefaction. Maceral composition as well as rank of a coal has got a tremendous significance in determining the quality of coal and suitability for specific applications. Coal is composed of microscopically distinct components of the organic matter, called macerals, derived from terrestrial, lacustrine and marine plant remains. They differ in their physico-chemical and

S. Bhattacharyya (✉)
Natural Energy Resources, Mission-IIB, Geological Survey of India, Bhu-Bijnan Bhawan, DK-6, Sector-II, Salt Lake City, Kolkata 700 091, India
e-mail: sudip.bhattacharyya@gsi.gov.in

U. K. Bhui (ed.), *Macromolecular Characterization of Hydrocarbons for Sustainable Future*, Green Energy and Technology,
https://doi.org/10.1007/978-981-33-6133-1_13

optical properties such as hydrogen content, aromaticity, reflectance, density. These are grouped as: (i) vitrinite, with intermediate reflectance and high oxygen-to-carbon ratios, derived from woody tissues; (ii) liptinite, with low reflectance and high hydrogen-to-carbon ratios, derived from spores, cuticles, resins and algal bodies and (iii) inertinite, with high reflectance and carbon contents, derived from fossil charcoal or decayed material. Hydrocarbon generation potential of coals are dependent on the amount, type and maturity of the inherent organic matter present within them [3, 4]. This organic matter within the coal changes from one lithotype to another with progressive changes of macerals composition [5–8]. Individual macerals and their associations in lithotypes may play different roles in generation and storage potentiality. Therefore, understanding the maceral compositions of lithotypes is important for gaining insights into their behaviour during hydrocarbon generation and storage.

In the present work, an attempt is made to understand the basic factors responsible for a favourable utilization of solid fossil fuel towards conversion to gas or oil. Based on the analysis of the available subsurface and surface data, either accrued through exploratory drilling by GSI or generated through researches all over the country, site-specific possibility has been explored for use of coal or lignite for UCG and CTL. The study for UCG was done with the physical, chemical, structural and petrographical data of the coal/lignite, while for CTL, petrography, thermogravimetric analysis and differential thermal analysis (TGA-DTA), Fourier transform infrared (FTIR) spectroscopy, UV–visible and fluorescence spectroscopy have been carried out to obtain the precise knowledge of macromolecular structure and composition for better site-specific and technology-specific utilization towards green and clean energy.

2 Underground Coal Gasification (UCG)

Underground coal gasification (UCG) is a technique of converting unrecoverable and unmineable coal into a combustible gas, which can be used for industrial heating, power generation or the manufacture of hydrogen, synthetic natural gas or diesel fuel. It allows to fully utilize the coal resource from otherwise unrecoverable coal deposits in an economically viable and environmentally safe way. UCG turns this resource into high-value products like clean power, liquid fuels, syngas, fertilisers and other chemical [3–6]. UCG has the potential to unlock vast coal energy resources that cannot be beneficiated economically through conventional mining methods. In the process coal, oxygen and water/steam are reacted at high temperatures through chemical reactions in the underground gasifier in situ to form synthesis gas (syngas), predominately being CO, H_2, CH_4 and CO_2. A UCG operation consists of a series of injection and production wells drilled into a coal seam where reactant gases, i.e. air, oxygen or steam, are injected to cause the gasification by partial combustion instead of complete combustion of coal (Fig. 1). Besides primary gas components, a number of organic and inorganic contaminants, such as phenols, benzene, sulphates, metal and metalloid elements, can be generated and released during the UCG process [9]. Simplicity of UCG is very attractive, but development of reliable working programme with acceptable cost benefit ratio seems difficult in practice. Technical feasibility depends on multidisciplinary aspects; thus, trials must be undertaken at pilot scale. For every geological set up, a tailor-made site-specific work plan has to be designed to exploit the targeted resource. UCG is an inherently clean coal technology as it reduces sulphur and nitrogen oxide emissions to very low levels and low volatile organic components, methane and greenhouse gases (GHG) emission to atmosphere. Total solid waste from UCG is typically half the volume generated by conventional coal plants and water use is substantially lower as well. UCG technology was first developed in the Former Soviet Union (FSU) in the 1920s. The USA and Europe also conducted several tests in the 1970s and 1980s. In the 1990s, China began UCG research and development and is continuing with this effort.

The first consideration is the selection of the site, which must take into account the condition and quality of the coal, the permeability of surrounding strata, the proximity to aquifers and the geological and hydrogeological conditions. This is a complex area, requiring detailed knowledge of site geology by exploratory drilling, seismic surveys, laboratory testing and modelling.

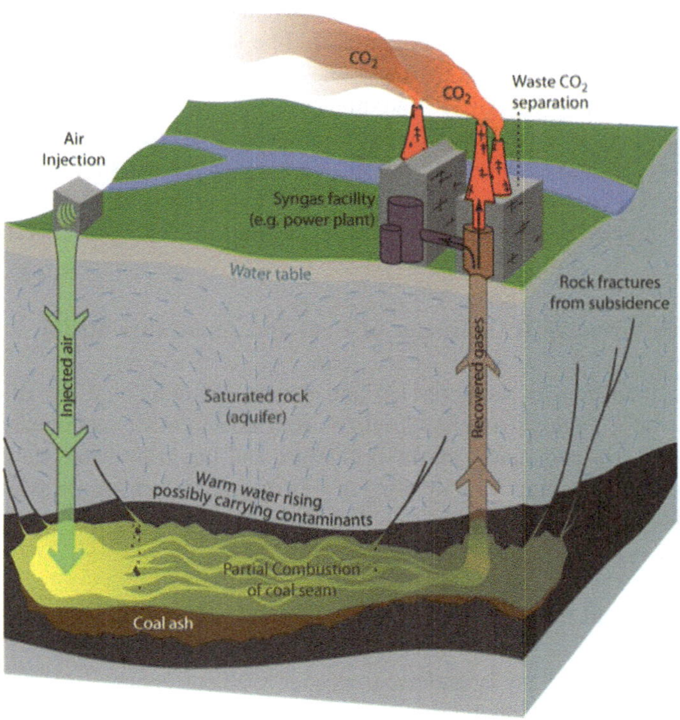

Fig. 1 Void formation during UCG process (*Source* Water Cartoon, Kissclipart.com)

2.1 *Desirable Coal Characters*

Coal with higher reactivity having significant amount of exinite/liptinite, high moisture content which facilitates reactions providing water, low-rank coal can be easily gasified. Caking and swelling index should be less; low mineral matter and ash fusion temperature and adequate permeability are important parameters for UCG treatment.

2.1.1 Physical Attributes of Target Coal Seam

(i) Depth of occurrence should be more than 100 m and preferably less than 300 m, (ii) thickness of coal seam should be more than 2.5 m with thickness variation less than 25%, and (iii) optimum permeability of the coal seam is desired as 50–150 md.

2.1.2 Chemical Properties of Target Coal Seam

(i) Ash content of coal seam should vary from 10 to 40% [heat value/calorific value (3.5–4 MJ/m^3] of product gas almost constant up to 40% of moisture content, beyond which steadily declines. (ii) Low-rank coal with relatively high moisture and porosity is preferred, desirable moisture content should be less than 15 wt%. (iii) Desirable sulphur content should be less than 1 wt%. (iv) Coal with low caking and swelling indices is preferable for UCG, and higher values result instability of gasifier. (v) Reactive macerals increase amount of hydrogen content in the product gas. Higher the liptinite content, higher will be the H/C ratio, which produces large amount of gas on heating. Low-rank coal with high inertinite and liptinite macerals are suitable because it suppresses excessive caking and swelling which are not desirable.

2.1.3 Geo-Mechanical Character of Coal and Associated Strata

Coal and strata density, compressive and tensile strength, modules of elasticity, Poisson's ratio, angle of internal friction, cohesion, porosity, permeability ratio, properties of clay and sands, plasticity number and liquid limit for clay and grain size analysis are required for modelling any UGC project, which are available from drill core study.

2.1.4 Structural Set-up

Higher dip of coal seam is preferred; steeper gradient enhances better pyrolysis producing more gas. Tars/liquids flow down in steep dipping beds favouring effective combustion.

Joints and faults are the discontinuities in coal seam which are identified based on surface and subsurface geological data. Coal seam targeted for UGC should be considerably away from identified major faults. Precise characterization of joints and faults as well as the mine-induced fractures within the coal/lignite seams is to be taken care of.

2.1.5 Overburden Characteristics

Impervious strata in roof and floor of seam is preferred, desirable thickness of the litho assemblage overlying targeted coal seam for UCG should not less 15 m which is to be devoid of any workable coal seam. Desirable permeability of overburden should be <5 md, having considerable mechanical strength to avoid roof collapse.

2.1.6 Hydrological Regime

No good aquifer should be in the vicinity of UGC project so that any quality groundwater supplies will not be put at risk. Thorough characterization of existing aquifer of the surrounding area is a must. Careful monitoring during operations, particularly maintenance of optimal pressure always less than hydrostatic pressure in gasifier is extremely necessary. Overlying aquifers must be sufficiently separated by an impermeable seal to prevent either excessive water inflow to the gasifier or gas escape to the overlying aquifer through any fractures that develop above gasifier. Similarly, underlying aquifers must be sufficiently distant to prevent excessive water inflow to the gasifier from deeper and higher pressure zones.

Groundwater pollution is considered to be one of the most serious environmental issues [10–13]. This may be caused by the dispersion and penetration of gaseous products into surrounding rock layers due to the development of higher pressure than hydrostatic pressure in gasification process to prevent the influx of groundwater, the emission and dispersion of contaminants with gas products and migration of gasification residue by leaching with the penetration of groundwater after gasification. The physicochemical properties of groundwater may be affected by the hazardous escaped gases. Phenolic compounds are major organic contaminants in groundwater, PAHs and heterocyclic compounds are minor organic contaminants of concern, and a wide array of ionic species, such as calcium, sulphate, bicarbonate, chloride, sodium and ammonium, etc., are also present in larger amounts [12]. Solid residue remained underground after UCG could be a source of contaminated groundwater [10, 14].

2.1.7 Land Use Considerations

Every geological setting is different, and the surface constraints vary due to land use, surface topography and other factors. The site of UGC project should be at outskirt with limited human activity, settlement or mine. No waterways (rivers/

lakes) should overly the site for UGC project. Field development plans and surface facilities layout should be studied to ensure the lowest lifecycle cost of the produced syngas.

2.2 Basics of UGC Process, Chemical Reactions and Void Formation

Higher coal quality, deeper coal formations targeted for UCG and lower water ingress results in improved oxygen utilization. Oxygen blown gasification is significantly more efficient in utilizing the injected oxygen as compared to air blown gasification and produces a syngas with a high-quality suitable for petrochemical applications. The gasification process requires high temperatures to convert the coal into syngas. In the gasification zone, temperatures exceed 1000 °C and may reach 1500–2000 °C when pure oxygen gasification is applied. The three essential types of char layer reactions which govern the UCG product gas composition are oxidation, reduction and pyrolysis (Fig. 2). In the oxidation zone, wet coal is heated and converted into dry coal by removing moisture attached to it. Upon further heating of the coal, the pyrolysis reactions begin at temperatures around 350–400 °C [11, 15] and coal loses its weight, generating volatile matters and the solid known as char. Finally, the char reacts with the injected/pyrolyzed gases to produce the syngas [15]. Empirical and exact mathematical tools and approaches can be used to provide insights into the underground processes through experimental cavity modelling [16–18]. The outflow channel is divided in small sections along the length, each consisting of rubble zone, void zone and roof at the top, which are well consistent with experimental models.

2.3 Environmental Impact

Environmental pollution in the form of generation and transport of contaminants via experimental and numerical analyses is the dominant issue. Leaching of organic substances from the gasifier, e.g. phenol, increased concentration of inorganic salts near the gasification zone dissolving of hazardous gases (H_2, CH_4, CO_2, H_2S and NH_3) in ground water, and leaching of heavy metals (Hg, As, Pb, Cr, Cd), emission of pollutants and greenhouse gases to the atmosphere are some of the possible contaminant during UCG operation. A spate of research work has been done in this respect, and possible measures have also been suggested [5, 12, 19–22] Several models with various parameters like modelling of contaminant transport and reactions [22, 23], general hydrogeological models were proposed for mitigation of the adverse effects of UGC.

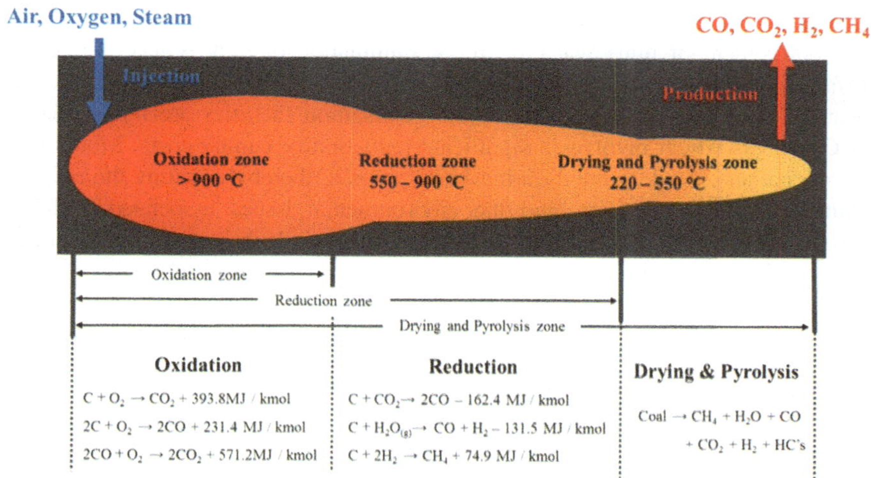

Fig. 2 Typical chemical reaction zone during UCG process [15]

2.4 Surface Subsidence and Effect on the Permeability of Overburden

Removing coal from the underground by mining, UCG can cause overburden to collapse, fracture and strain in the form of surface subsidence, which may likely to change the permeability field. In large-scale UCG operations, it is more likely that higher permeability would be created far above the seam. This can make groundwater contamination worse. To isolate the UCG operation environmentally from shallower sensitive contaminant receptors, the zone of enhanced permeability must stay below the impermeable barriers that are counted on to protect shallower aquifers. Surface subsidence can be mitigated by appropriate site selection, including depth and strength of rock volume; spacing of UCG reactors to leave walls and pillars between produced zones, and identification and avoidance of structural weaknesses as pre-existing faults. Gasification of thick seams may result in a greater vertical extent of roof collapse (goafing). This can open up gas connectivity far up into the fractures of a sagging overburden, making it more likely for product gas to escape up into shallow strata.

2.5 Advantages of UCG

Lower emissions of particulates, reduced methane emissions, SO_2 and NOx, no noise and visual impact on the surface and disposal at mine sites and power and station sites, less transport cost, reduction of operating costs and surface impacts, occupation of smaller land area, lower risk of surface water pollution, non-requirement of mine water recovery, elimination of mine safety issues, and surface hazard liabilities on mine abandonment are some of the advantages of UCG over coal mining. Besides, conventional gasification facilities are not needed in UCG process which involves a significant reduction of capital costs. Most of the ash in coal stays underground during UCG process, thereby avoiding the need for additional syngas clean-up, and the environmental issues associated with ash storage. Commercial UCG applications use substantially less water than conventional gasification technologies.

2.6 UCG, Greenhouse Gases and CO_2 Capture Storage (CCS)

Efforts were made globally effort to understanding UCG's advantages and disadvantages with respect to greenhouse gas (GHG) emissions. Methane comprises a significant fraction of UCG product gas. Methane is a much more potent greenhouse gas than carbon dioxide. Potential leaks of methane containing product gas from UCG operations need to be included in analyses. Gasification is technically amenable to efficient separation and CCS that could then be sequestered (Fig. 3). This has been advanced for surface gasifiers in recent developments and demonstrations. UCG could work similarly, although the methane in UCG's product gas cannot easily be water-gas shifted, limiting the extent of CCS that can be achieved. More researches about carbon capture from UCG syngas and environmental protection are needed with the objective of its application in terms of comparison of global GHG emissions with the fossil fuel and renewable alternatives [20]. There may be sites where the coal is adjacent to a suitable CO_2 store, like a depleted gas reservoir or saline aquifer.

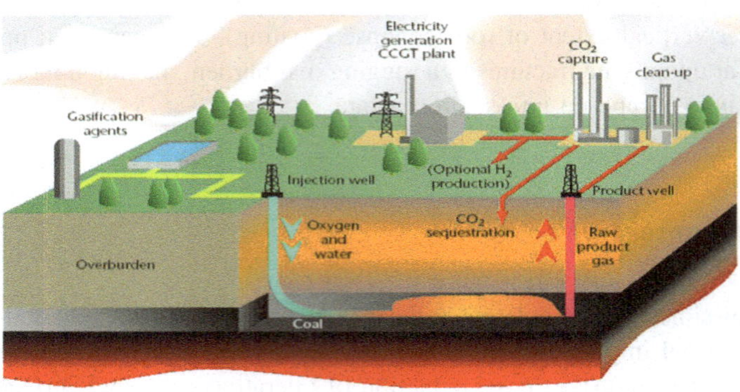

Fig. 3 Site characterization and selection of pilot plant for UCG (*Source* Online published in a blog 'Mining methods in India'; on June 04, 2011

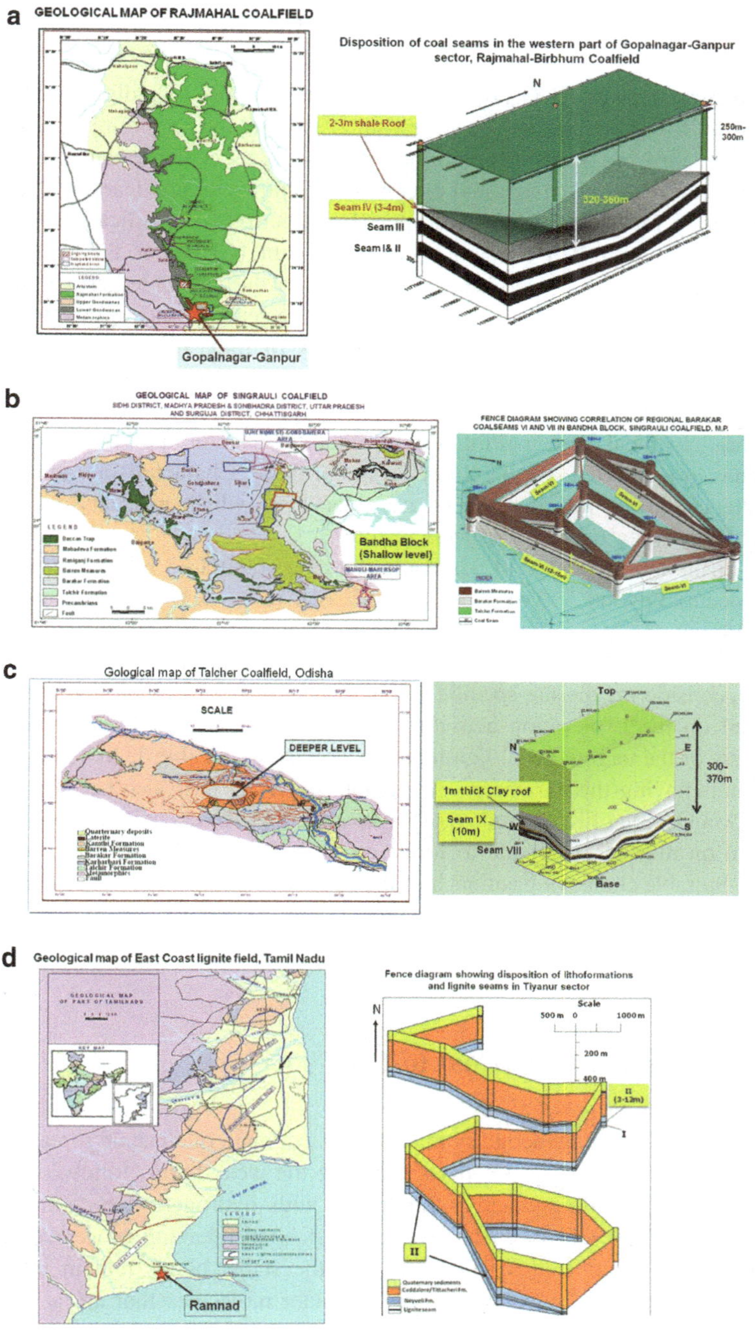

Fig. 4 **a** Gopalnagar–Ganpur block, Rajmahal coalfield and stratigraphic disposition of seam-IV. **b** Bandha block, Singrauli coalfield and fence diagram of seam-VI. **c** Jalatap block, Talcher coalfield and stratigraphic disposition of seam-IX. **d** Tiyanur block, Ramnad Sub-basin and fence diagram of seam II

2.7 *Future Potentiality of UCG in India*

The strategic case for UCG lies with the plentiful supply of indigenous coal, the high cost of fossil fuel alternatives like natural gas and the opportunities it presents for new industry. An instrumented pilot test in the target coal is essential in order to obtain the data to plan, design and operate the commercial application. The choice of technology for UCG is still wide open. Long periods of gasification have been achieved with controlled retracting injection point (CRIP) or the Chinese and Russian methods of multiservice wells. Based on the available surface and subsurface data in and around the targeted coal/lignite seam, choice of technology for UCG should be site specific. Geological Survey of India has created a vast subsurface database of the coal and lignite seams in all the Gondwana and Tertiary coal and lignite fields by exploratory drilling through decades. In-depth analysis of structural, chemical and coal/lignite petrographic data of individual coal/lignite seams has imparted UCG suitability of some of these. Characteristics and essential parameters required for such seams are given here, which warrant specific technologies for taking up UGC projects in future.

Out of the total resource of 32.6 billion tonne, about 13.8 billion tonne of coal (42%) occurs beyond 300 m depth and the total resource of lignite stands as 4.5 billion tonne (as per Inventory of Indian Coal and Lignite Resources as on 01-04-2019), which indicates that there are ample potentiality of UGC projects in the country, considering the site-specific favourable parameters for their implementation.

Based on the exploration activities carried out by GSI in India, some of the characters of the targeted coal/lignite seams are given in Table 1. In addition, some coal seams having the parameters favourable for UCG may also be targeted from Hasdo-Arand, East of Kothagudem of Godavari and Ib-River coalfields, where either embargo on mining exists due to forest cover and are not suitable for immediate mining. Suitable choice of gasifiers and site-specific implementation of pilot projects are mandatory for amplifying this unconventional green energy resource to occupy major space of 'Energy Basket' of India keeping pace with the outer world.

3 Coal Liquefaction (Coal to Oil CTL)

Liquefaction of coal to generate synthetic oil is a domain of constant research, and it is generally presumed to proceed via breaking the weak bonds followed by stabilization of the resulting reactive fragments with hydrogen. Depending on the structural bonds present in coal, a sequence of bond breaking process eventually produces reactive molecules while many of the stable bonds remain intact. The liquid product of the liquefaction should contain molecules that are a vestige of structural arrangements present in the original coal. Coals may contain significant quantities of the hydrogen-rich and hydrocarbon-generating maceral group, liptinite. When heated, coals produce petroleum-like pyrolysis products. Bituminous

Table 1 List of some of coal/lignite seams recorded in areas explored by GSI warranting attention for UCG

Coal/ lignite basin	Rajmahal Basin	Singrauli Basin, Son valley	Talcher Basin, Mahanadi valley	Ramnad Sub-basin, East coast lignite
Area	Gopalnagar–Ganpur [24] (Fig. 4a)	Bandha [25] (Fig. 4b)	Jalatap [26] (Fig. 4c)	Tiyanur [27] (Fig. 4d)
Seam no.	IV	VI	IX	II
Thickness (m)	3–4	12–15	10	3–12
Depth (m)	320–360	125–200	300–370	340–460
Roof lithology	Sh: 2–3 m, T. O. > 20 m	Sst: 20 m	Clay: 1 m, T. O. > 15 m	Clay > 20 m
Dip	20–40	20–30	30	20–30
Rank	HVB	HVB	HVB	Lignite
Ash (%)	20–49	23–27	15–45	5–15
Moisture (%)	4–6	6–9	2–13	45
Maceral (%)	L(4–16), I(22–44)	L(16–19), I(32–42)	L(21), I(26)	L(2–10), I(5–10)
Resource (Sq km)	20(4)	235(4)	125(4)	150(28)

Sh Shale, *Sst* Sandstone, *T.O.* Total Overburden, *HVB* High Volatile Bituminous, *L* liptinite, *I* Inertinite

coals contain bitumen that can be easily released by extraction using common solvents. Some coals are oil prone, whereas others are not potential for generating liquid hydrocarbons. Role of microconstituents of coal and corresponding changes affecting its macromolecular arrangement influences the potentiality of generation and expulsion of oil from coal.

3.1 *Coal Petrographic Input in Favour of CTL*

3.1.1 Role of Reactive Macerals

Consideration of liquefaction requires that the components of coals are commonly divided into reactive groups (vitrinite and liptinites) and inerts (inertinites and mineral matter). Semifusinite is considered to be partly reactive. It is clear that enhanced conversion will be related to greater amounts of reactives and lower contents of inerts in coal samples. Liptinite group of macerals is rich in hydrogen and are suitable for hydrogenation. Coal with higher liptinite content has higher volatile matter. Generally, higher liptinite content is found within coal with high ash content, and often, the true liptinite content is masked by the mineral matter. During thermal maturation, the kerogen (Type I, II, III and IV) involves hydrogen disproportionate reactions in which the coal and kerogen loses hydrogen in succession. These hydrogen depleted coal and kerogen condenses and aromatizes, eventually to form graphite [28, 29].

Critical importance of liptinites in estimating the potential of coal for oil generation has been assessed, and attempts were made for estimation of the minimum volume of liptinite required for capable of generating oil by many workers [29–31]. A minimum of 15–20% [28, 32, 33] and 20–25% [34] liptinite were estimated for potentiality of coal as oil prone. Assessment of potentiality of hydrocarbon formation of individual liptinite maceral reveals the order as alginite, resinite, cutinite, sporinite and suberinite [35, 36].

The liptinite and vitrinite macerals are transformed to inertinite by virtue of higher temperature and time with increasing rank [37, 38]. The higher temperature and time changes the maceral composition of coal and kerogen thus changing the physical, chemical and optical properties from a sapropelic coal to humic coal and from humic coal towards more humic leading to graphite at last.

Vitrinites typically contain about 70% aromatic carbon atoms in hard brown coals and over 90% in anthracites. Liptinites initially have a very low aromatic content, but in low-rank bituminous coal, 50% carbon atoms are found in aromatic rings. Thereafter, the aromaticity rises rapidly to join that of vitrinite in anthracites. The inertinite maceral group has 90–100% aromatic carbon at all ranks. This invariably supports that hydrogenation is the basic process by which aliphatic-rich liptinite maceral may be derived from more aromatized vitrinite having a large range of aromaticity, while it is quite impossible to revert back from the maximum aromatized inertinite maceral in any way.

Some vitrinite-rich coals have hydrogen content (4.6–5.4 wt%), higher than the normal vitrinites, and show reddish-brown fluorescence under UV light. These hydrogen-rich vitrinites are called perhydrous vitrinite, found particularly in marine-influenced coals with increased wax–resin component [37]. The visible fluorescence of these vitrinites is commonly taken to be a strong indicator of a hydrogen-rich composition when chemical analyses are not available. The maceral composition of perhydrous vitrinite rich and liptinite poor New Zealand coals [39–42] confirms the potentiality of perhydrous vitrinite as an important source of liquid hydrocarbons in oil-prone coals. Similar observation was also noted from Cretaceous coals of the San Juan Basin, New Mexico [43], Gippsland Basin, Australia [44, 45]. In humic coals, high liptinite contents are commonly associated with perhydrous vitrinite compositions [46–48].

3.1.2 Vitrinite Reflectance (VR) and Effect of Its Suppression

Vitrinite reflectance (VR) is the proportion of incident light reflected from a polished vitrinite surface. Vitrinite is a maceral (an organic component) of coal. As vitrinite is heated in the earth during burial, it systematically changes its reflective properties. These changes are caused by increasing condensation of aromatic (carbon ring) structures in the coal matrix [30, 49]. Vitrinite reflection is a standard method for calculating the relative degree of coalification to reveal the rank and maturation that a coal has undergone. Coals with VRo% less than 0.45 lie within the zone of early stage of oil generation, whereas the values of 0.5–0.57% indicate the beginning of commercial oil generation from coal within the diagenetic stage [28].

Suppression of vitrinite reflectance is a common phenomenon in which reduced reflectance values are accounted lower than those produced by low reflecting, non-suppressed vitrinite. Suppressed reflectance values are intermediate between those for liptinite and those for non-suppressed normal vitrinites with hydrogen-poor lignocellulosic structures. Such suppression may occur through the modification of vitrinite structure due to interplay of a number of factors like the high hydrogen contents produced by reducing environment of deposition, impregnation of high bitumen contents produced during early stages of diagenesis along with the formation of perhydrous vitrinites, highly aliphatic components formed due some specific palaeoecological factors or presence of high liptinite content and type of liptinite maceral present in the maceral assemblage [50]. Quantitative assessment of liptinite required for suppression of vitrinite reflectance varies from at least 20% [51] to 35% having 24% resinite component [50]. Suppression effect is primarily attributed to variation in the hydrogen content of vitrinite, while anomalously low reflectance is exhibited by perhydrous vitrinites [52, 53].

3.2 *Expulsion of Liquid Hydrocarbon*

Extent of expulsion of generated liquid hydrocarbons from coal-bearing sequences is probably the most critical factor in determining whether or not a coal has given rise to liquid petroleum products. Evidences are there for movement of petroleum molecules from their structural site of origin and redistribution of generated bitumen in the form of exsudatinite in fractures and fluorescing cell fillings in macerals within coal seams [54, 55]. Poor expulsion is attributed to the adsorption of oil on the organic macromolecule.

3.3 *CTL Processes*

Coal liquefaction is a general term referring to a family of processes for producing liquid fuels from coal. Specific liquefaction technologies generally fall into two categories: direct (DCL) and indirect liquefaction (ICL) processes. Indirect liquefaction processes generally involve gasification of coal to a mixture of carbon monoxide and hydrogen (syngas) and then using a process such as Fischer–Tropsch (FT) to convert the syngas mixture into liquid hydrocarbons. By contrast, direct liquefaction processes convert coal into liquids directly, without the intermediate step of gasification, by breaking down its organic structure with application of solvents or catalysts in a high pressure and temperature environment. Since liquid hydrocarbons generally have a higher hydrogen–carbon molar ratio than coals, either hydrogenation or carbon-rejection processes must be employed in both ICL and DCL technologies. South Africa developed CTL technology at Sasol in the 1950s during an oil blockade, and CTL now plays a vital part in South Africa's national economy, providing over 30% of their fuel demand. DCL works by dissolving the coal in a solvent at high temperature and pressure. This process is highly efficient, but the liquid products require further refining to achieve high-grade fuel characteristics. ICL processes operate in two stages. In the first stage, coal is converted into syngas (a purified mixture of CO and H_2 gas). In the second stage, the syngas is converted into light hydrocarbons using one of three main processes: Fischer–Tropsch synthesis, methanol synthesis with subsequent conversion to gasoline or petrochemicals and methanation. Carbon capture and storage, which is still the subject of much research, would alleviate the environmental impact of carbon dioxide being released into the environment, the main argument against CTL by critics.

3.4 Indian Scenario, Recent Advancement and Future Potential

Permian coal of India is composed of variable proportion of vitrinite and inertinite (17–70%) with subordinate amount of liptinite (<20%) [56]. The average maceral composition of Permian coal of India indicates lower fusible or reactive (vitrinite, exinite and semivitrinite) and higher 'inert' constituents and mineral matter similar to Permian coal of western Australia and South Africa [57]. As a whole in all the Gondwana basins of India, vitrinite content increases upward along the stratigraphic succession. General value of vitrinite content in Karharbari coal seams is <20% (fusic-vitrofusic), in lower Barakar coal seams 25–40%, (trimaceritic), in upper Barakar Member 50–70% and in Raniganj coal seams up to 70–80% (vitric-fusovitric). Coking property of coal is also a function of reactive constituent (vitrinite + liptinite) content. In Indian Gondwana basins, high liptinite value up to 25% is found within the coals of Mahanadi and Son Valley although isolated high values are reported from almost all the coalfields. GSI is having a comprehensive inventory of petrographic composition and proximate and ultimate analytical data of Indian coal, generated through decades through exploration activities of coal. Coal petrographic study of some regional coal seams of Ib-River and Talchir Coalfields of Odisha and Mand-Raigarh Coalfield of Chhattisgarh of Mahanadi Valley Gondwana master basin, so far, has already established their suitability for potentiality of oil generation [58, 59]. They meet the basic petrographic criteria for potentiality of oil generation, i.e. Vitrinite + Liptinite > 60%, vitrinite reflectance value (Ro%) < 0.80 and H/C atomic ratio > 0.75. Along with petrographic character of the coal seams, their thickness and depth of occurrence are also favourable for CTL process.

3.4.1 Talcher Coalfield, Mahanadi Valley Basin

Detailed coal petrographic study carried out in different exploration blocks by GSI revealed that coals from Talcher Coalfield have vitrinite (20–65%), liptinie (12–54%) and inertinite (7–28%), the total reactive contents (vitrinie + liptinite) of most of the coal seams being quite high (up to 74%), which is very much desirable for coal liquefaction process. Occurrence of increased liptinite content may be due to decomposition of tissue cells during fusinitization process and redistribution of the bitumen within the fusinite [58]. Liptinite macerals are composed of sporinite, sporangium, cutinite, resinite, alginate, suberinite and liptodetrinite. Coal seams are of subbituminous to high volatile bituminous 'B' type with VRo% varying from 0.30 to 61. During bituminization, perhydrous vitrinite has also taken an active part to form petroleum-like substances. Depth (208–279 m) and thickness (0.58–47.35 m) of the ten regional Barakar coal seam zones (II to XI) have mineable prospect through suitable CTL technology.

3.4.2 Ib-River Coalfield, Mahanadi Valley Basin

Regionally co-relatable subbituminous R-I and R-III Raniganj coal seam zones were found to exhibit high percentage of vitrinite (38.53–53.92%) group of macerals followed by consistently higher liptinite content (12.92–27.45%) with VRo% ranging from 0.31 to 0.40 [59]. Again sub-bituminous to high volatile bituminous C-type regional Barakar Coal seam zones of Belpahar, Parkhani, Lajkura and Rampur show moderate to high vitrinite (12.48–67.60%) followed by liptinite (12.43–35.92%) with VRo mean% ranging from 0.30 to 0.53 [59]. Total vitrinite present in the coal may act as reactive maceral along with liptinite indicate in the process of coal liquefaction. In general, considerable persistent thickness of the coal seams occurring within 700 m may open up a new vista for the future potentiality inviting modern CTL process.

3.4.3 Mand-Raigarh Coalfield, Son Valley Basin

Sub-bituminous to high volatile bituminous 'C'-type regional Barakar coal seams (Nos. IX, VII, VI, IV, III, II and I) show a moderate to high vitrinite (11.14–72.39%) and liptinite (2.68–11.73%), resinite being most abundant, which show higher value (12.46–25.92%) with VRo ranges from 0.35% to 0.57% [59]. Thickness and depth of occurrence within 650 m of the coal seams render a good prospect for CTL.

3.4.4 Godavari Coalfield

In the north-western part of the coalfield from Belampalli area, the middle to upper part of thick 'Bottom seam' of high volatile bituminous 'C' type is comprised of vitrinite (25–44%), liptinite (24–32%) and inertinite (11–26%) group of macerals with the Ro mean of 0.50% stage [60]. Petrographic analysis of four economic coal seams (I, II, III and IV) from Ramagundam north-western part of the coalfield indicates that these coals are dominated by inertinite followed by vitrinite and liptinite [61]. The vitrinite group of macerals ranges from 20 to 64% in Seam IIIA, 31 to 57% in Seam II and 24 to 63% in Seam III. The liptinite macerals vary between 10 and 18% in Seam IIIA, 12 and 19% in Seam II, and 10 and 17% in Seam III. In the Koyagudem area of the south–eastern continuity of Mulug Coal belt located in the central part of Godavari Coalfield also reveals favourable petrographic characters for CTL. The 'Queen Seam' occurring between 163.00 m and 185.50 m depth range has shown vitrinite as the dominant maceral group (40–59%), followed by inertinite (21–35%) and liptinite (9–23%). The basal part of Queen seam has dominance of liptinite (33–38%). In general, Ro mean% ranges between 0.48 and 0.64, which shows that this coal seam has reached high volatile bituminous C rank [60]. Domainal enrichment of reactive macerals (>60%) of some selected thick coal seams throughout Godavari Coalfield from NW to SE warrants its introspection for their CTL potentiality.

3.4.5 Lignites of Cambay Basin

Petrographic study of the three lignite fields, i.e. Vastan, Rajpardi and Tadkeshwar of the Cambay Basin of Gujarat was studied in detail [62]. The lower and upper Vastan lignite seam consists of huminite (58.0–83.1%), followed by liptinite (6.9–16.6%) and inertinite (1–17.9%). In Tadkeshwar upper and lower lignite seam, huminite (52–75.9%) is the dominant maceral group followed by liptinite (6.8–24.4%) and inertinite (1.4–14.8%). Rajpardi lignite comprises of huminite (67.9–80.0%), liptinite (8.7–16.3%), followed by inertinite (2.8–9.3%). The huminite reflectance of the lignites of Cambay Basin shows that Ro mean varies from 0.27% to 0.34% (average 0.31%). All the lignite petrographic data show favourable character for CTL.

Photomicrographs of some Barakar and Raniganj coal seams enriched in different varieties of liptinite macerals from Singrauli Coalfield, Madhya Pradesh; Ib-River Coalfield, Odisha; and Raniganj Coalfield, West Bengal, are given here (Fig. 5a–j). Petrographically, all the coal seams indicate their oil-prone nature [63].

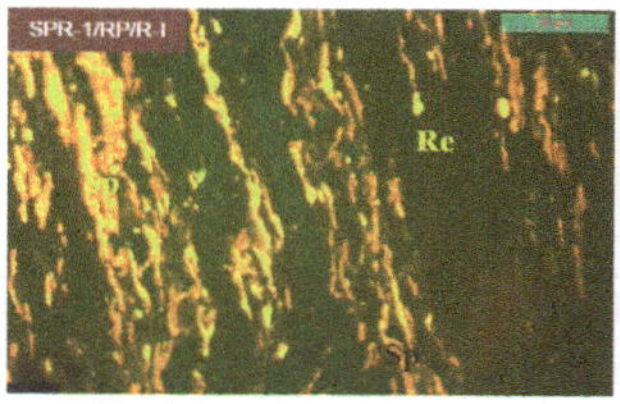

a. Thin elongated soprinites (Sp) and small globules of resinite (Re). Raniganj Formation, Singrauli Coalfield.

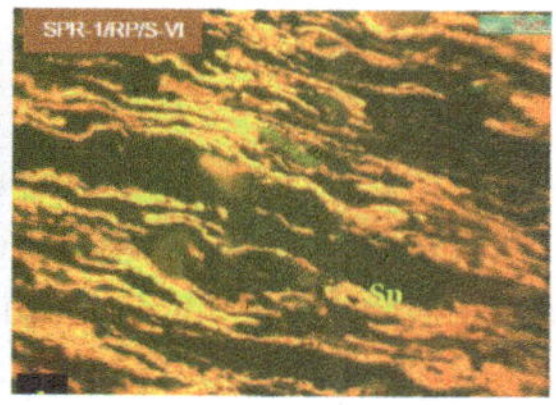

b. Liptinite rich bands with thin elongated soprinites (Sp) Barakar Formation, Singrauli Coalfield.

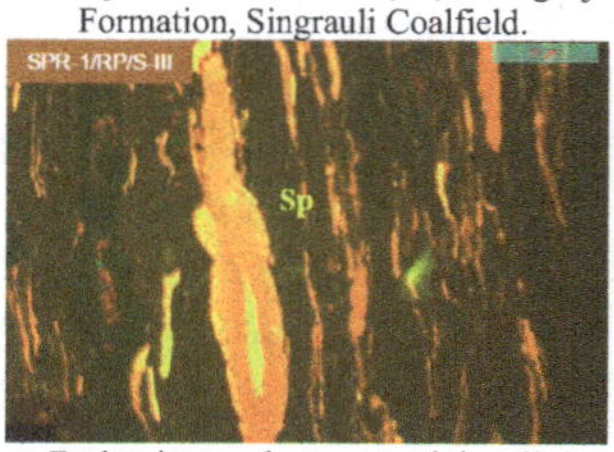

c. Both micro and mega sporinites (Sp), Barakar Formation, Singrauli Coalfield.

d. Thick resinite (Re), Barakar Formation, Singrauli Coalfield.

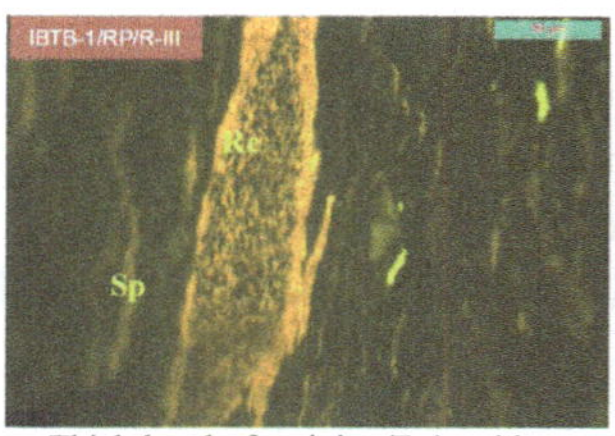

e. Thick band of resinite (Re), with sporinite (Sp), Raniganj Formation, Ib-River Coalfield.

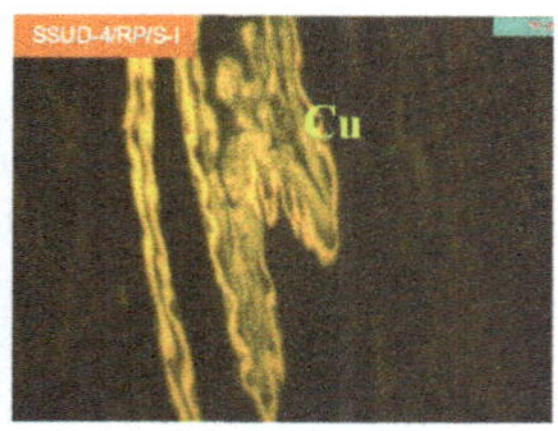

f. Cutinite (Cu) in association with collodetrinite, Raniganj Formation, Ib-River Coalfield.

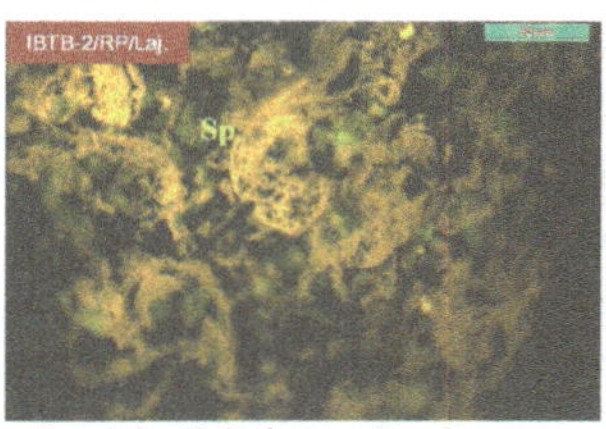

g. Sporinite (Sp) clusters, Barakar Formation, Ib-River Coalfield.

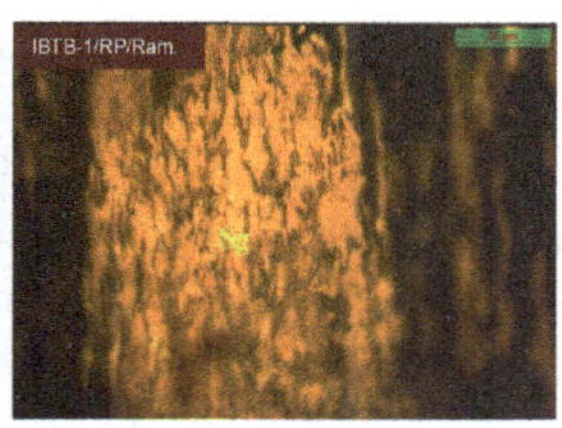

h. Large sporangium, Barakar Formation, Ib-River Coalfield.

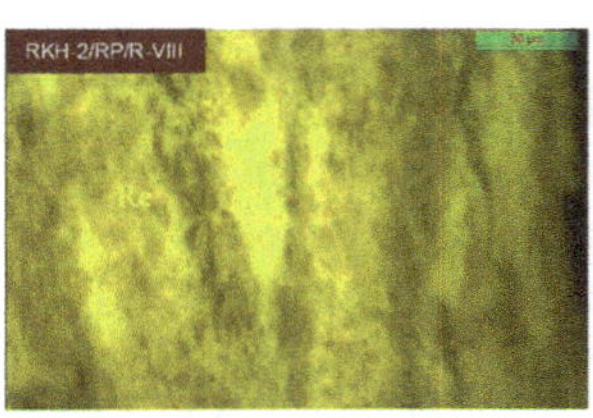

i. Thick band of resinite (Re), Raniganj Formation, Raniganj Coalfield.

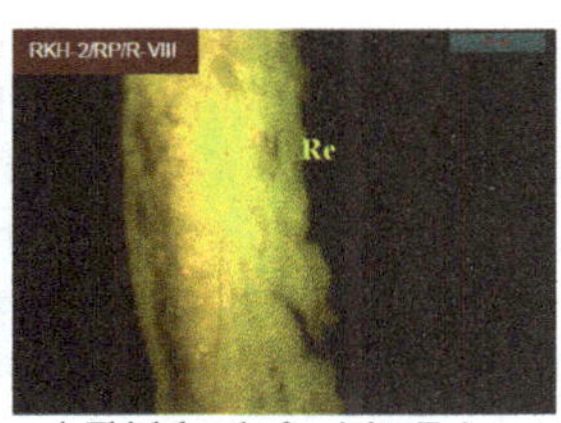

j. Thick band of resinite (Re), Barakar Formation, Raniganj Coalfield.

Fig. 5 a–j Photomicrographs of some Barakar and Raniganj coal seams from different coalfields enriched in different varieties of liptinite macerals

3.5 *Recent Advances*

In this context, a thorough petrographic scan of the low- to medium-rank Gondawana coal and Tertiary lignite and low-rank Tertiary coal of North East India is required with special reference to suitability for CTL. Multiparametric approach including petrography, thermal, ultimate and proximate analyses, spectroscopy using UV and IR may be used to identify the oil-prone nature. Coal oxidation at moderate temperatures (≤ 400 °C) is a complicated process affected by several factors, including coal compositions, temperature and oxygen concentration in the environment and alters the coal properties as fuel, thereby making pronounced impact on its subsequent utilization [64, 65]. Recent researches are going on in this direction with different spectroscopic techniques on coals and their soluble extracts to understand the macromolecular structures, which is essential for evaluate the oil producing and expelling potentiality of them.

3.5.1 TGA-DTA Analysis

Thermogravimetric analysis (TGA) and derivative thermogravimetry analysis (DTA) studies are useful tools for the determination of combustion characteristics (e.g. peak temperature, burnout temperature and activation energy) of coals during heating in a controlled system [66–68]. The procedure is strongly affected by particle size, ash content, sample amount, heating rate and gas flow rate [66]. High heating rates will cause simultaneous evolution and ignition of volatiles, whereas with low heating rates devolatilization will occur prior to ignition and combustion. The burning profile of coal can be an instrumental analysis in distinguishing between different coals. TGA curves show the change in weight of coal as a function of temperature and corresponding DTA curves shows the thermal events like exothermic and endothermic reaction, decomposition, melting, crystallization and change in crystal structure, etc. Analysis have been carried out for low-rank liptinite-rich non-coking coal of Ib-River Coalfield, Odisha, high-rank coking coal of Jharia Coalfield, Jharkhand, non-coking coal of Raniganj seams of Raniganj Coalfield and Tertiary coal of Meghalaya which gives some important inference regarding the energy release and decomposition behaviour of coal under a programmed and controlled environment (Fig. 6).

The thermal properties of coal strongly depend on coal rank. The ignition temperature of coals increases with decreasing volatile matter content. The ignition temperatures of maceral groups increase in the order of liptinite–vitrinite–inertinite [69]. Some recent studies have also shown that the combustion behaviour of coal cannot be merely correlated to petrographic composition. Maceral interactions can occur, and depending on the rank of coal, liptinite has a more important role in the maceral interaction [67]. The rank or parameter comprising rank and petrographic composition has more effect on combustion behaviour than merely maceral composition does [70, 71].

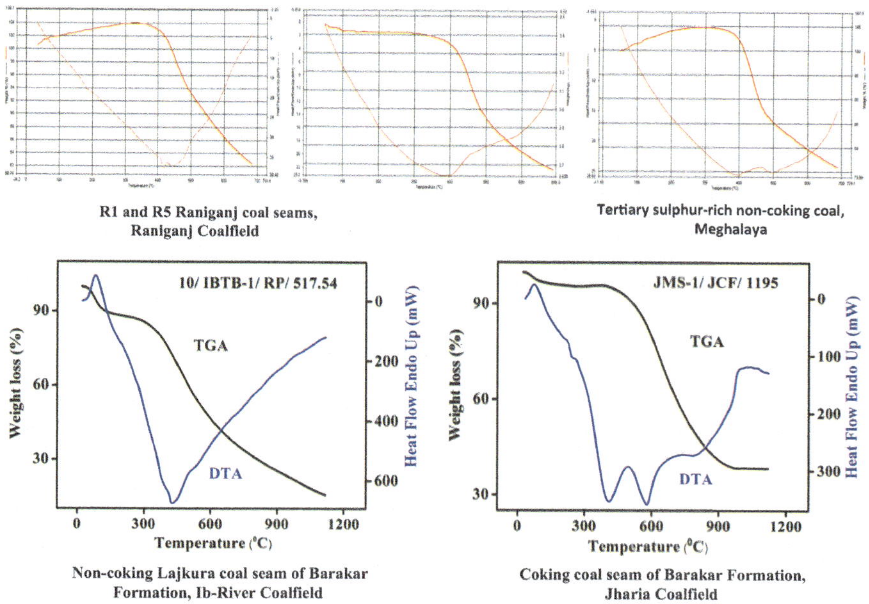

Fig. 6 TGA-DTA profiles of coal seams from different coalfields

3.5.2 FTIR Spectral Analysis

FTIR spectra of the non-coking Gondwana coals from R-I and R-V Raniganj seams of Raniganj Coalfield, West Bengal, Lajkura and Parkhani seams of Barakar Formation of Ib-River Coalfield, Odisha and Tertiary coal from Meghalaya indicate distinctly different signals indicating different macromolecular structures (Fig. 3) The peak ranging from 2800 to 3700 cm^{-1} is due to the major water and presence of C–H stretching (2800–3100 cm^{-1}) and C–H bending (700–900 cm^{-1}) of hydrocarbon components. A distinct peak around 1600 cm^{-1} region is due to the C=C ring stretch mode arising from irregular aromatic units as polycyclic aromatic hydrocarbons (PAH) containing ortho-substituted end rings or substituted with aliphatic chains (Fig. 7). Enriched aliphatic signatures of coals from Ib-River Coalfield indicate its oil-prone nature.

3.5.3 Ultraviolet (UV) Visible Spectroscopy

Analysis of UV-visible spectra of the water-soluble condensate of Lajkura coal seam of Barakar Formation of Ib-River Coalfield derived from destructive distillation up to 160 °C reveals that UV excitation enhances the visualization of the

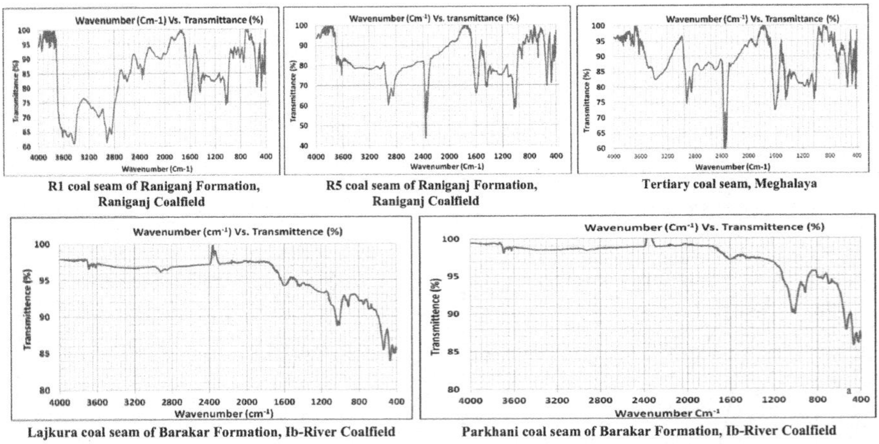

Fig. 7 FTIR spectra of coal samples from different seams of Gondwana and Tertiary coalfields

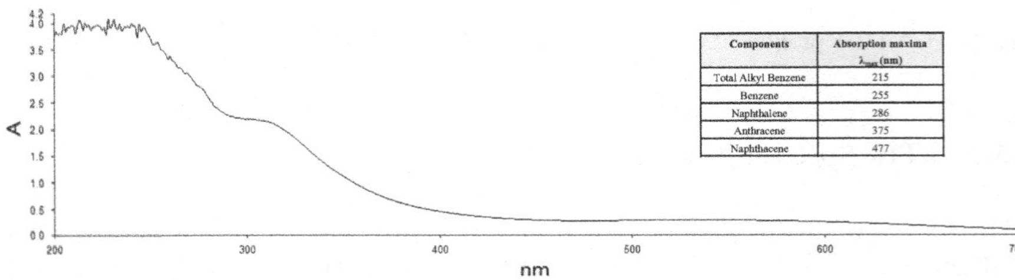

Components	Absorption maxima λ_{max} (nm)
Total Alkyl Benzene	215
Benzene	255
Naphthalene	286
Anthracene	375
Naphthacene	477

Fig. 8 The wavelength (nm) versus absorption (A) spectra of condensate dissolved in water

structures of chromophores, the light incorporating group of the molecules. The absorption spectra of condensate clearly demonstrate the presence of dominantly aromatic ring hydrocarbon of composition of alkyl benzene, benzene and naphthalene with lesser presence of higher ring polycyclic aromatic hydrocarbons (Fig. 8).

3.5.4 Ultraviolet (UV) Fluorescence Spectroscopy

Fluorescence spectroscopy, in the form of steady-state emission (SSE) and synchronous fluorescence scanning (SFS), is a diagnostic tool for investigating polycyclic aromatic hydrocarbon (PAH) structures in terms of the size of the fused aromatic ring (FAR) present in the condensate sample. It has been found from the

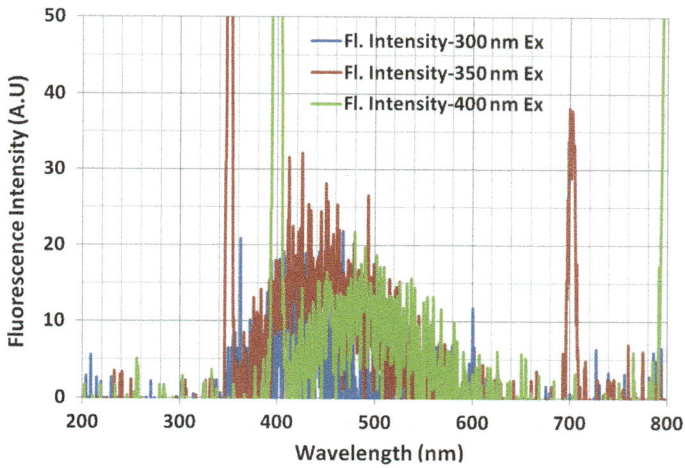

Fig. 9 Steady-state emission spectra (at 300 nm, 350 nm and 400 nm excitation wavelength) of condensate dissolved in water

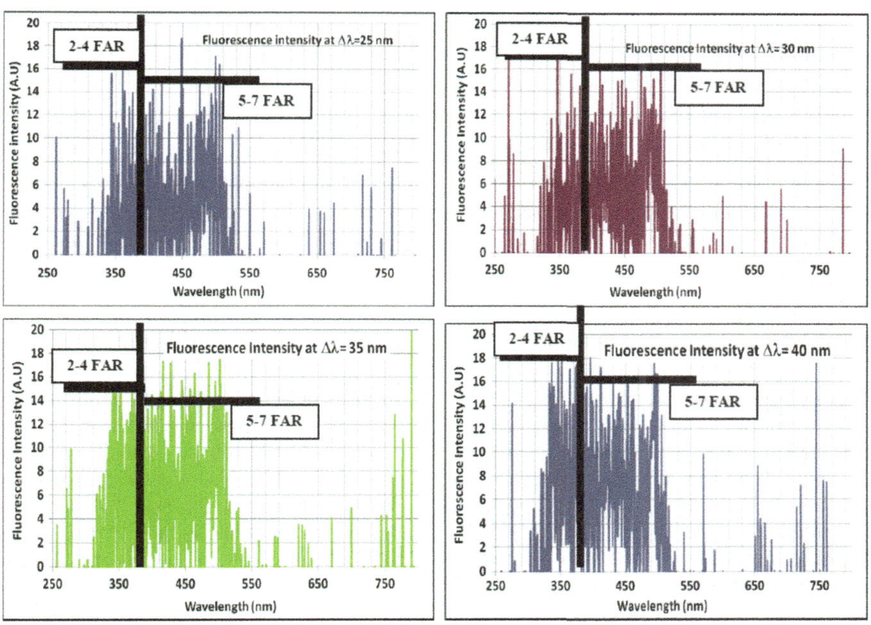

Fig. 10 Synchronous fluorescence spectra of the water dissolved condensate at **a** $\Delta\lambda = 25$ nm, **b** $\Delta\lambda = 30$ nm, **c** $\Delta\lambda = 35$ nm and **d** $\Delta\lambda = 40$

above figure that the SSE spectra at different excitation wavelengths are exhibiting a characteristic broad range of emission pattern from 360 to 650 nm. This observation can be correlated with the fact that the condensate sample contains high molecular weight polycyclic aromatic ring structures, i.e. asphaltenes (Fig. 9).

For the detailed investigation of the ring size (fused aromatic ring, FAR) associated with the high molecular PAH structures in the condensate, SFS spectra of the water-soluble condensate are captured which shows the presence of both small (2–4) FAR and larger (5–7) FAR (Fig. 10).

4 Conclusions

Studies undertaken so far from coal and lignite of non-coking and coking coal of both Raniganj and Barakar formations of Gondwana coalfields and Tertiary Coalfields of North-eastern region as well as lignites from west and coast lignite fields, unravel the intrinsic potentiality of some of them for unconventional greenfield energies like gas and oil. A huge data accrued through exploratory drilling in different coal and lignite fields are available, which are to be introspected in detail to unlock their potentiality. Favourable factors with respect to physic-mechanical, chemical, structural and petrographical characters are present for Indian solid fossil fuel, which may be utilized with advanced site-specific and composition-specific technology in near future. Maximum thrust is to be given on the different spectroscopic studies of coal and lignite to achieve the fundamental knowledge of their macromolecular structure so that they may be used more effectively for specific purposes of UCG or CTL. Understanding the size of various polyaromatic components released during heating of coal will help to design customized systems for coal gasification/liquefaction as well as residual materials which could be used as carbon feedstock. Understanding of this specific knowledge of aromatic components in coal has direct impact to determine the internal macromolecular structures to ascertain their suitability for coal utilization process.

Acknowledgements SB thankfully acknowledges Director General, GSI, for giving opportunity to represent GSI in the 'one-day conference' on 'Macromolecular characterization of coal and hydrocarbon components for future' jointly organized by School of Petroleum Technology, Pandit Deendayal Petroleum University (PDPU), Gujarat, with S. N. Bose National Centre for Basic Sciences, Kolkata, on 125th Birth Anniversary of Prof. Satyendra Nath Bose held on 19th November, 2018, in PDPU Campus, Gandhinagar. The present paper is an outcome of the deliberation of SB in this conference. He also thankfully acknowledges the support of Petroleum Engineering Laboratory of PDPU for the spectroscopic analysis and characterization of data. Data generated by GSI, which have been used in this paper are taken from the different unpublished technical reports of investigations carried out as approved FSP of GSI.

References

1. Niekerk VD, Mathews JP (2010) Molecular representations of Permian-aged vitrinite-rich and inertinite-rich South African coals. Fuel 89(1):73–82
2. Lievense MEH, Lucht M, Peppas NA (1985) Macromolecular structure of coals. V. Molecular weight distribution of extracted coal chains. Macromol Mater Eng 134 (1):73–95
3. Bhutto AW, Bazmi AA, Zahedi G (2013) Underground coal gasification: from fundamentals to applications. Prog Energy Combust Sci 39:189–214
4. Friedmann SJ, Upadhye R, Kong F (2009) Prospects for underground coal gasification in carbon constrained world. Energy Procedia 1:4551–4557
5. Imran M, Kumar D, Kumar N, Qayyum A, Saeed A, Bhatti MS (2014) Environmental concerns of underground coal gasification. Renew Sustain Energy Rev 31:600–610
6. Shafirovich E, Varma A (2009) Underground coal gasification: a brief review of current status. Ind Eng Chem Res 48:7865–7875
7. Liu S, Liu J, Yu L (2002) Environmental benefits of underground coal gasification. J Environ Sci 14:284–288
8. Stanczyk K, Kapusta K, Wiatowski M, Swidrowski J, Smolinski A, Rogut J (2012) Experimental simulation of hard coal underground gasification for hydrogen production. Fuel 91:40–50
9. Sury M, White M, Kirton J, Carr P, Woodbridge R (2004) Review of environmental issues of underground coal gasification. WS Atkins Consultants Ltd., University of Liège Belgium, FWS Consultants Ltd 126:532, pp 1–126
10. Kapusta K, Stanczyk K, Wiatowski M, Chećko J (2013) Environmental aspects of a field-scale underground coal gasification trial in a shallow coal seam at the experimental mine Barbara in Poland. Fuel 113:196–208
11. Kapusta K, Stanczyk K (2011) Pollution of water during underground coal gasification of hard coal and lignite. Fuel 90:1927–1934
12. Liu S, Li J, Mei M, Dong D (2007) Groundwater pollution from underground coal gasification. J China Univ Min Technol 17:467–472
13. Pankiewicz-Sperka M, Stanczyk K, Płaza GA, Kwasniewska J, NałeRcz-Jawecki G (2014) Assessment of the chemical, microbiological and toxicological aspects of post-processing water from underground coal gasification. Ecotoxicol Environ Saf 108:294–301
14. Xu B, Chen LJ, Xing BL, Yi GY, Li L (2013) Experimental study on the environmental behaviors of poisonous trace elements in post-gasified residues, Guangzhou, China. Adv Mater Res 805–806:1478–1483. https://doi.org/10.4028/www.scientific.net/AMR.805-806.1478
15. Hamanaka A, Su F, Itakura K, Takahashi K, Kodama J, Deguchi G (2017) Effect of injection flow rate on product gas quality in underground coal gasification (UCG) based on laboratory scale experiment: development of co-axial UCG system. Energies 10:238. https://doi.org/10.3390/en10020238
16. Khan M, Mmbaga J, Shirazi A (2015) Modelling underground coal gasification—a review. Energies 8:12603–12668. https://doi.org/10.3390/en81112331
17. Samdani G, Aghalayan P, Ganesh A (2016) A process model for UCG—part 1. J Fuel 181:690–703, part 2 181:587–599
18. Duan T, Lu C, Xiong S (2016) Pyrolysis and gasification modelling of underground coal gasification and the optimisation of CO_2 as a gasification agent. J Fuel 183:557–567
19. Green M (2018) Recent developments and current position of underground coal gasification. Proc Inst Mech Eng, Part A: J Power and Energy 232(1):39–46
20. Soukup K, Hejtmanek V, Capek P, Stanczyk K, Solcova O (2015) Modeling of contaminant migration through porous media after underground coal gasification in shallow coal seam. Fuel Process Technol 140:188–197
21. Jiang L, Chen Z, Ali SF (2018) General hydro-geological impact of cleats on underground coal gasification. Fuel 224:128–137

22. Kostúr K, Laciak M, Durdan M (2018) Some influences of underground coal gasification on the environment. Sustainability 10(5):1512
23. Yang L, Zhang X (2008) Modeling of contaminant transport in underground coal gasification. Energy Fuels 23(1):193–201
24. Ray A, Chatterjee DP, Chattopadhyay G, Datta A, Khangar RG (2005) Final report on the regional exploration for coal in Gopalnagar—Ganpur sector, Birbhum coalfield, Birbhum district, West Bengal (Unpublished progress report for the field seasons 2001–02 to 2003–04 of Geological Survey of India), Code No. MIE/CW/CW/2001/005
25. Roy A, Raychaudhuri, AK (2006) Final report on regional exploration for coal in Bandha block, Singrauli coalfield, Sidhi district, Madhya Pradesh (Unpublished progress report for the field seasons 2001–2002 to 2002–2003 of Geological Survey of India), Code No. MIE/CW/CW/2001/01
26. Chakraborty BK, Nanda PK, Behera SN, Pal S (2010) Final report on regional exploration for coal in Jalatap block, Talcher coalfield, Angul district, Orissa (Unpublished progress report for the field seasons 2003–04 to 2007–08 of Geological Survey of India), Code No. MIE/C/CW/CW/2003/008
27. Giritharan TS, Balukkarasu A, Baskaran K (2010) Report on exploration for lignite by scout drilling in Tiyanur sector, Ramnad sub basin, Ramanathapuram district, Tamil Nadu. (Unpublished progress report for the field seasons 2004–05 to 2006–08 of Geological Survey of India), Code No. MIE /C/CW/CW/2004/024
28. Hunt JM (1996) Petroleum geochemistry and geology. W.H. Freeman and Company, New York, p 743
29. Tissot B, Welte DN (1984) Petroleum formation and occurrence, 2nd edn. Springer Verlag, Heidelberg, p 699
30. Mukhopadhyay PK, Hatcher PG (1993) Composition of coal. In: Law BE, Rice DD (eds) Hydrocarbons in coal. American Association of Petroleum Geologists Studies in Geology, 38, pp 79–118
31. Curry DJ, Emmett JK, Hunt JW (1994) Geochemistry of aliphatic-rich rocks in the Cooper Basin, Australia, and Taranaki Basin, New Zealand: implications for the occurrence of potentially oil-generative coal. In: Scott AC, Fleet AJ (eds) Coal and coal-bearing strata as oil-prone source rocks?, vol 77. Geological Society Special Publication, pp 149–182
32. Hendrix MS, Brassell SC, Carroll AR, Graham SA (1995) Sedimentology, organic geochemistry, and petroleum potential of Jurassic coal measures: Tarim, Junggar, and Turpan Basins, Northwest China. Am Asso Petrol Geol Bull 79:929–959
33. Snowdon LR (1991) Oil from type III organic matter: resinite revisited. Org Geochem 17:743–747
34. Mukhopadhyay PK, Hatcher PG, Calder JH (1991) Hydrocarbon generation from deltaic and intermontane fluviodeltaic coal and coaly shale from the Tertiary of Texas and Carboniferous of Nova Scotia. Org Geochem 17:765–783
35. Powell TG, Boreham CJ (1991) Petroleum generation and source rock assessment in terrigenous sequences: an update. Aust Petrol Explor Assoc J 31(1):297–311
36. Qin K, Huang D, Li L, Guo S (1993) Oil and gas potential of macerals as viewed by C-13 NMR spectroscopy. organic geochemistry. In: Poster sessions from the 16th international meeting on organic geochemistry. Stavanger, Falch, Oslo, pp 758–762
37. Stach E, Mackowesky M, Teichmuller M, Taylor GH, Chandra D, Teichmuller R (1982) Stach's textbook of coal petrology, 3rd edn. Gebruder Borntraeger, Berlin, 535p
38. Taylor GH, Teichmuller M, Davis A, Diessel CFK, Littke R, Robert P (1998) Organic petrology. Gebruder Borntraeger, Berlin, p 704p
39. Newman J, Newman NA (1982) Reflectance anomalies in Pike River coals: evidence for variability in vitrinite type, with implications for maturation studies and 'Suggate Rank'. NZ J Geol Geophys 25:233–243
40. Killops SD, Funnell RH, Suggate RP, Sykes R, Peters KE, Walters C, Woolhouse AD, Weston RJ, Boudou JP (1998) Predicting generation and expulsion of paraffinic oil from vitrinite-rich coals. Org Geochem 29:1–21

41. Newman J, Price LC, Johnston JH (1997) Hydrocarbon source potential and maturation in Eocene New Zealand vitrinite-rich coals. J Pet Geol 20:137–163
42. Norgate CM, Boreham CJ, Kamp PJJ, Newman J (1997) Relationships between hydrocarbon generation, coal type and rank for Middle Eocene coals, Buller Coalfield, New Zealand. J Pet Geol 20:427–458
43. Clayton JL, Rice DD, Michael GE (1991) Oil-generating coals of the San Juan Basin, New Mexico and Colorado, U.S.A. Org Geochem 17:735–742
44. Smith GC, Cook AC (1984) Petroleum occurrence in the Gippsland basin and its relationship to rank and organic matter type. Aust Petrol Explor Assoc J 24(1):196–216
45. Moore PS, Burns BJ, Emmett JK, Guthrie DA (1992) Integrated source, maturation and migration analysis, Gippsland basin, Australia. Aust Petrol Explor Assoc J 32(1):313–324
46. Price LC, Barker CE (1985) Suppression of vitrinite reflectance in amorphous rich kerogen—a major unrecognised problem. J Pet Geol 8:59–84
47. Kalkreuth WD (1982) Rank and petrographic composition of selected Jurassic-Lower Cretaceous coals of British Columbia, Canada. Bull Can Pet Geol 30:112–139
48. Raymond AC, Murchison DG (1991) Influence of exinitic macerals on the reflectance of vitrinite in carboniferous sediments of the midland valley of Scotland. Fuel 70:155–161
49. van Krevelen DW (1993) Coal typology–physics–chemistry–constitution, 3rd edn. Elsevier, Amsterdam, p 979
50. Petersen HI, Vosgerau H (1999) Composition and organic maturity of Middle Jurassic coals, North-East Greenland: evidence for liptinite-induced suppression of huminite reflectance. Int Journ Coal Geol 41:257–274
51. Murchison DG, Pearson J, Raymond A (1991) Anomalies in vitrinite reflectance gradients. Bull Soc Geol France 162:183–191
52. Wenger LM, Baker DR (1987) Variations in vitrinite reflectance with organic facies-examples from Pennsylvanian cyclothems of the midcontinent, U.S.A. Org Geochem 11:411–416
53. Hao F, Chen J (1992) The cause and mechanism of vitrinite reflectance anomalies. J Pet Geol 15:419–434
54. Littke R, ten Haven HL (1989) Palaeoecologic trends and petroleum potential of Upper Carboniferous coal seams of Western Germany as revealed by their petrographic and organic geochemical characteristics. Int J Coal Geol 13:529–574
55. Levine JR (1993) Coalification: the evolution of coal as a source rock and reservoir rock for oil and gas. In: Law BE, Rice DD (eds) Hydrocarbons in coal. American Association of Petroleum Geologists Studies in Geology, 38, pp 39–77
56. Chakrabarti NC (1996) Rank and petrographic characters of Gondwana (Lower Permian) coal of peninsular India. Indian Miner 50(1–2):41–52
57. Mishra HK, Chandra TK, Verma RP (1990) Petrology of some Permian coals of India. Int J Coal Geol 16(1–3): 10, 47–71
58. Chaudhuri SN (2015) Petrographic approach on suitability of a few coal seams for liquefaction in Talcher Coalfield, Odisha, India. Indian J Sci 69(2):103–116
59. Geethumol KG, Ramteke CP, Chaudhuri SN (2019) Petrographic studies to delineate suitable coal seams for coal liquefaction and their palaeo-environmental analysis: a case study from the Ib river and Mand-Raigarh coalfields of the Mahanadi valley coalfields India. J Earth Syst Sci 128:165
60. Sarate OS (2017) Petrography of coal seams from Belampalli coalfield, Godavari valley, Telangana state, India. J Geosci Res 2(1):29–36
61. Singh PK, Singh GP, Singh MP (2011) Characterization of coal of seams II, III, and IIIA from Ramagundam coalfield, Godavari valley, Andhra Pradesh. In: India energy sources 2011, Part A 33, pp 1863–1870, ISSN: 1556-7036, print: 1556-7230. https://doi.org/10.1080/15567030903468536
62. Singh PK, Singh VK, Singh MP, Rajak PK (2017) Petrographic characteristics and paleoenvironmental history of Eocene lignites of Cambay basin, Western India. Int J Coal Sci Technol 4(3):214–233. https://doi.org/10.1007/s40789-017-0173-2

63. Geethumol KG, Ramteke CP (2018) Final report on a comprehensive study on the petrological and depositional characteristics of early Permian coals (Barakar Formation) and late Permian coals (Raniganj formation) from selected gondwana basins of india. (Unpublished progress report for the field seasons 2016–2017 to 2017–2018 of Geological Survey of India), 2018, Code No. M2BRP/NC/CHQ/2017/12338
64. Wang H, Dlugogorski BZ, Kennedy EM (2003) Role of inherent water in low temperature oxidation of coal. Combust Sci Tech 175:253–270
65. Clemens AH, Matheson TW, Rogers DE (1991) Low temperature oxidation studies of dried New Zealand coals. Fuel 70:215
66. Blazek A (1973) Thermal Analysis. In: Tyson JF (ed) Chapter 2. Van Nostrand Reinhold, London
67. Milligan JM, Thomas KM, Crelling JC (1997) Temperature-programmed combustion studies of coal and maceral group concentrate. Fuel 76:1249–1255
68. Sun CI, Kozinski JA (2000) Ignition behavior of pulp and paper combustible wastes. Fuel 79:1587–1593
69. Shibaoka M (1969) Combustion of coal in thin sections. Fuel 47:285–295
70. Pregermain S (1988) Rank and maceral effects on coal combustion characteristics. Fuel Process Tech 20:297–305
71. Furimsky E, Palmer AD, Kalkreuth WD, Cameron AR, Kovacik G (1990) Prediction of coal reactivity during combustion and gasification by using petrographic data. Fuel Process Tech 25:135–151

Maturity Assessment of Cambay Shale Formation (CSF) for Hydrocarbon Prospect: A Molecular Structure Approach with Optical Spectroscopy Study

Uttam K. Bhui, Ravinder Ariketi, Archchi Sarkar, Saheli Sanyal, and Rincy Anto

1 Introduction

Sedimentary organic matter (SOM) when preserved in subsurface conditions undergoes irreversible physical and chemical reordering owing to the increase in temperature in time duration of geological time-scale, hence increase in thermal maturity. The thermal evolution of SOM is the gradual transformation under the effect of a given thermal history, while the whole organic content evolves progressively towards a carbon residue (during metagenesis stage) with liberation of oxygen bearing gaseous materials like CO, CO_2, H_2O in early stage (diagenesis) followed by hydrocarbon liquid or gas in the later stage (catagenesis). Thermal evaluation of kerogen leads to relative condensation and aromatization in the residual portion, whereas increasing thermal cracking of the kerogen leads to higher concentration of lighter molecules in the liberated products (bitumen). This causes organic matter to form clusters of aromatic compounds which eventually transforms into polycyclic aromatic hydrocarbon (PAH) [1, 2]. With increase in temperature

U. K. Bhui (✉) · A. Sarkar · S. Sanyal · R. Anto
School of Petroleum Technology, Pandit Deendayal Energy University (PDEU, Formerly Pandit Deendayal Petroleum University-PDPU), Raisan, Gandhinagar, Gujarat 382007, India
e-mail: Uttam.bhui@spt.pdpu.ac.in

A. Sarkar
e-mail: sarkar.archchi@gmail.com

S. Sanyal
e-mail: sahelisanyal87@gmail.com

R. Anto
e-mail: rincyanto19@gmail.com

R. Ariketi
Gujarat State Petroleum Corporation Ltd., Gandhinagar, Gujarat 382010, India
e-mail: ravinder@gspc.in

U. K. Bhui (ed.), *Macromolecular Characterization of Hydrocarbons for Sustainable Future*, Green Energy and Technology,
https://doi.org/10.1007/978-981-33-6133-1_14

and pressure, it has been observed that these PAHs respond predictably towards progressive alkalization to form thermally more stable isomers. The asphaltenes are PAHs, containing fused aromatic ring (FAR) up to maximum ten in number, where the increasing ring numbers indicate increasing aromaticity of the structural component. Asphaltenes and kerogen pyrolysates would be similar as asphaltenes are basically subunits of kerogen formed by breaking enough cross-link of their original structures. The bitumen becomes more aromatic as thermal severity increases [3] asphaltenes separated from the aromatic part shown similar changes in chemical composition [4]. The asphaltene content of extracted organic matter generally increases with thermal exposure. PAHs present in hydrocarbon are primarily resistant to biodegradation and can span through a wide range of thermal maturity, making them useful as maturity indicators. Asphaltenes do not have unique chemical composition and structure where their average molecular formula changes with source type and level of maturity. So, chemical structure of asphaltenes is diagnostic of both original kerogen structure and their thermal exposure. Analysis and integration of spectroscopic characterization of molecular structure present in the bitumen and residual kerogen portion of the resource rock samples with their population distribution could be potential parameter for indicating thermal maturity.

Optical properties of organic matter such as the reflectance of light in polished sections, the colour of transmitted light in thin section and fluorescence in either transmitted or reflected light are characteristics of thermal maturity for generation of hydrocarbon as liberated by-product, from the organic matter. The direct measurement of these optical properties of different organic constituents records the thermal evolution level in order to produce by-products like hydrocarbon oil and gas. Various geochemical scales, such as vitrinite reflectance (VRo), Rock-Eval pyrolysis, thermal alteration index (TAI), biomarker population are presently being used to indicate the level of thermal maturity of organic matter in source rocks. It is important to note that VRo is considered highly valuable and important as it can show uniform change during maturation of hydrocarbon [5, 6]. Moreover, vitrinite is present in most sedimentary kerogens, and hence, it has widespread availability. But most of the hydrocarbon generation occurs from liptinite macerals having higher H/C ratio compared to vitrinite. Therefore, the reaction kinetics of hydrocarbon generation and maturation of liptinites must be similar like vitrinite to use VRo as maturity indicator. The advantage of Rock-Eval pyrolysis over vitrinite reflectance study of disseminated kerogen is that it is conducted on whole rock rather than on isolated/separated kerogen.

Complex aromatic (fluorophores) bearing organic molecules, kerogen is optically active which absorb as well fluoresces when irradiated optically. Tremendous advanced in spectroscopic characterization made it possible to determine molecular structures of the chemical components, but the structure of components during thermal cracking of kerogen is sufficiently complex and varied that only qualitative picture has emerged so far. Researchers used fluorescence study for fingerprinting, concentration quantification as well as maturity assessment of petroleum crudes [7, 8]. The population distribution of small and large PAH with FAR system in different component (bitumen and residual kerogen) of prospective source rock samples in comparison with already established maturity parameters like VRo, Rock-Eval data and TAI may help to understand the thermal cracking processes in natural system. With the help of optical spectroscopic study particularly UV–visible and fluorescence (steady-state and synchronous) spectroscopic study, Bhui et al. [9, 10] characterized PAH with FAR structures present in solvent-extracted shale samples and their demineralized counterpart for evaluation of thermal maturity for hydrocarbon prospect. Optical spectroscopic approach was undertaken in the present work to understand population distribution of macromolecular organic structure in the samples of Cambay shale formation (CSF), for assessment of thermal maturity for hydrocarbon prospect.

2 Background Geology and Sample for Study

The Cambay basin is located in the western flank of Peninsular India, between the latitudes 21° N to 25° N and longitudes 71°30′ E to 73°30′ E. Geographically, it extends from Surat in the south to Sanchor in the north, divided into five tectonic blocs (Fig. 1) comprising an area of about 53,500 sq.km. The Cambay basin forms a significant petroleum province with established hydrocarbon source potential. Earlier studies indicate proven occurrence of hydrocarbon in the Cambay-Kadi/Kalol petroleum system in north and Cambay-Hazad petroleum system in south of Cambay basin [11].

The present study area lies in the Narmada-Tapti block and is delineated in the north by the major Narmada fault. The thickness of the sedimentary sequence reaches up to 2000 m in the western part of the block, with underlying Deccan trap basalts exposed in the eastern margin. This block is an eastern continuation of ENE-WSW Satpura/Narmada–Son lineament. Owing to this, nearly all structures and faults in the Narmada-Tapti block are aligned in the ENE-WSW direction which further continues west into the Gulf of Cambay. The study area covers 160 sq.km in which the thickness of Cambay shale formation (henceforth CSF)

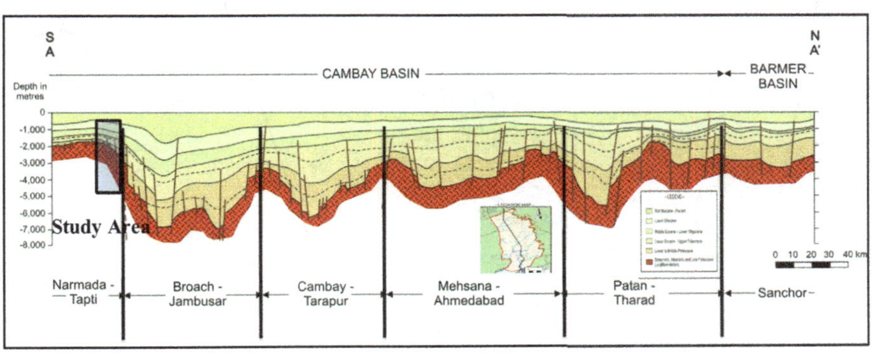

Fig. 1 Regional N-S cross section of the Cambay basin showing the sediment thickness in different depocenters [12]. The study area of the present work lies in the south of the Broach depocenter of the Narmada-Tapti block

varies from 25 m in the east to 700 m in the west. The depth of CSF varies from 1100 to 2500 m. Oil to source correlation studies indicate that the Cambay shale is the main source rock with middle to lower shale horizon is in catagenesis to metagenesis stage of thermal maturity while upper Cambay shale still in the hydrocarbon generation phase. In the southern blocks, there occurs subpetroleum system in relation to the lower and upper Cambay shale with huge thickness, high TOC and thermal maturity (Fig. 2) [11, 13, 14].

Five vertical wells viz. Well#1, Well#2, Well#3, Well#4 and Well#5 were drilled through CSF in the study area (Fig. 3). The correlation and distribution of Cambay shale in different wells in the study area are given in Fig. 4. Drill cutting

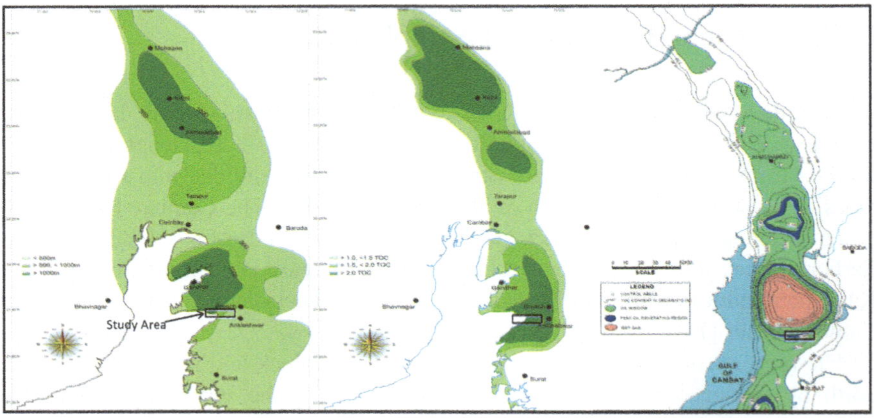

Fig. 2 Source rock distribution and maturity map of in Cambay shale in different depocenters of the Cambay basin with the location of the study area [11]

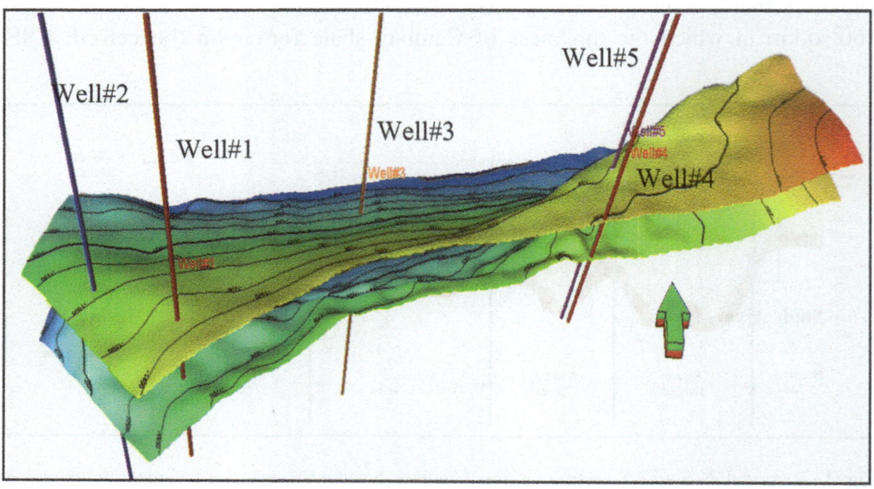

Fig. 3 Location map of the drilled wells in the study area (Schlumberger® Petrel, version 2014). Arrow in this map indicates north direction [15]

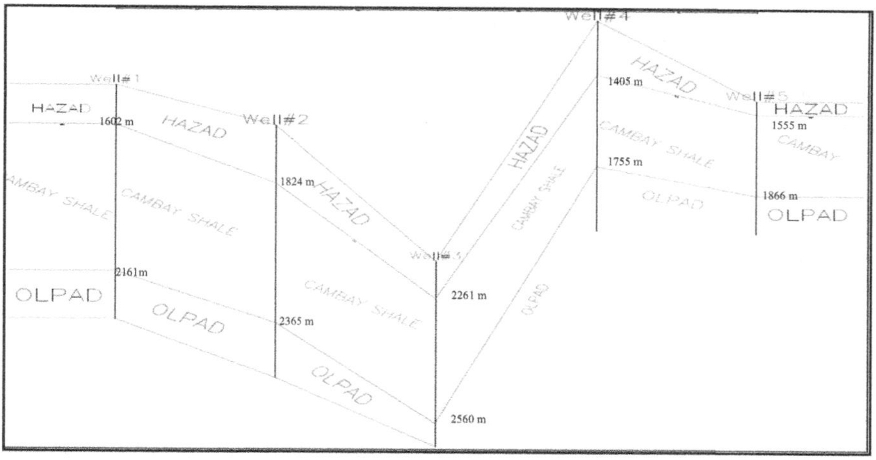

Fig. 4 Schematic representation of Cambay shale distributions in different wells of the study area with Cambay shale top and Cambay shale bottom from the drilling data [15]

samples from different depths of Well#2 and Well#3 were collected and presented in Table 1. Three samples from Well#2 and four samples are used in the present study (Bold emphasized in Table 1).

Table 1 Details of the drill cuttings collected of the Cambay shale from different wells and the samples considered in the present study

Well No. #2			Well No. #3		
Cambay shale	Top	1824 m	Cambay shale	Top	2261 m
	Bottom	2365 m		Bottom	2560 m
Sample No.		Drill cutting depth (m)	Sample No.		Drill cutting depth (m)
1		1826–1829	1		2260–2265
S-2		**1829–1832**	2		2263–66
3		1838–1841	3		2266–69
4		1847–1850	4		2275–78
5		1880–1885	**CS-5**		**2281–2284**
6		1885–1890	**CS-6**		**2284–2287**
7		1905–1910	**CS-7**		**2287–2290**
S-6		**1915–1920**	8		2290–2293
9		1935–1940	9		2302–2305
10		1940–1945	10		2305–2310
11		1947–1950	11		2325–2330
12		1955–1960	12		2330–2335
13		1965–1970	13		2335–2340
S-12		**1980–1985**	**CS-14**		**2345–2350**

Bold emphasized samples are used in this study

3 Methodology and Instrumentation

The selected samples from the collected drill cuttings are thoroughly washed with water, air dried and examined under binocular microscope for complete removal of any sort of contamination. The dried and powdered samples were homogenized in spectrophotometric grade KBr in an agate mortar and pressed at 3 mm pellets with a hand press for FTIR analysis. The infrared spectra were acquired using a PerkinElmer Spectrum Version 10.4.2, at a resolution of 4 cm^{-1} and presented in Fig. 5. The UV–visible and fluorescence spectroscopic analyses were carried out at room temperature for the solvent (toluene)-extracted portion of the powder drill cutting samples. For UV–visible spectroscopy, the instrument used was Perkin Elmer Lambda 35 UV–visible double beam spectrometer. Auto-correction of the sample was done prior to the analysis, while absorption spectrographs were recorded along with baseline correction for all the samples. The acquisition of fluorescence spectroscopy data was carried out using a Perkin Elmer LS 55 spectrophotometer. The emission spectra were acquired from 200 to 800 nm keeping the excitation wavelength at constant values (250 and 300 nm) with fixed slit width for the solvent-extracted fraction. Representative UV–visible and fluorescence graphs are presented in Figs. 6 and 7, respectively, for analysis.

4 Results

4.1 FTIR

FTIR for each sample was done in order to account for the type of bonds which were easily detected from the vibrations, bending and stretching frequencies of the bonds and possible band assignments were made. The infrared spectra of the bulk rock samples show presence of characteristic band groups in the high-frequency region (3700–3600 cm^{-1}), in the medium frequency region (1700–1000 cm^{-1}) and in the low-frequency region (1000–600 cm^{-1}). The O–H absorption peaks present at 3694, 3650 and 3622 cm^{-1}. Near 1640 and 1411 cm^{-1}, the presence of olefin C=C absorption peaks is marked, whereas near 1002 cm^{-1} C–O, absorption peaks are noticed, between 534 and 913 cm^{-1}, the heteroatom absorption peaks are prominent.

4.2 UV–Visible Spectroscopy

The solvent-extracted component from the shale samples of both the wells (Well#2—Fig. 6a and Well#3—Fig. 6b) is showing more or less similar absorption behaviour. Saturated organic groups with non-bonding electrons can give rise to

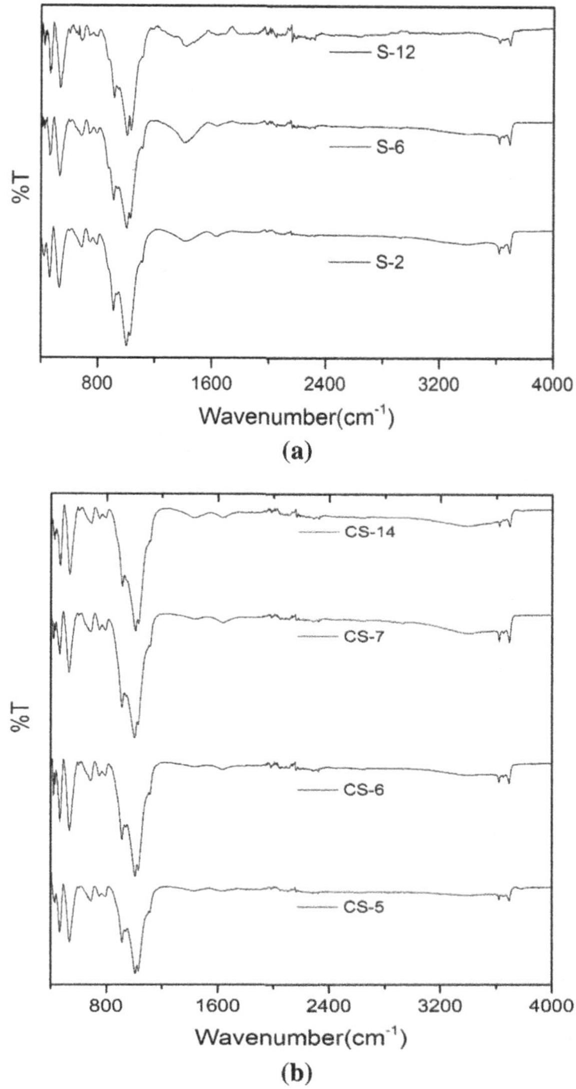

Fig. 5 FTIR spectra of powder shale samples from specific depths of Well#2 (**a**) and Well#3 (**b**) of the study area

transitions in the UV and near-UV wavelength range while unsaturated groups which give rise to absorptions involving π or π^* orbitals in the near-UV/visible wavelength range. Absorption peaks obtained in the wavelength range from 250 to 270 nm indicate the presence of carbonyl compounds with mainly $n \rightarrow \pi^*$ transitions. Some chromophores with bonding C=O functional group, $n \rightarrow \pi^*$ transitions gave rise to absorption (λ_{max}) at 285 nm and N=O ($n \rightarrow \pi^*$) transition at 275 nm. Absorption peak around 310 nm indicates n to π^* transitions with presence of nitro-compounds.

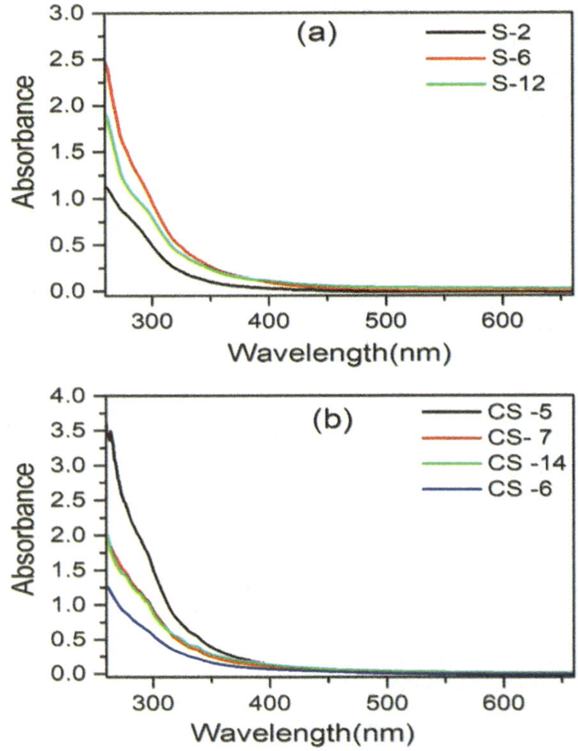

Fig. 6 Absorption spectra of solvent-extracted shale powder samples from different depths of **a** Well#2 (S-2, S-6, S-12) and **b** Well#3 (CS-5, CS-6, CS-7, CS-14)

Most of the absorptions observed in the study samples involve σ to σ^* (alkenes), $n \rightarrow \sigma^*$ (N, S, O), $n \rightarrow \pi^*$ (Carbonyl) and $\pi \rightarrow \pi^*$ (aromatic) transitions (Table 2).

The UV–visible spectra of shale (solvent extracts) samples from Well#2 and Well#3 exhibit high absorbance values in the ultraviolet wavelength (UV) range with gradual decrease up to 500 nm and showing constant absorption after 500 nm wavelength suggesting presence of both the fine structures of a series of individual ortho-fused end ring PAHs [16]. The absorption spectra of all the samples of Well#2 show deflections up to 320 nm, whereas the absorption spectra of shale samples of Well#3 exhibit broad peaks up to 360 nm (300, 323, and 340 nm). With increasing number of aromatic rings in case of polycyclic aromatic hydrocarbon, a shift of absorption maxima shifts towards higher wavelength. This reflects the presence of larger or heavier aromatic species (larger PAH) in the samples of Well#3 than Well#2.

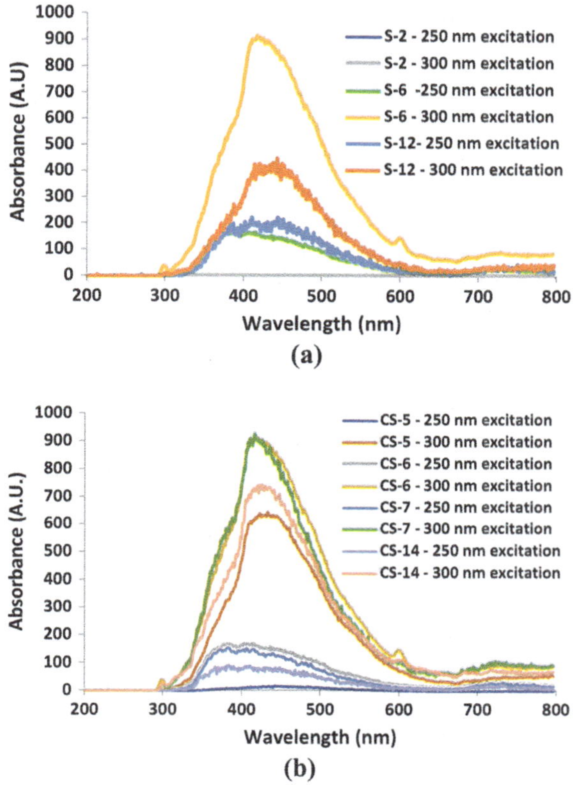

Fig. 7 Fluorescence spectra of solvent-extracted portion of shale samples at 250 and 300 nm excitation wavelengths for Well#2 (**a**) and Well#3 (**b**)

Table 2 Range of maximum absorption in the UV–visible spectra with possible transition and possible compounds in the solvent-extracted shale samples of Well#2 and Well#3

	Max. λ_{max} range	Max. possible transition	Possible compounds
Well#2			
S-2	247–290	$n \rightarrow \pi^*$	Alkenes, Carbonyl, NSO with PAH
S-6	247–290	$n \rightarrow \pi^*$	Alkenes, Carbonyl, NSO with PAH
S-12	241–259	$n \rightarrow \pi^*$	Alkenes, Carbonyl, NSO with PAH
Well#3			
CS-5	240–270	$n \rightarrow \pi^*$	Alkenes and Carbonyl compounds PAH
CS-6	242–296	$n \rightarrow \pi^*$	Alkenes, Carbonyl, NSO with PAH
CS-7	240–290	$n \rightarrow \pi^*$	Alkenes, Carbonyl, NSO with PAH
CS-14	240–290; 339	$n \rightarrow \pi^*$; $\pi \rightarrow \pi^*$	Alkenes, Carbonyl, alkynes, NSO with PAH/ asphaltene porphyrins

4.3 Fluorescence Spectroscopy

Fluorescence spectroscopy is commonly used as selective diagnostic tool for the analysis of polycyclic aromatic hydrocarbon (PAH) molecules in a complex aromatic system [17]. The high λ_{max} obtained between 320 and 400 nm is probably associated with 1–3 FAR aromatic structures, while λ_{max} obtained between 400 and 550 nm is probably associated with PAH with 4–6 FAR structures (Table 3) [18].

The steady-state emission spectra of solvent-extracted portion of the shale samples from specific depths of Well#2 and Well#3 are captured at fixed excitation wavelengths of 250 and 300 nm and presented in Fig. 7. Samples S-6 and S-12 from Well#2 are characterized by broad emission pattern extending from 300 to 600 nm range (Fig. 7a), while sample S-2 does not have any emission at both the excitation wavelengths. The fluorescence intensity at 250 nm excitation wavelength is less compared to 300 nm excitation wavelength. Small deflections present at or near 366, 387, 410, 440 nm within the broad spectra are indicative of presence of PAH with 3–6 FAR structures [18–21]. All the samples from Well#3 have broad emission pattern and extending from 300 to 650 nm with considerable absorption intensity at 300 nm, while at 250 excitation wavelengths, the absorption intensity is very less for all the samples. This is very much indicative of presence of slightly larger PAH with FAR in Well#3 compared to Well#2.

4.4 Mud Logging Data of Well#2 and Well#3

Drilling parameters like ROP, fluorescence cut, total gas integrated with shale density, D-exponent, shale factor, etc., give indications of subsurface geology,

Table 3 Range of peak wavelength (λ_{max}) values of fluorescence spectra for polycyclic aromatic hydrocarbon (PAH) with different fused aromatic rings (FAR)

PAH with FAR structure	Names	Peak emission wavelength (λ_{max}) range (nm)
Single Benzene ring structure	Benzene	250–300
2-Benzene ring FAR structure	Naphthalene	300–365
3-Benzene ring FAR structure	Fluorene	302–370
	Anthracene	372–460
	Phenanthrene	348–407
4-Benzene ring FAR structure	Napthacene	460–580
	Chrysene	380–400
	Pyrene	370–400
5-Benzene ring FAR structure	Pentacene	620–750
	Perylene	450–510

while systematic study of fluorescence, background gas, cut in drill cutting is very much helpful in identification of hydrocarbon reservoirs as well as source rock potential. In Well#2, from depth 2027–2267 m, the deeper section shows golden yellow fluorescence and mild cut with presence of gas (Fig. 8 and Table 4). Similarly, the deeper section (from 2395–2418 m) of Well#3 also shows reasonably good golden yellow fluorescence with very good cut and high total gas (Fig. 9 and Table 5).

5 Discussion and Conclusions

The thermal maturation of CSF in the study area was performed with the combination of methods like Rock-Eval pyrolysis, optical methods, well log data for Δlog *R* techniques and 1D modelling for Well#1, drilled as a parametric well for resource estimation [15, 22]. The Rock-Eval data shows the max TOC is 5.14 wt% while the Δlog *R* techniques show the max TOC is 4.85 wt%. The log-derived VRo is 0.86% at 1935 m, and the core analysis VRo is 0.84% at 1941 m indicating the Oil window. The cross-plots between T_{max} Versus HI, VRo Versus LOM and 1D modelling show that the CSF in the study area falls in the mature zone of oil to wet gas generation window. The integrated results of the different methods indicate that the thermal maturity of CSF in the study area is falling within the oil generation window (shallow depths) to gas generation window (deeper depth) in Well#1 (Fig. 10).

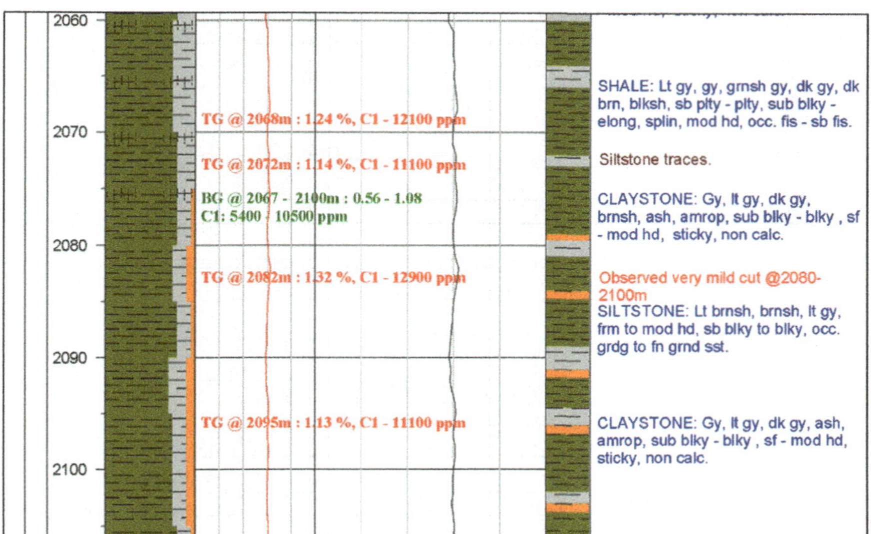

Fig. 8 Fluorescence cut in drill cutting of Well#2 in different depth (extreme left values are in metre)

Table 4 Hydrocarbon shows observed during drilling in Well#2

Well No.	Drill cutting depth (m MD)	Total Gas (%)	Cut/Fluorescence
Well#2	2027	1.26	Few grains show golden yellow fluorescence and mild cut
	2068	1.24	Few grains show golden yellow fluorescence and mild cut
	2082	1.32	Few grains show golden yellow fluorescence and mild cut
	2152	2.21	1–3% grains show golden yellow fluorescence and mild cut
	2267	1.43	1–5% grains show golden yellow fluorescence and mild cut

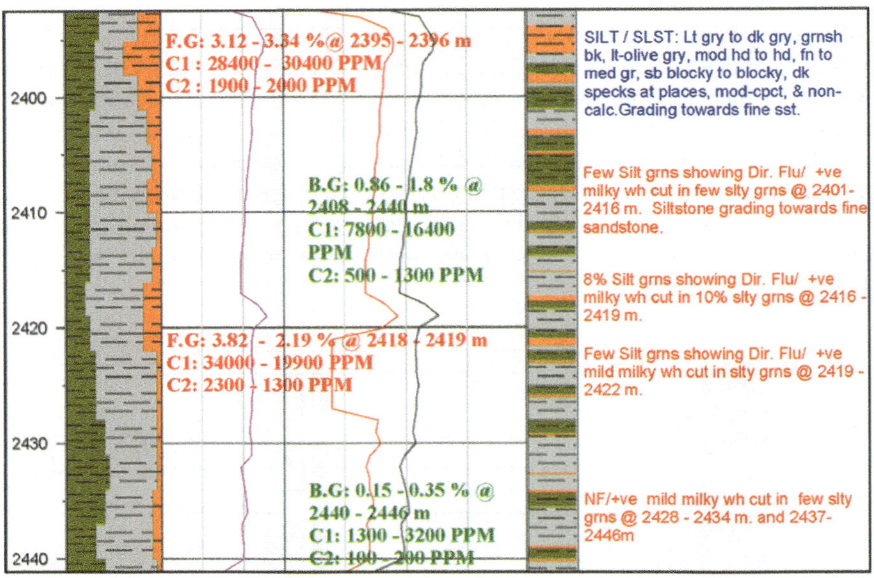

Fig. 9 Fluorescence cut in drill cutting of Well#3 in different depth (extreme left values in metre)

Table 5 Hydrocarbon shows observed during drilling in Well#3

Well No.	Drill cutting depth (m MD)	Total gas (%)	Cut/Fluorescence
Well#3	2395	3.12	10–15% grains show golden yellow fluorescence and very good cut
	2418	3.82	5–10% grains show golden yellow fluorescence and very good cut

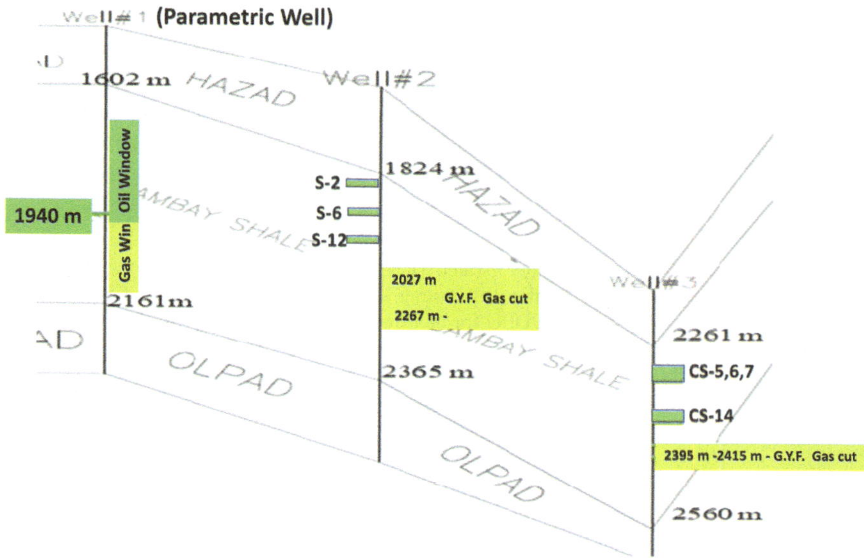

Fig. 10 Schematic correlation of thermal maturity parameters of CSF from different wells of the study area [G.Y.F = Golden Yellow Fluorescence]

Mud logging information from the deeper zones specifically from 2027 to 2267 m of Well#2 shows golden yellow fluorescence with mild gas clearly indicates hydrocarbon generation (Fig. 10). Depth interval from 2395 to 2418 m of Well#3 also showed golden yellow fluorescence and very good gas cut, clear indicative of hydrocarbon generation in this depth zone (Fig. 10).

Thermal maturity information of nearby well drilled through CSF (Well#1) and mud logging parameters from the deeper depths of the study wells (Well#2 and Well#3) clearly showed that the study samples lie within the zone of thermal maturation for generation of hydrocarbons. As known, thermal evolution of kerogen leads to relative condensation and aromatization in the residual portion with higher concentration of relatively smaller molecules in the generated hydrocarbon product which are generally present (as soluble components) in rock extracts with solvent of different nature. With the help of modern spectroscopic techniques, Kelemen et al. [23] pointed out the role of nitrogen, sulphur and oxygen functionality as a function of thermal maturity for all kerogen types. Carbonyl (ketone, carboxyl, ester) and ether groups are at roughly equal abundance at low maturation level of kerogen with low aromaticity. With increasing thermal maturity, aromaticity increases with the increase of ether groups and decrease of carboxyl groups, while abundance of carboxyl group is negligible at higher maturation level. Optical characterization of molecular moieties through FTIR, UV–visible and fluorescence spectroscopic study of the present CSF samples from two wells clearly demonstrated the presence of different PAH with FAR structures with functional group in solvent-extracted portion. Depth-wise variation of the characteristics of the molecular moieties and their attached functional group may lead to understand thermal evolution of the organic matter. Detail molecular characterization and their quantification in the source rocks through optical spectroscopic approach in comparison with already established maturity parameters may help to understand the thermal cracking processes in the natural system and to develop a new scale of thermal maturity on the basis of prevalent molecular structure present in a prospective source rock.

Acknowledgements UKB and Ravider Ariketi (RA) acknowledged the support of GSPC for providing the official permission for the drill cutting samples for doctoral study of RA. AS, SS, Rincy Anto thankfully acknowledge PDPU for their research fellowships. Petroleum Engineering (PE) Laboratory, Reservoir Characterization Laboratory (RCL) of SPT, and Chemistry Laboratory of PDPU are thankfully acknowledged for the instrumentation support of optical characterization of the study samples.

References

1. Mukhopadhyay PK (1992) Maturation of organic matter as revealed by microscopic methods: applications and limitations of vitrinite reflectance, and continuous spectral and pulsed lased fluorescence spectroscopy. In: Wolf KH, Chillingarian CV (eds) Diagenesis, III, Developments in sedimentology, vol 47. Elsevier Science Publishers
2. Mukhopadhyay PK (1994) Vitrinite reflectance as maturity parameter petrographic and molecular characterization and its applications to basin modeling. In: Mukhopadhyay and Dow (eds) Vitrinite reflectance as a maturity parameter ACS symposium series. American Chemical Society, Washington, DC
3. Feng Y, Le Doan TV, Pomerantz AE (2013) The chemical composition of bitumen in pyrolyzed Green river oil shale: characterization by 13C NMR spectroscopy. Energy Fuels 27:7314–7323
4. Pomerantz AE, Le Doan TV, Craddock PR, Bake KD, Kleinberg RL, Burnham AK, Wu Q, Zare RN, Brodnik G, Lo WC-H, Grayson MB, Mitra-Kirtley S, Bolin TD, Wu T (2016) Impact of laboratory-induced thermal maturity on asphaltene molecular structure. Energy Fuels. https://doi.org/10.1021/acs.energyfuels.6b01238
5. Hunt JM (1979) Petroleum geochemistry and geology, 1st edn. Freeman
6. Tissot BP, Welte DH (1984) Petroleum formation and occurrence. Springer-Verlag, New York
7. Wang C, Li W, Luan X, Liu Q, Zhang J, Zheng R (2010) Species identification and concentration quantification of crude oil samples in petroleum exploration using the concentration-synchronous-matrix-fluorescence spectroscopy. Talanta 81:684–691
8. Ryder AG, Glynn TG, Feely M, Barwise AJG (2002) Characterization of crude oils using fluorescence lifetime. Spectrochim. Acta Part A 58:1025–1037; Ryder AG (2002) Applied spectroscopy 56(1):107–116
9. Bhui UK, Basu A, Sanyal S (2018) Maturity assessment of Shale from Kachchh basin of India for hydrocarbon prospect: an optical spectroscopic approach. In: Tu P4 07, 80th EAGE conference and exhibition 2018
10. Bhui UK, Sarkar A, Basu A, Sanyal S (2018) A molecular structure approach for evaluating thermal maturity during hydrocarbon generation: a study on Jhuran shale of Kachchh basin, India. In: Abstract volume of 2nd national conference and field workshop on 'recent studies on the geology of Kachchh Basin, p 24

11. Mishra S, Patel BK (2011) Gas shale potential of Cambay formation, Cambay basin, India. Geo-India, New Delhi, pp 1–6
12. HIS Report (2005). Basin monitor Cambay basin, India. pp 1–16
13. Banerjee A, Pahari S, Jha M, Sinha AK, Jain AK, Kumar N, Thomas NJ, Mishra KN, Chandra K (2002) The effective source rocks in the Cambay basin, India. AAPG Bull 86: 433–456
14. Banerjee A, Sinha AK, Jain AK, Thomas NJ, Misra KN, Chandra K (1998) A mathematical representation of rock-eval hydrogen index versuss Tmax profiles. Org Geochem 28:43–55
15. Ariketi R (2016) Integrated reservoir characterization of Cambay shale for shale gas resource estimation in Ankleswer area, Cambay basin, Gujarat, India. Unpublished Ph.D. Thesis, PDPU
16. Fetzer JC (2000) Large (C > = 24) polycyclic aromatic hydrocarbons: chemistry and analysis. Wiley
17. Berlman I (1971) Handbook of fluorescence spectra of aromatic molecules, 2nd edn. Academic Press; Apicella B, Ciajolo A, Tregrossi A (2004) Fluorescence spectroscopy of complex aromatic mixtures. Anal Chem 76:2138–2143
18. Merola MC, Carotenuto C, Gargiulo V, Stanzione F, Ciajolo A, Minale M (2016) Chemical physical analysis of rheologically different samples of a heavy crude oil. Fuel Process Technol 148:236–247
19. Groenzin H, Mullins OC (2000) Molecular size and structure of asphaltenes from various sources. Energy Fuels 14:677–684
20. Kasha M (1960) Paths of molecular excitation. In: Proceedings of a symposium sponsored by the U.S. atomic energy commission, Newyork
21. Pampanin D, Kemppainen E, Skogland K, Jorgensen K, Sydnes M (2016) Investigation of fixed wavelength fluorescence results for biliary metabolites of polycyclic aromatic hydrocarbons formed in Atlantic cod (Gadus morhua). Chemosphere 144:1372–1376
22. Ariketi R, Bhui UK, Chandra S, Biswal S (2017) Evaluation of organic richness and thermal maturity of Cambay shale formation for Shale gas exploration—a study from Ankleshwar area, Cambay basin, India. Geohorizons, V-Jan 2017:17–27
23. Kelemen SR, Afeworki M, Gorbaty ML, Sansone M, Kwiatek PJ, Walters CC, Freund H, Siskin M, Bence AE, Curry DG, Solum M, Pugmire RJ, Vandenbroucke M, Leblond M, Behar F (2007) Direct characterization of kerogen by X-ray and solid-state 13C nuclear magnetic resonance methods. Energy Fuels 21:1548–1561

Radio Frequency (RF) Plasma Treatment of Coal: Preliminary Results and Future Projections

Joydeep Ghosh, Archchi Sarkar, Uttam Sharma, Sachin Singh Chouhan, Jayshree Sharma, and Uttam K. Bhui

1 Introduction

With the decline of oil production from the existing oil fields and limited discovery of large- to medium-scale new oil fields, coal raises as a potential candidate for energy production in more environmentally friendly manner. Coal represents a substantial fraction of the estimated non-renewable fossil fuel resources of the world and may become an increasingly important source of hydrocarbon fuels and chemical feedstock. Coal is the residual organic matter derived mainly from plant materials through thermal modification within the Earth's interior during their burial. Coal has a complex macro-molecular network, consisting of polyaromatic clusters linked with side chains and functional groups on peripheral positions by covalent and non-covalent bonds [1]. The structure of coal encompasses a system of

J. Ghosh (✉)
Institute for Plasma Research, Gandhinagar, Gujarat, India
e-mail: jghosh@ipr.res.in

A. Sarkar · U. K. Bhui
School of Petroleum Technology, Pandit Deendayal Energy University (PDEU, Formerly Pandit Deendayal Petroleum University-PDPU), Raisan, Gandhinagar, Gujarat 382007, India
e-mail: archchi.sphd18@spt.pdpu.ac.in

U. K. Bhui
e-mail: Uttam.bhui@spt.pdpu.ac.in

U. Sharma · S. S. Chouhan
Shri Vaishnav Institute of Science, Indore, Madhya Pradesh, India
e-mail: druttamsharma1971@gmail.com

S. S. Chouhan
e-mail: sachinrajsingh2012@gmail.com

J. Sharma
Shri Vaishnav Institute of Management, Indore, Madhya Pradesh, India
e-mail: jayshreesharma71@gmail.com

U. K. Bhui (ed.), *Macromolecular Characterization of Hydrocarbons for Sustainable Future*, Green Energy and Technology,
https://doi.org/10.1007/978-981-33-6133-1_15

aromatic and hydroaromatic–hydrocarbon clusters linked by flexible aliphatic, sulfidic and etheric bridges of various types, and the structure hosts various smaller molecules. Coal is a very complex in nature where the main bonding structure varies systematically with their carbon percentage. The estimated relative importance of various forces prevailing in coal structure holding the coal clusters together modelled by Aida, where it has been clearly shown that the type of chemical forces holding the component clusters together evolve with maturity from dominantly oxygen related at low rank to covalent hydrocarbon and carbon–carbon (π-system) interactions at higher ranks [2]. Studies have shown the prevalence of Van der Waals and hydrogen bonding in coal with very low bonding energies (<1 eV), whereas the covalent bonds existing in relatively lower percentage have higher energies (4–6 eV) [3, 4] (Fig. 1).

Thermal disintegration of coal is important since it is the initial step in most coal conversion processes and depends mostly on the coal properties. The term 'pyrolysis' is usually applied to a process which involves an extensive thermal decomposition of coal or organic matter where above a threshold temperature; the complicated structure of coal is dissociated into aromatic ring units by rupture of bridge carbons with formation of 'free radicals'. Recombination of the smaller radicals results in the formation of small aliphatic gas molecules, and water and tar components of medium molecular weight. The volatile compounds diffuse from the interior of the coal/char particle into the bulk gas phase and react with the surrounding atmosphere. The diffusion can also happen from the surrounding atmosphere to the interior of coal. Recent works have carefully emphasized on the rapid heating or 'flash pyrolysis' of coal samples, with the introduction of plasma as an efficient mean for pyrolysis. Behind this emphasis, there has been the desire to elucidate the role of plasma in coal combustion and gasification under different thermodynamic conditions and hence the possibility of successful conversion of coal to liquid fuels and other chemicals such as acetylene. Previous studies have confirmed the possibility of a direct reaction of coal with hydrogen plasma [5].

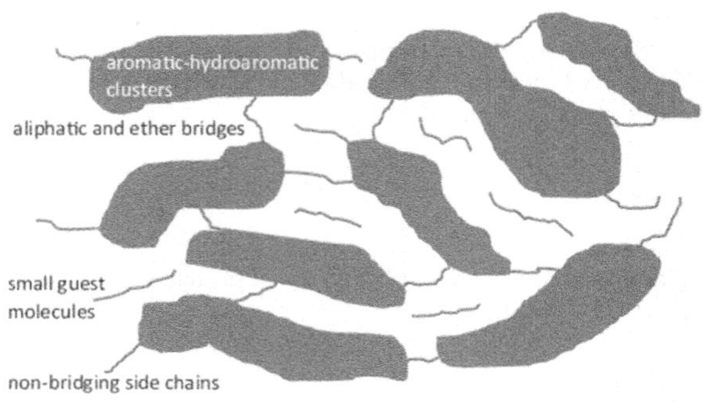

Fig. 1 Schematic diagram of coal structure showing a system of aromatic–hydroaromatic clusters linked by flexible aliphatic, sulfidic and etheric bridges. The clusters also have aliphatic and O_2-containing side groups of various types, and the structure hosts various smaller 'guest' molecules [2]. Reprinted with permission ©Springer International Publishing AG 2017

Desypris et al. [6] had observed that the degree of polycondensation decreases with increasing pyrolysis temperature, and at higher temperature, the product distribution is skewed towards components having lower molecular weight. Despite their diversity, the products of coal pyrolysis under plasma conditions typically include char, volatiles and CH species, which depending upon the temperature and reaction speed can react among themselves to produce similar compounds and other secondary products [7].

It is a topic of research for the last century that production of distillate liquid fuel is possible by high pressure hydrogenation from residual crude oil and coal where the residue of the hydrogenated coal has microstructure resembling those of metallurgical coke [8–10]. Thermal hydrogenation is considered as an important process where non-coking coal could be converted to coking coal [9, 11–13] or coking property of the coal could be improved or restored [14].

2 Materials and Methods

In the present attempt to observe the influence of hydrogen plasma on three different coal samples, the samples were placed inside a plasma reactor. Plasma is complex energy source known for tailoring the decomposition of diverse materials in controlled fashion. Plasmas are chemically active media, representing uniquely large variety of components in the form of excited and ionized particles—both atomic and molecular, photons and radicals. All these species are capable of carrying significant energy to induce chemical reactions in the plasma volume and initiate different processes like adsorption, desorption, chemical and physical sputtering, etc., at its interface with any solid surface present. In addition to it, various types of plasmas—cold, warm and hot can be generated with different energies of the constituent particles to bring in desired changes in the contact material. Depending upon the energies of the different species present in the plasma, molecular bonds of the materials present are broken, and release of reactive free radicals among other smaller composition present in the material is facilitated. The plasma particles are also known to diffuse inside the material to influence its composition.

Inside the plasma reactor [15], plasma is produced in the chamber by a capacitive discharge initiated between two parallel plates with the application of radio frequency (RF) voltage at 13.56 MHz in hydrogen gas. The RF power can be varied from 10 to 600 W in the reactor. The schematic of the plasma reactor along with its subsystems is shown in Fig. 2. The reactor is made up of a cylindrical stainless-steel (grade SS304) vacuum chamber of 36 cm diameter and 30 cm height with its axis perpendicular to the base support surface. The top and bottom flanges are vacuum sealed with large 'O' rings and contain ports at the centre of the flanges for electrode insertion. The top flange also contains five more 35CF ports for diagnostics. The azimuthal periphery of the cylinder contains eight 63CF ports for gas-feed, electrical feedthrough, probes and optical diagnostics. The volume

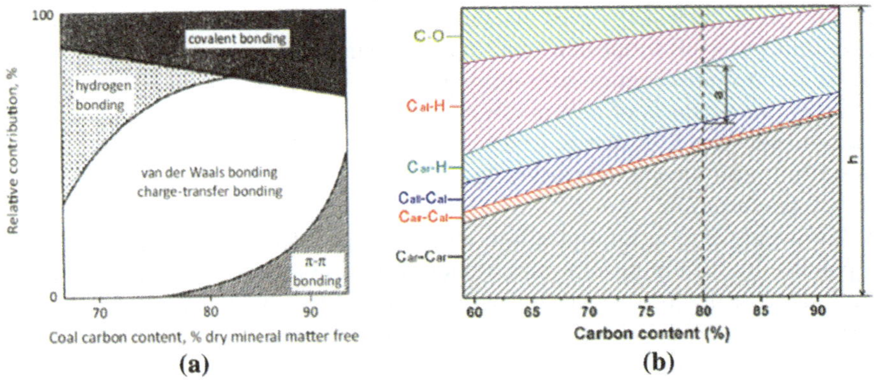

Fig. 2 **a** Model of Aida, estimating the relative importance of various forces holding coal clusters together [2, 3]. **b** Typical bonding forces of coal with % carbon content [4]. Reprinted with permission ©2016 Elsevier

~0.5 m^3 is pumped to a base vacuum of 2 × 10^{-6} mbar using a turbo molecular pump (TMP), backed by a suitable rotary pump. The TMP is having a pumping speed of 260litre/s with nitrogen, and the backing rotary pump is having pumping speed of 12 m^{3}h^{-1}. The TMP is connected to the chamber with a gate valve through a peripheral port. Rough vacuum on the system is monitored by a Pirani gauge (IHV-made) and the high vacuum is measured with a cold-cathode Pirani gauge (Pfeiffer-made). The gas-feed system is equipped with remote-controlled mass flow controllers calibrated for hydrogen. The gas-feed line is connected to the peripheral ports at a vertical midplane of the chamber. A high precision residual gas analyser (RGA) is connected to the differential section of the chamber to monitor the presence of radicals and compound formation during and after the plasma treatment. Maximum mass range detection limit of the RGA is 300amu with a mass resolution of 0.5amu.

Two electrically isolated stainless-steel circular discs of diameter ~150 mm and thickness ~10 mm are placed inside the plasma chamber from the top and bottom flanges. These two parallel discs work as electrodes for the plasma production, and the distance between them can be adjusted using movable vacuum feedthrough. The electrodes are powered using a RF source (Serene-made) having frequency of 13.56 MHz and maximum power ~600 W with an appropriate auto-matching network which can be operated in both manual and automatic modes. The electrical connections are made with appropriate high voltage RF feedthrough. A picture of the complete arrangement is shown in Fig. 3.

As stated earlier, plasma in the chamber is produced by a capacitive discharge initiated between the two parallel plates with the application of radio frequency (RF) voltage at 13.56 MHz in hydrogen gas. The application of the RF voltage between the electrodes allows free electrons to oscillate and collide with neutral gas molecules which cause ionization and produce sustainable plasma in the chamber. Such plasmas can be maintained without secondary emission from the electrode

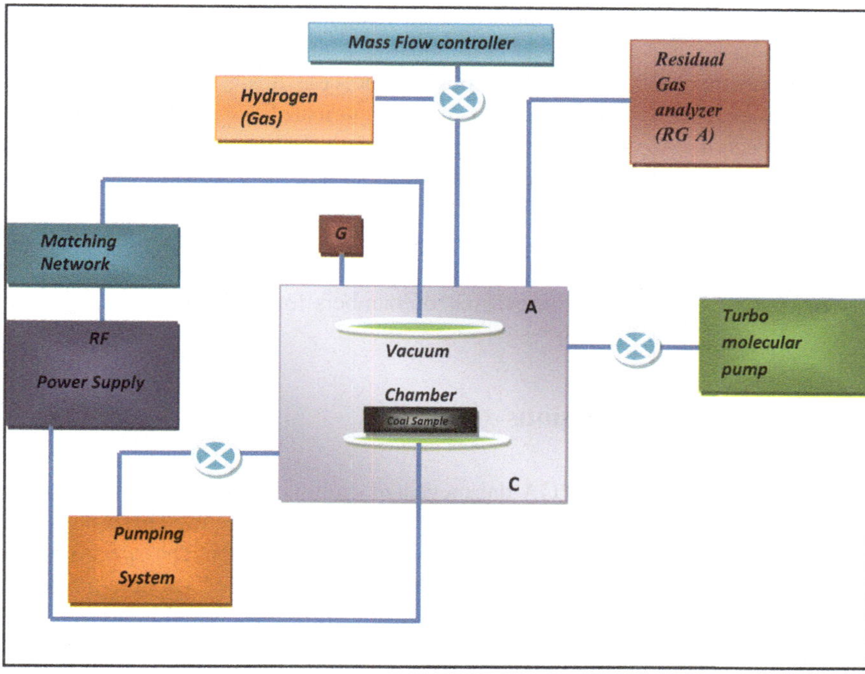

Fig. 3 Schematic of the RF plasma treatment reactor at SVVV

Fig. 4 RF plasma treatment reactor at SVVV

surfaces, and ionization is caused using the energy gained by the electrons during oscillation under influence of the RF field. For the results presented in this paper, the plasma is produced with 60 W RF power (see Fig. 4).

During the experiment, the coal samples are placed on the bottom electrode, which is put at the electrical ground and exposed to the plasma. The plasma reactor is equipped with several diagnostics for plasma characterisation, i.e. for measuring the density and temperature of its species [15, 16]. A low-resolution spectrometer in the visible region is present to identify the components in the plasma. Most importantly, the reactor is equipped with a residual gas analyser (RGA) with a maximum mass range detection limit of 300amu with a mass resolution of 0.5amu.

The experiments are initiated by placing the coal inside the chamber of the reactor and evacuating it to a pressure of $\sim 10^{-5}$ mbar. Several scans of the RGA have been recorded to obtain the mass spectra before the treatment of the coal. After that, hydrogen plasma is generated in the chamber and the coal samples have been soaked in the plasma for 3–4 h. The plasma is then switched off, and again several scans of RGA are recorded and compared to the earlier made preplasma treatment recordings. The experiments are repeated for different sets of coal samples. The treated coal samples are kept in evacuated chambers for further analysis.

3 Results and Discussions

After plasma treatment, the RGA data revealed multiple prominent peaks near the lower end and clusters near the higher end of the mass spectra. These peaks are identified as saturates, aromatics and non-organic compounds. The spectra reflect the presence of both saturated and unsaturated hydrocarbons of variable molecular weight in the coal sample which were released during the plasma treatment. This preliminary data indicate that plasma is indeed able to sever the weak bonds in coal, separating most of the side chains and possibly some of the heteroatoms. It is also possible that some compounds are chemically modified after reacting with hydrogen and thus generated the clusters of peaks near the higher end of the mass spectrum. In this regard, the ratios of H/C in the treated samples are also under investigation to study the possibility of hydrogenation of coal using the plasma (see Fig. 5; Table 1).

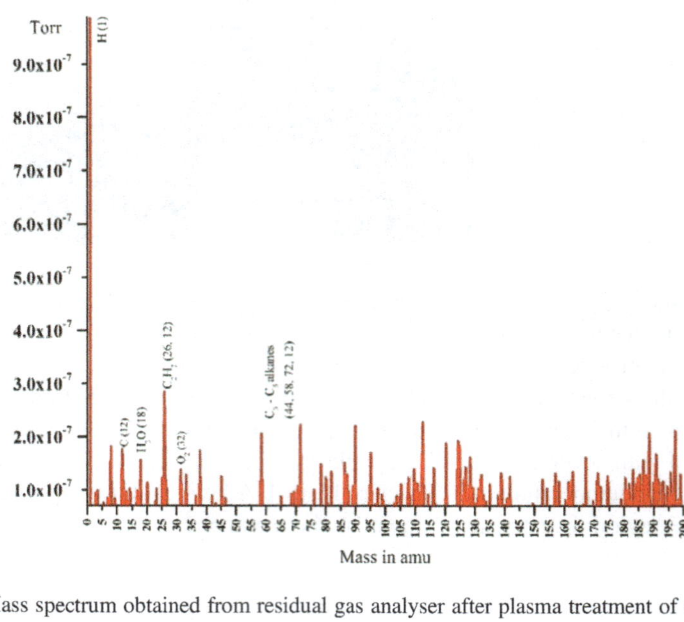

Fig. 5 Mass spectrum obtained from residual gas analyser after plasma treatment of coal sample

Table 1 Data from RGA after plasma treatment of coal sample

Mass	Possible chemical species present	Relative abundance (relative intensity)
1	Hydrogen (H)	Very high (parent peak)
12	Carbon (C)	High
18, 17	Water vapour (H_2O)	Moderate
26, 12	Acetylene (C_2H_2)	High
32	Oxygen (O_2)	Low
44, 12	Propane (C_3H_8)	Low
58, 12	Butane (C_4H_{10})	High
72, 12	Pentane (C_5H_{12})	High
>78	Higher HCs or secondary reaction products	High

The largest and most prominent peak present in the spectrum is at 1amu, and it is characteristic of hydrogen. Other peaks are representative of different compounds, fragment ions and radicals. It is possible that some peaks have overlapping mass signatures which can arise due to different compounds having different ionizations and cracking energies. Compounds having mass >78amu, which were not observed on the mass spectrum taken before the coal treatment, are possibly aromatic in nature or can be secondary products generated by hydrogenation of primary volatile species. It is evident from the mass spectra that most of the representative fraction with mass >78amu are heavier HCs with one or more benzene ring in the structures. The data also reveal clusters of peaks near the higher end of the mass spectrum which is supportive of the fact that both saturated and unsaturated heavy HCs are present.

Comparing the data with literatures and previous works done on plasma treatment of hydrocarbons and coal-related products [17–19], it was inferred that the major peaks present in the spectrum are characteristic peaks of acetylene (C_2H_2) at 26amu, Propane (C_3H_8) at 44amu, Butane (C_4H_{10}) at 58amu. Minor peaks indicating the present of certain chemical species in smaller concentration are characteristic of H_2O at 18amu, CO_2 at 44amu, O_2 at 32amu. In some places, radicals and ionized species might be present. It is also possible that isomers of some compounds are also present which bear the same mass but are different in structure. Therefore, it requires further analysis using spectroscopy to determine the structures of the compounds in order to make a more concrete conclusive statement.

4 Conclusions

In the present study, it was possible to successfully implement plasma treatment on coal and identify chemical species present in the residual volatile matter. Several lighter hydrocarbon fractions including C3 to C5 alkanes, the major pyrolysis product of acetylene (C_2H_2) along with elemental carbon, water vapour and traces of oxygen were present in the residual gas. A possible reaction of hydrogen with primary pyrolysis products producing secondary HC components or heavier aromatics was observed in the mass spectrum. To conclude in a more obvious note, it is required to implement spectroscopic and total mass spectrometric methods to identify the structural moieties present in the components.

Acknowledgements The authors would like to thank all the technical staff of the Centre of Excellence in Plasma Physics, SVVV, Indore, for their support during the experiments. Furthermore, the authors would like to thank Dr. Upinder Dhar, Vice Chancellor, SVVV, Indore, for his patronage and encouragement towards the work. Also, thanks are due to the Board of Research in Nuclear Science (BRNS), Department of Atomic Energy, Government of India, for funding the construction of the plasma reactor. AS thankfully acknowledges PDPU for providing her research fellowship.

References

1. Van Niekark D, Mathews JP (2010) Molecular representations of Permian-aged vitrinite-rich and inertinite-rich South African coals. Fuel 89:73–82
2. Burnham AK (2017) Global chemical kinetics of fossil fuels. Springer International Publishing AG, pp 75–105
3. Aida T (1991) Solvent swelling dynamics as a probe of coal structure. J Fuel Soc Jpn 70:820–826
4. Zhou B, Shi L, Liu Q, Liu Z (2016) Examination of structural models and bonding characteristics of coals. Fuel 184:799–807
5. Baumann H, Bittner D, Beiers H, Klein J, Jüntgen H (1988) Pyrolysis of coal in hydrogen and helium plasmas. Fuel 67:1120–1123
6. Desypris J, Murdoch P, Williams A (1982) Investigation of the flash pyrolysis of some coals. Fuel 61:807–816
7. Xie K-C (2015) Structure and reactivity of coal. Springer-Verlag, Berlin Heidelberg, pp 77–413
8. Nandi BN, Montgome DS (1967) Thermal behaviour of massive and granular micrinite. Fuel 46(4–5):394
9. Nandi BN, Ternan M, Belinko K (1981) Conversion of non-coking coals to coking coals by thermal hydrogenation. Fuel 60(4):347–354
10. Orchin M, Golumbic C, Anderson JE, Storch HH (1951) Studies of the extraction and coking of coal and their significance in relation to its structure. US Bur Mines Bull 505 (BM-Bull-505)
11. Lander C, Sinnatt FS, King JG, Crawford A (1928) Improvements in and relating to the treatment of coal and like carbonaceous material. Brit Pat Spec 301720
12. Crawford A, Williams FA, King JG, Sinnatt FS (1931) The action of hydrogen upon coal. Fuel research technical paper no. 29, Department of Scientific and Industrial Research, Her majesty's stationary office, London
13. Horton L, King JC, Williams FA (1933) Progressive action of hydrogen on coal [Using compounds of Ge, Sn, and Pb as catalysts]. J Inst Fuel 7:85
14. Ahuja LD, Sharma JN, Kini KA, Lahiri A (1958) Conversion of a noncoking coal into coking type by partial hydrogenation. J Sci Ind Res 17:27–29

15. Chauhan SS, Sharma U, Sharma J, Sanyasi AK, Ghosh J, Yadava N, Choudhary KK, Ghosh SK (2018) Development of RF based capacitively coupled plasma system for tungsten nano layer deposition on graphite. Mat Res Express 5:115020
16. Chauhan SS, Sharma U, Sharma J, Sanyasi AK, Ghosh J, Choudhary KK, Ghosh SK (2019) Radial density profile measurement at different RF power in Argon plasma using RF compensated Langmuir probe. AIP Conf Proc 2100:020187
17. Kukharenko TA, Matveeva II (1950) *Izuchenie prirody nespekayushchikhsya uglei metodom gidrirovaniya ikh nizhe temperatury razlozheniya. Zh Prikl Khim 23(7):732–736
18. Li M, Fan Y, Bao W, Guan Y, Li S (2006) The reactions of carbon in a hydrogen plasma. Energy Sources, Part A: Recovery, Utilization, And Environ Effects 28(9):829–835
19. Beiers H, Baumann H, Bittner D, Klein J, Jüntgen H (1988) Pyrolysis of some gaseous and liquid hydrocarbons in hydrogen plasma. Fuel 67:1012–1016

Biomass System, Carbon Materials and Carbon Capture and Storage (CCS)

Recent Advances in Biomass Gasification: A Review

Sukumar Mandal, Sateesh Daguppati, Rajib Bandyopadhyay, and Asit Kumar Das

1 Introduction

Over the last few decades, global energy consumption has been nearly doubled. With the industrialization, there was a decline in urban air quality with the serious and severe degradation of soils and water. In addition, the use of fossil fuels still represents a large percentage as an energy source, and this results in increased greenhouse gas emissions (GHG). World energy consumption is projected to increase by 48% over the next three decades, led by strong increases in the developing world—especially in Asia [1].

The use of non-renewable energy resources has been widely studied with the aim of increasing the efficiency of conversion, reliability and economic. It is also equally important to understand energy resources and their limitations, as well as the environmental concern of their usage. Biomass is one of the most widely available potential sources of energy as it is carbon neutral. Therefore, it is green energy source as it gives energy security for sustainable development with no or minimal negative impact on environmental. Most of biomass is burnt either for heating purpose or simply burn in the open field causing environmental pollution. Since biomass is a potential element of energy of different forms and renewable,

S. Mandal · S. Daguppati · A. K. Das (✉)
Refining Research and Development, Reliance Industries Limited, Jamnagar, Gujarat, India
e-mail: asit.das@ril.com

S. Mandal
e-mail: sukumar.k.mandal@ril.com

S. Daguppati
e-mail: Sateesh.Daggupati@ril.com

S. Mandal · R. Bandyopadhyay
Pandit Deendayal Energy University (PDEU, Formerly Pandit Deendayal Petroleum University-PDPU), Knowledge Corridor, Raisan, Gandhinagar 382007, Gujarat, India
e-mail: Rajib.Bandyopadhyay@sot.pdpu.ac.in

U. K. Bhui (ed.), *Macromolecular Characterization of Hydrocarbons for Sustainable Future*, Green Energy and Technology,
https://doi.org/10.1007/978-981-33-6133-1_16

energy recovery from biomass can be achieved through several routes of thermochemical conversion technologies. These thermochemical conversions allow the transformation of biomass into various energy products, such as liquids (bio-oil), gases and char [2]. One of the thermochemical conversion routes is gasification that has several advantages such as conversion of entire biomass to syngas which is building block for making several energy forms and chemicals. Biomass gasification technology is not yet matured, and hence, researchers are engaged to make this process more reliable and economical. This paper is attempted to summarize the latest development and future needs for further development to make biomass energy source is one of attractive sources over fossil fuel.

2 Biomass Feedstock

Lignocellulosic biomass is predominantly composed of five major components; three of them are oxygenated solid polymers, accounts for the bulk of biomass, i.e. cellulose (30–50 wt%), hemicellulose (10–40 wt%) and lignin (15–40 wt%). Small amounts of inorganic ash and low-molecular weight extractives make up the reminder. The relative amounts of these components are dependent on the type of biomass [3–5]. As per Jiang et al. [6], the chemical composition of biomass plays an important role in the distribution of pyrolysis products and hence gasification. Each material exhibits a particular characteristic when it is pyrolyzed, due to the proportion of the constituent components. The ultimate and proximate analysis of few biomass that are available in India in plenty, are given in Tables 1 and 2 [7].

Biomass consists basically of carbon, hydrogen and oxygen. Nitrogen and sulphur usually present in small proportions. Some biomasses also contain considerable amounts of inorganic compounds, being the main elements are Si, Ca, K, Fe, P, Al, Na and Mg. The ash concentration of these inorganic compounds can range from less than 1% in softwoods to 15% in herbaceous biomass and agro industrial residues [8]. The minerals present in ashes are constituted of inorganic matter found between the carbonic chains whose structure is different for each feedstock. In agricultural residues, it is about 23% ash in rice husk, less than 3% in sugarcane

Table 1 Ultimate analysis of different biomass

Biomass name	Composition (wt%)				
	C	H	N	S	O
Castor stalk (CS)	43.27	5.89	0.65	0.19	49.99
Castor DOC (CDOC)	44.14	6.08	8.85	0.60	40.33
Wood sawdust (WSD)	48.12	6.24	0.32	0.04	45.29
Tur stalk (TS)	44.24	5.96	0.82	0.05	48.93
Cotton stalk (CTS)	43.69	5.86	0.69	0.06	49.71
Sugarcane stalk (SCS)	38.73	5.75	0.60	0.13	54.79

Table 2 Proximate analysis of different biomass samples

Biomass	Proximate analysis (wt%)				
	Moisture	Volatile	Ash	Fixed carbon	Char + ash
Castor stalk (CS)	10.5	69.5	5.76	14.24	20
Castor DOC (CDOC)	10.88	64.12	6.93	18.07	25
Wood sawdust (WSD)	10.36	65.64	3.82	20.18	24
Tur stalk (TS)	10.26	71.74	5.2	12.8	18
Cotton stalk (CTS)	13.44	63.56	0.55	22.45	23
Sugarcane stalk (SCS)	11.05	62.95	12.89	13.11	26

bagasse or only 0.7% in coconut husk [9]. Very little information is available about the effect of ash on product yields during the thermochemical conversion. The catalytic role played by some of these constituents may be important. Some studies revealed that effect of mineral content in biomass components in the pyrolysis process inhibits the formation of certain compounds and alters product distribution during decomposition [10]. According to Gianesella [11], in a thermochemical conversion process, the chemical composition of the ashes can create serious operational problems, i.e. a liquid phase can be formed at elevated temperatures and can deposit on downstream equipment, resulting in high operating costs.

3 Biomass Gasification

In pyrolysis technology, the main objective is to obtain products, called bio-oil, with higher energy density and better properties than those of the initial biomass. However, bio-oil (rich with oxygenates) does not meet the quality of transportation fuel [12, 13], and it is essential to upgrade further (de-hydro oxidation) which demands additional cost-effective processing units for removal of oxygenates. In the case of gasification, the technology allows the conversion of solid biomass into a mixture of fuel gases (predominantly CO and H_2) called synthesis gas [14, 15] which is having versatile applications. In general, gasification is a thermo-conversion process that transforms solid or liquid fuels into a combustible gas mixture through partial oxidation and high temperature [16]. The resulting gas is a mixture of carbon monoxide, hydrogen, methane, carbon dioxide and nitrogen, the proportions of which vary depending on the process conditions, particularly if it is air or pure oxygen and/or steam being used in the oxidation. Air-based gasifiers typically produce a product gas containing a relatively high concentration of nitrogen resulting low heating value between 4 and 6 MJ/m^3; however, pure oxygen and steam-based gasifiers produce a product gas containing a relatively high concentration of hydrogen and CO with a heating value between 10 and 20 MJ/m^3.

3.1 Gasifier Types and Operating Parameters

Gasifiers can be broadly classified in four categories such as fixed-bed updraft, fixed-bed downdraft, bubbling fluidized bed and circulating fluidized bed [17, 18], depending on biomass bed in the reactor vessel, the direction of flow of both the biomass and oxidant and the mode of energy supply for meeting endothermic heat demand of steam gasification to the reactor.

1. In fixed-bed updraft (counter flow) gasifier, biomass flows down the reactors, and a grate at the bottom of the reactor supports the reacting bed. The oxidant like air or pure oxygen and/or steam is introduced below the grate and flow up through the beds. Due to counter current flow, syngas leaves from the top of the reactor, has lower temperature of ~750 °C as it transfers heat to incoming biomass, resulting very high tar content (10–20%) [18].
2. All feed and product material and oxidant flows downward in case of fixed-bed downdraft in a co-current manner. Unlike fixed-bed updraft, the product gases are not heat exchanging with incoming biomass feedstock, passing through hot char bed at the bottom, and hence, it remains hot resulting less tar formation (5 g/nm^3) [19, 20].
3. A bubbling fluidized bed (BFB) gasifier consists of fine, inert particles of sand or alumina as bed material, operates in bubbling fluidization regime, that ensure good heat transfer throughout the reactor without loss of any particle from the bed. The feed biomass entered into the fluidized bed closes the bottom zone which will be fluidized by introduction of oxidizing/gasifying agents (air, oxygen and/or steam) injected from the bottom of the gasifier and product gases come out from the top of gasifier.
4. Circulating fluidized bed gasifiers consist of cyclone which is attached to main reactor wherein reactor has bed materials that operate in entrainment bed fluidization regime and resulting in entrainment of the particles along with product gas stream. The entrained particles are separated in a cyclone and returned to the reactor through cyclone dip leg. The biomass is fed to the bottom part of the bed for ensuring enough residence time in the bed.
5. Dual bed gasifier performs gasification and combustion reaction in separate circulating fluidized beds that prevent mixing of syngas and combustion gas, resulting production of nitrogen-free syngas having higher calorific value. The steam gasification biomass takes place in gasification bed wherein hot inert/catalyst bed material received from combustion vessel. Usually, steam is used as fluidizing medium as well as gasifying agent which is introduced to the gasifier at the bottom. The char produced in gasifying bed is transferred to combustion vessel where it is burned in presence of air that heats the bed material. By continuous circulating this material between gasification and combustion vessels, endothermic heat demand in gasifier is met by combustion heat produced in combustion.

6. Other Gasification technique
   Special gasifiers are also developed to handle specific feedstock such as toxic organic wastes.

   (1) Plasma Gasification—Plasma is one of the four fundamental states of matter. It consists of a gas of ions. The plasma is formed by passing an electrical discharge through a gas such as air or oxygen (O_2). The interaction of the gas with the electric arc dissociates the gas into electrons and ions and causes its temperature to increase significantly, often exceeding 6000 °C. There are two types of plasma—thermal plasma and cold plasma. Thermal plasma is created at ambient pressure, while cold plasma is produced in a vacuum. The present major application of plasma technology is used to treat toxic waste biomass. However, the technology has also gained interest for syngas production and electricity generation in the recent years as the costs of this technology have significantly improved to be competitive commercially. Lang et al. [21] have given a comprehensive review on the development of fundamental researches on plasma pyrolysis/gasification systems including direct current (DC) arc plasma system and radio frequency (RF) plasma system with an emphasis on reactor design such as plasma fixed/moving bed reactor system, plasma entrained-flow bed reactor system and plasma spout-fluid bed reactor system. The syngas produced from plasma technology contains mainly hydrogen and CO with low CO_2 and no tar, resulting high calorific value. Eco-Valley® became one of the first waste-to-energy facilities in the world to utilize Plasma Gasification technology on a commercial basis in 2003 [22]. Eco-Valley® is located in Japan and has been successfully processing up to 220 tonnes per day of MSW or up to 165 tonnes per day of a 50/50 mixture of MSW and auto shredder residue.
   (2) Supercritical water gasification (SCWG) for wet biomass
   Normal water becomes supercritical water at 22.12 MPa pressure and 374.12 °C temperature. Supercritical water is a poor solvent for electrolytes, which tend to form ion-pairs. Thus, supercritical water has unique properties as a solvent. With a lower dielectric constant, the solubility of organic polar substances increases, and the solubility of inorganic ionic substances is reduced. In the recent past, supercritical water gasification (SCWG) of wet biomass ($\sim$80% water) has received significant attention as an alternative process, since it does not need high energy consuming drying process for wet biomass [23]. SCWG is typically performed at temperatures between 600 and 650 °C and at a pressure of about 30 MPa. Tar and coke formation is inhibited by fast solution of the formed gas components in the supercritical water. Methane is the main component in the produced gas at reaction temperatures below 450 °C, whereas at reaction temperatures above 600 °C, hydrogen is dominant. The reaction temperature can be decreased if suitable catalyst is used to make this process very commercially completive. Hence, catalytic gasification of biomass in supercritical water

has received much attention. Several researchers [23] have investigated the application of both homogeneous and heterogeneous catalyst in supercritical waster gasification. Several catalyst such as Ni and Ru, activated carbon, Pt-based catalysts and alkali metal-based materials such as $Na_3(CO_3)(HCO_3)\_2H_2O$ (trona), KOH, NaOH, K_2CO_3 have been tested. It has been reported [23] that the mixture of Raney-Ni and NaOH shows great synergistic effect in the process of biomass gasification in supercritical water. Even though considerable improvement is made from its inception, further improvement is essential for commercial deployment.

4 Theoretical Aspects of Conversion Processes

In the pyrolysis process, thermal decomposition occurs between 280 and 450 °C. Drying and pyrolysis of biomass with release of water, volatiles and tar occur in three phases [24, 25].

- Evaporation of water.
- Decomposition of carbohydrates (hemicellulose, cellulose, lignin) (250–300 °C).
- Production of tar and light acids (350–430 °C).

The products obtained are

- Non-condensable gases (CO, CO_2, H_2, CH_4, O_2).
- Char.
- Condensate: tar and acids.

The chemical reactions that occur in the combustion zone are basically the combination of oxygen in the air with carbon and hydrogen. A simplified pyrolysis scheme of the solid [25] can be described as

$$\text{Biomass (s)} + \text{heat} \rightarrow \text{coke (s)} + \text{gases (g)} + \text{tar (1)} + \text{condensate (g)} \qquad (1)$$

The gasification process

During the gasification processes, exothermic reactions of oxidation (combustion) and endothermic reduction reactions involving solid and gaseous phases occur [26].

$$\begin{aligned}\text{Biomass}_{(S)} + (O_2(g) + \text{with/without } H_2O(1)) \rightarrow & CO(g) + CO_2(g) + H_2O(1) \\ & + H_2(g) + CH_4(g) + \text{Tag}(1) \\ & + \text{Char}(s) + \text{ash}(s) + HCl(1) \\ & + NH_3(g) + HCN(g) \\ & + H_2S(g) + \text{others}\end{aligned} \quad (2)$$

Different reactions that are occurring in different stages are given below:

I. Pyrolysis

$$\text{Biomass (s)} + \text{heat} \rightarrow \text{coke (1)} + \text{gases (g)} + \text{tar (1)} + \text{condensate (g)} \quad (3)$$

In this stage, cracking of chemical bonds takes place with the formation of lower molecular weight specially gaseous and liquid fraction and solid char. These products are formed through very complex phenomenon which include heat transfer, product diffusion from biomass towards gas bulk phase and reactions in series. At low temperature, kinetic of the reactions may be limiting step, while at high temperature, heat transfer limits or product diffusion are limiting step.

II. Oxidation

$$C(S) + 1/2\ O_2(g) \rightarrow CO(g) \quad \Delta H = -110.6\ \text{KJ.mol}^{-1} \quad (4)$$

$$C(S) + O_2(g) \rightarrow CO_2(g) \quad \Delta H = -393.8\ \text{KJ.mol}^{-1} \quad (5)$$

$$H_2(g) + O_2(g) \rightarrow H_2O(1) \quad \Delta H = -242\ \text{kJ.mol}^{-1} \quad (6)$$

A small fractions of biomass are oxidized, especially when heat is not supplied through external source, to maintain the reaction temperature while supplying heat for endothermic reactions such as pyrolysis, gasification and tar cracking reactions. The products are CO, CO_2 and water. In case, the source of oxygen is air, then the product gas contains nitrogen also. Otherwise, no nitrogen will be present in combustion products.

III. Gasification
(Reverse Boudouard reaction)

$$C(S) + CO_2(g) \leftrightarrow 2CO(g) \quad \Delta H = 171.6\ \text{KJ.mol}^{-1} \quad (7)$$

$$C(S) + H_2O(l) \leftrightarrow CO(g) + H_2(g) \quad \Delta H = 131.4\ \text{KJ.mol}^{-1} \quad (8)$$

$$C(S) + 2H_2(g) \leftrightarrow 2CH_4(g) \quad \Delta H = -74.9\ \text{KJ.mol}^{-1} \quad (9)$$

Homogeneous reactions

$$CO(g) + H_2O(1) \leftrightarrow CO_2(g) + H_2(g) \quad \Delta H = -41.2\ KJ.mol^{-1} \tag{10}$$

$$CH_4(g) + 2H_2O(1) \leftrightarrow CO(g) + 3H_2(g) \quad \Delta H = 201.9\ KJ.mol^{-1} \tag{11}$$

These reactions are occurred in reducing atmosphere where all products formed in oxidation and pyrolysis reactions are involved. In fact, all gas/liquid products and char are reacted each other to form final products.

IV. Cracking Tar

$$Tar(1) + Vapour(g) + heat \rightarrow CO(g) + CO_2(g) + CH_4(g) \tag{12}$$

V. Partial Oxidation of Pyrolysis Products

$$CO(g) + H_2(g) + CH_4(g) + O_2(g) \rightarrow CO_2(g) + H_2(g) \tag{13}$$

Even though water gas shift reaction (10) and methanation reaction (9) are exothermic, but the degree of endothermic nature of Boudouard reaction (7) and char reforming reaction (8) are such that the overall reduction reactions are endothermic which is supplied by oxidation reaction. The reactions from (7) to (11) are in equilibrium, and concentration of each products and reactants is governed by Laws of Thermodynamic. In other word, reduction zone temperature and pressure are the both factor that determine the concentration of reactants and products. However, higher temperature increases the oxidation of tar resulting reduction of tar/char. This may end up to high risk of sintering and reduction of caloric value of syngas particularly at the higher equivalence ratio.

Operating Parameters

There are various parameters that affect the operation of gasification and hence product quality/yield. These parameters are equivalence ratio (ER), steam to biomass ratio (SBR), gasification temperature, composition and physical properties of biomass, inlet temperature and composition of the oxidant and tar content. The effect of equivalence ratio(ER), steam to biomass ratio (SBR), gasification temperature, residence time and biomass composition, etc., are explained elsewhere [27–32] (Table 3).

Table 3 Advantages and technical challenges of different gasifier designs, gasifying agents and operations for syngas production [17, 33]

Gasifier design	Main advantages	Main technical challenges
Fixed-bed updraft	• Simple and reliable design • Can handle high moisture and ash • Favourable economics on a small scale	• High tar contents requiring expensive tar removal system • Low CGE • Low productivity
Fixed-bed downdraft	• Simple and reliable design • Favourable economics on a small scale • Very low tar formation	• Need low moisture in feed • Exit syngas has high 'T' • Need heat recovery system
Bubbling fluidized bed (BFB)	• Uniform temperature • Uniform syngas composition • No limit on particle size • High conversion • Low tar	• Large bubble size
Fluidized bed	• Short residence time • High productivity • Uniform 'T' distribution • Low char or/and tar contents • High CGE • Min issues with ash	• High particulates in dust in syngas • Favourable economics on a medium to large scale • Size of fuel particles determine min.transport velocity; high velocities may lead to erosion
Dual bed gasifier	• High CV of syngas as gasification and combustion take place in a two separate vessels • Other merit as in fluidized bed reactor	• High inert material circulation required for satisfying heat demand • Other demerit as in case of fluid bed
Plasma gasification	• Decomposition of any organic matter • Treatment of hazardous waste	• High energy requirement and high cost • Low efficiency
Supercritical water gasification	• Biomass with high moisture is treated	• High energy requirement and high cost
Gasifying agent	**Main advantages**	**Main technical challenges**
Air	• No need of air separation unit • Low cost for gasification • Combustion takes place in separate vessel	• Low heating value syngas (4–7 MJ/Nm^3)
Steam	• High heating value syngas (13–20 MJ/Nm^3) • H_2-rich syngas (e.g. >50% by volume)	• Require indirect or external heat supply for gasification • High tar content in syngas • Require catalytic tar reforming
Caron dioxide	• High heating value syngas • High H_2 and CO and low CO_2 in syngas	• Require indirect or external heat supply • Required catalytic tar reforming

(continued)

Table 3 (continued)

Gasifier design	Main advantages	Main technical challenges
Gasifier operation	**Main advantages**	**Main technical challenges**
Increase of temperature	• Decreased char and tar content • Decreased methane in syngas • Increased carbon conversion • Increased heating value	• Decreased energy efficiency • Increased ash-related problems
Increase of pressure	• Low char and tar content • No costly syngas compression required for downstream utilization of syngas	• Limited design and operational experience • Higher costs of a gasifier at a small scale
Increase of ER	• Low char and tar content	• Decreased heating value of syngas

5 Uses of Syngas

Gasification processes are widely applied for over last 50 years in the production of fuels and chemicals. Current trends in the manufacture of chemicals and in the petroleum industries indicate growth in the use of gasification for the production of synthesis gas (primarily composed of CO and H_2), mainly due to the production of a consistent and high quality gas and the possibility of using a wide variety of materials as feed for the system. The main product of gasification is the synthesis gas, and it can be used in several ways [16, 34, 35] as can be seen in Fig. 1.

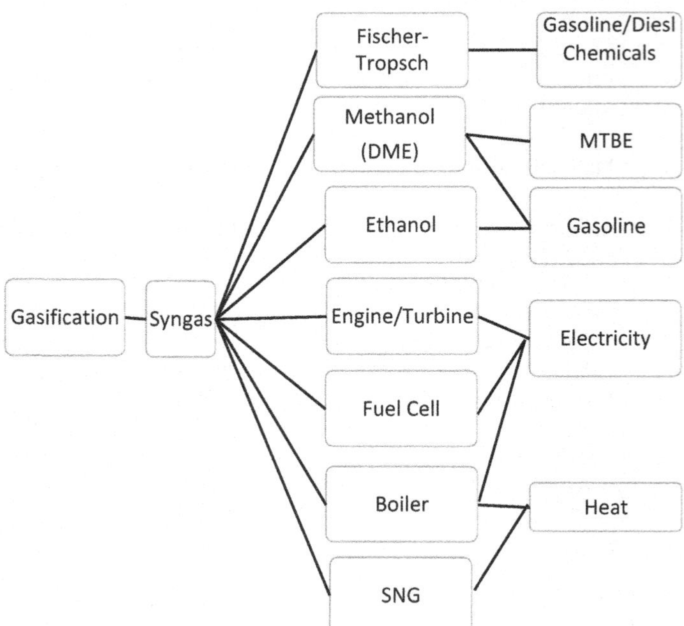

Fig. 1 Possible commercial applications of synthesis gas

Usually, low calorific value syngas is used in engine, turbine or boiler to produce electricity/heat. However, higher caloric value syngas is for production of fuel or different chemicals.

However, it is important to note that depending on the process to be adopted to convert syngas to chemicals or power, syngas contaminant should be within the specified limits that unit can handle. These contaminates are particulates, condensable organic compounds collectively known as tar, nitrogenous compounds like ammonia (NH_3) and hydrogen cyanide (HCN), sulphur containing inorganic compounds like hydrogen sulphide (H_2S), carbonyl sulphide (COS) and carbon disulphide (CS_2), hydrogen halides and halogens such as hydrogen chloride (HCl) and chlorine (Cl_2), and trace metals like sodium (Na) and potassium (K) [36]. For example, the upper limit (in ppmv) of tar, sulphur, nitrogen, alkali and halides in syngas suitable for FT synthesis are 0.1–1, 0.01, 0.02 and 0.01 for both alkali and halides, respectively. In case of methanol synthesis, tar and sulphur are <1 mg/Nm_3, and nitrogen and halides are 0.1 mg/Nm_3. These values are little less stringent for gas turbine application. Among all these impurities, tar is the most abundant per unit of biomass gasified and has been the focus of the majority of contaminant remediation studies. This tar is usually removed by cold gas clean-up technique. Since gasification is carried out at high temperature (>800 °C), the main disadvantage of this technique is the efficiency loss due to syngas cooling and additional cost incurred to treat or dispose of contaminant streams. Hence, warm and hot gas clean-up approaches are of great interest in contaminant removal from syngas at high temperature (>300 °C) [37]. These approaches address the efficiency loss as experienced in cold gas clean-up process. Whether syngas is treated in cold or hot conditions, it needs extra cost for treating. Hence, mush progress has been made to use catalytic material in gasification bed so that no extra hardware or loss of efficiency is envisaged.

6 Catalytic Gasification of Biomass

Some catalysts have been used as active bed additives as a heat carrier material inside the gasifier during gasification. There is a great potential of in-bed additives in terms of tar reduction and thus avoidance of complex downstream tar removal methods. These bed additives act as in situ catalysts, promoting several chemical reactions in the same gasifier. The presence of additives influences not only the gas composition, but also the heating value of the product gas. The use of catalysts during biomass gasification promotes char gasification, changes the product gas composition and reduces the tar yield. The addition of active bed materials also prevents the solid agglomeration tendencies and subsequent choking of the bed. These are the reasons why the major part of ongoing biomass gasification research deals with the development of new catalysts or the improvement of existing active materials to produce high quality tar-free syngas and/or hydrogen. A significant number of studies have been carried out using dolomite, olivine, alkali and noble metals catalysts for this purpose.

6.1 Dolomite, Olivine and Alkali Metal-Based Catalysts

Dolomite, a magnesium ore with the general formula $MgCO_3{\cdot}CaCO_3$, has been considered as a catalyst in biomass gasification. It has attracted much attention as it is a cheap disposable material that can significantly reduce the tar content of the product gas. The main issue with these materials is their fragility as they are soft and quickly attrite in fluidized beds under the prevalent high turbulence conditions.

Dalai et al. [38] studied the performance of a CaO catalyst by varying the catalyst loading from 0 to 8.9 wt% during temperature programmed gasification (TPG) and constant temperature gasification (CTG) processes. The experiments showed that the use of CaO as a catalyst reduced the gasification temperature by maximum 150 °C. Also, total fuel yields (H_2) and carbon yields were significantly increased with the impregnation of CaO in cellulose, cedar and aspen. The rate of production and cumulative production of H_2 from cedar and aspen were significantly higher than those from cellulose for catalytic and non-catalytic TPG and CTG processes.

A few studies have been done recently about the catalytic activity of olivine for tar elimination. Hu et al. [39] tested calcined olivine and dolomite as downstream catalysts in a fixed-bed reactor. The results show that the catalytic activities of calcined catalysts are higher than the activities of the natural catalysts. A similar system was used by Lopamudra et al. [40] who observed that tar conversion increases with a temperature rise from 800 to 900 °C. They found that water-soluble heterocyclic compounds can be 100% converted at 900 °C. Additionally, the conversion of heavy polyaromatics can increase from 48 to 71% with the use of 17 wt% olivine in sand at 900 °C, compared to a conversion of up to 90% with 17 wt% of calcined dolomite. A total tar amount of 4.0 g/m^3 could be reduced to 1.5 and 2.2 g/m^3 using dolomite and olivine, respectively.

Aznar et al. [41] conducted parametric studies using dolomite as a tar cracking catalyst. The feedstock was composed of blends of plastic waste with pine wood sawdust and coal, at flow rates of 1–4 kg/h. The operating variables that were studied were gasifier bed temperature (750–880 °C), equivalence ratio (0.30–0.46), feedstock composition and the influence of secondary air insertion in freeboard. As a result, a gas with medium hydrogen content (up to 15% dry basis) and low tar content (less than 0.5 g/m^3) was obtained. Additionally, these authors found that the injection of secondary air in the freeboard reduces tar content by 50%. Tar content obtained was less than 0.5 g/m^3, and as a result, an essentially clean gas was obtained under these conditions.

Xu et al. [42] demonstrated that for atmospheric gasification of biomass, CaO could also be an effective on-site CO_2 acceptor, provided the reaction temperature is controlled at appropriately low values, such as 973 °K. It was shown that at temperatures <1000 °K, the acceptor captured CO_2 with fuel gasification, lowering the CO_2 content in the product gas to a few percent (<10 vol.%), and increasing the gas's heating value considerably. The addition of CaO into the fuel increased the H_2 content of the gas, while decreasing its CO concentration, irrespective of the reaction temperature. This result corroborates the commonly known catalytic effect of CaO on CO shift and tar reforming/cracking reactions.

Monovalent alkali metals such as lithium (Li), sodium (Na), potassium (K), rubidium (Rb), cesium (Cs) and francium (Fr) belong to group 1A of the periodic table are all highly reactive and electropositive. Alkali metals, principally K and to a lesser extent Na, exist naturally in biomass and accumulate in the gasifier ashes. Furthermore, the use of ash itself, as a catalyst, solves the problem of ash waste handling and gives an added value to the gasification by increasing the gasification rate and reducing the tar content in the produced gas. However, the major disadvantage of these ash-based catalysts is the loss of its activity due to particle agglomeration. Sutton et al. [43] reported several disadvantages related to the direct addition of alkali metals, such as formation of water insoluble alkali alumino silicates and the difficult and the expensive recovery of the catalyst, increased char content after gasification, and ash disposal problems. On the other hand, Lee et al. [44] found that the addition of Na_2CO_3 enhances the catalytic gasification of rice straw over a nickel catalyst and significantly increases the formation of permanent gases. The same authors found that the formation of permanent gases depends on the nature of the alkali metal carbonate with the following reactivity order being proposed as $Na \geq K > Cs > Li$.

The use of activated alumina as a secondary catalyst for tar reduction comes from its high catalytic activity, comparable to dolomite [45], although it deactivates due to coke deposition faster than dolomite. Juutilainen et al. [46] tested its activity in the selective oxidation of tar and ammonia using catalysts containing zirconia and alumina. Their performance was compared with that of nickel and dolomite catalysts. Synthesis gas with toluene as a tar model compound was used as feed in a fixed-bed tube reactor. In the presence of oxygen, zirconia and alumina-doped zirconia yielded high toluene and ammonia conversions, below 600 °C. These catalysts were the most active catalysts for toluene oxidation below 700 °C and for ammonia oxidation below 650 °C. At higher temperatures, these impregnated ZrO_2/Al_2O_3 catalysts performed better. Oxidation selectivity was improved, and toluene and ammonia conversions were higher. These authors concluded that the both zirconia and alumina promoted toluene and ammonia conversions at lower temperatures. This shows the enhanced oxidation activity of zirconia with alumina improving oxidation selectivity. H_2S had little effect on the activity of alumina-doped zirconia.

6.2 *Nickel-Based Catalysts*

Among the transition metals (group VIII), nickel is the most widely used in the industry to catalyse steam reforming and dry reforming reactions [47]. Commercially available nickel reforming catalysts have been used extensively for biomass gasification [48–53]. Heavy hydrocarbon steam reforming catalysts are according to Aznar et al. [49] more active than light hydrocarbon steam reforming catalysts for tar removal. These catalysts promote steam and dry reforming reactions including water gas shift reactions. They are very effective in tar conversion and in adjusting the gas composition to syngas desired H_2/CO ratios. According to Olivares et al. [54], nickel reforming catalysts display 8–10 times more reactivity than calcined dolomite.

The formulation of nickel catalysts involves (i) an active phase (i.e. Ni), (ii) promoters and (iii) a support phase. Generally, higher nickel content results in lower tar yield and higher H_2 and CO yields. On the other hand, according to Bartholomew et al. [58], nickel content has a significant effect on the catalyst deactivation by coking. They suggested that lower metal crystallite concentration results in stronger interaction with the support phase. This normally yields higher metal dispersion and therefore more resistance to deactivation caused by carbon fouling. Metal dispersion may be improved by addition of promoters and thus minimizing the coking tendency. It has been proven that the activity and life (deactivation) of nickel-based catalysts depend greatly on the type of support and the presence of additives and promoters.

The support phase gives the catalyst mechanical strength and protection against severe conditions such as attrition and heat [55]. The pore structure of the support, the metal-support interactions and the acidity-basicity of the support significantly influence the metal dispersion, metal crystallite size and carbon deposition on the catalyst surface, thus affecting the overall catalytic performance and catalyst coking resistance [61]. Baker et al. [50] also reported that the acidity of the support affects coke deposition and catalyst deactivation. For instance, higher acidity of support materials favours tar cracking reactions causing higher carbon build-up on the catalyst surface.

Alumina-based materials are considered to be the primary support material for most reforming catalysts. Gadalla and Bower [62] investigated performance of α-Al_2O_3 and γ-Al_2O_3 supported Ni catalysts for the reforming of methane with CO_2. They reported that Ni/α-Al_2O_3 catalyst provided lower methane conversion than Ni/γ-Al_2O_3, given the stable allotropic form and smaller surface area of α-Al_2O_3. They also reported that Al_2O_3 supports with MgO/CaO addition were more stable than with silica, which favour rapid Ni/Al_2O_3 catalyst deactivation. Wang and Lu [61] also reported higher conversion and lower deactivation rate for Ni/γ-Al_2O_3 over the Ni/α-Al_2O_3 catalysts. They found that nickel aluminate (Ni Al_2O_4) was formed due to phase transformation of γ-Al_2O_3 supported Ni catalyst during calcinations. While this aluminate is hard to reduce at lower temperatures, reduced Ni/Al_2O_4 appears active for reforming reactions, being quite resistant to coking.

On the other hand, deactivation of nickel catalyst is a primary concern, it was reported that nickel-based catalysts are more prone to several deactivations such as poisoning by sulphur, chlorine and alkali metals, sintering of Ni particles and coke formation [55]. Ni-based catalysts deactivate rapidly due to coke formation and catalyst attrition. Coke formation is inherent in the steam reforming processes and in the high temperatures associated with reforming. These conditions promote higher hydrogen and carbon yields. Coking of Ni-based steam reforming catalysts is reasonably well understood [56]. High temperatures promote dissociation of tars, light and unsaturated hydrocarbons both in the gas phase and on the catalyst surface producing carbon deposits. They can block the access to the catalyst pore network resulting in catalyst activity loss. The formed carbon may be gasified, encapsulated on the surface or diffused through the nickel crystallites. Carbon may at a later reaction stage nucleate and/or precipitate lead to the formation of carbon whiskers. Formation of carbon whiskers lifts nickel crystallite from the surface resulting in catalysts sintering. Therefore, nickel-based catalysts deactivate by carbon in two ways: (1) encapsulation of nickel crystallites by layers of inactive carbonaceous material and (2) formation of inactive bulk nickel carbide phases [56–59]. Furthermore, there is a tendency for coke to be formed with the increase of unsaturation, molecular weight and aromaticity of the feed.

Regarding coke formation, it can be minimized through the use of excess steam as required by gasification stoichiometry. In this respect, it is possible to estimate a minimum steam/carbon ratio required to avoid coke formation [60]. This provides a very useful guideline to establish the desired operating conditions. However, the practical negative effect is that it increases the overall energy costs for plant operation. Therefore, given the above-mentioned consideration, it is crucial to maintain as low steam/C ratio as possible [56]. However, if coke deposits on the catalyst surface at the same rate as it is removed by combustion, the catalyst surface remains clean. Thus, the catalyst remains effective in cracking newly formed tar and/or preventing char formation yielding gas products with increased ability [50]. This is the ideal scenario that may happen in autothermal gasification of biomass where the air fluidizing the bed (catalyst and biomass) may contribute to keeping the catalyst free of coke. Otherwise, it is essential to ensure the coke removal in a systematic manner, otherwise, it can lead to poor catalyst activity, selectivity and limited catalyst life.

6.3 Dual Bed Gasifier with Catalyst

Dual bed gasifier with catalyst approach gives substantial benefits over in-bed catalyst application as it employs catalytic gasification and combustion in separate vessels thereby it gives very high calorific value syngas, no need of pure oxygen for combustion, lower gasification temperature due to the presence of catalyst, no char yield as it gets combusted in combustion vessel and maintains constant catalyst activity by continuous addition of catalyst on-line.

Mandal et al. [7] revealed catalytic dual bed process wherein the heated catalyst from combustor in which some portion of feedstock are burned with air goes to the bottom of gasification bed. In the gasifier, steam and biomass of size 180–300 μ, prepared by dry milling of biomass, are injected. The steam acts as pneumatic transport medium as well as reactant for gasification reaction. Gasifier bed is operated in bubbling or turbulent regime where reaction mixture is allowed more reaction time for completion of gasification reaction utilizing the heat carried by the catalyst. The product syngas gets separated by cyclone, heat exchanged and pass through filter for separating fine catalyst particle <10 μ as cyclone is not enough to separate particle less than 10 μ. The catalyst from gasification reactor is withdrawn in catalyst withdrawal vessel and then goes to combustor where required amount of biomass is burnt to heat the catalyst which then goes to gasification bed. The flue gas from combustor goes to stack after separating solid particles in cyclone and extracting heat through power recovery and waste heat recovery train. The rate of catalyst circulation is controlled by combination of proper pressure balance and heat demand in gasifier. Maintaining the higher delta temperature between combustor and gasifier is very important and essential to supply endothermic heat in gasifier. The lower is the delta temperature, higher will be catalyst circulation demand for supplying gasifier heat and vice versa. The main advantage of using a supported catalyst, i.e. catalyst as separate solid particles is that the active catalyst particles remain within the bed without losing their activity, so that the catalyst loss, recovery and reimpregnation issues are completely eliminated, and hence, the cost of operation is expected to be much lower than the commercially available gasifiers of today. Due to catalytic gasification reaction, syngas produced is completely tar free. The catalyst employed for this process is K_2CO_3 impregnated on high pore alumina support. The mechanical strength of this catalyst is sufficient enough to use in circulating fluid bed system. It was also reported that the product syngas is completely tar free, i.e. catalyst is capable to destruct the tar completely into permanent gases.

7 Conclusion

Biomass gasification known for more than 50 years gets renewed attention globally by researchers and Government legislation bodies as biomass is carbon neutral and abundantly available in most part of globe, unlike fossil fuel that is responsible for GHG effect and environment degradation. Direct heating gasification configuration such as fixed-bed updraft or down draft, BFB, CFB with air using as oxidant are well established, which gives very low calorific syngas and suitable for power generation. However, these configurations with oxygen as oxidant produce high calorific syngas that are suitable for chemical production. The syngas produced by any of the reactor configurations has lots of contaminants specially tars that needs to be cleaned before end use. Hence, different sorbent like dolomite, olivine, Cao or nickel-based catalyst has been developed and being used in situ gasifier to reduce tar and improvement of syngas composition with high H_2 and CO. However, deactivation of these sorbents at gasification conditions is major issue. There are significant progress on plasma and supercritical water gasification for gasifying toxic solid waste and high water containing biomass, respectively. These gasifiers are energy intensive and hence costly. Duel bed catalyst gasification is the new area of research which has multiple advantages over conventional gasification as it gives high calorific value syngas with no need for use of oxygen as oxidant. The catalyst deactivation and ash removal are managed by continuous addition and withdrawal of catalyst. The major challenge is to manage catalyst circulation to supply heat for endothermic gasification reactions and catalyst development that has high mechanical strength of catalyst and low deactivation. This technology is yet to be demonstrated in commercial scale.

References

1. https://www.eia.gov/todayinenergy/detail.php?id=26212
2. Bridgwater AV, Czernick S, Piskorz J (2001) An overview of fast pyrolysis. In: (ed) Progress in thermochemical biomass conversion. [S. l.:] IEA Bioenergy; Blackwell Sciences, pp 977–997
3. Lee DK, Owens VN, Bow A, Jeranyama P (2007) Composition of herbaceous biomass feedstocks. South Dakota State University
4. Yaman S (2004) Pyrolysis of biomass to produce fuels and chemical feedstocks. Energy Convers Manage 45:651–671
5. Kumar P, Barrett DM, Delwiche MJ, Stroeve P (2009) Ind Eng Chem Res 48:3713–3729
6. Jiang G, Nowakowski DJ, Bridgwater AV (2010) A systematic study of the kinetics of lignin pyrolysis. Thermochim Acta 498(1):61–66
7. Mandal S, Daggupati S, Majhi S, Thakur S, Rajib Bandyopadhyay R, Das AK (2019) Catalytic gasification of biomass in dual-bed gasifier for producing tar-free syngas. Energy Fuels 33(3):2453–2466
8. Vassilev SV, Baxter D, Andersen LK, Vassilev CG (2010) An overview of the chemical composition of biomass. Fuel 89:913–933
9. Raveendran K, Ganesh A, Khilar KC (1996) Pyrolysis characteristics of biomass and biomass components. Fuel 75(8):987–998
10. Antal MJ, Morten G (2003) The art, science, and technology of charcoal production. Ind Eng Chem Res 42(8):1619–1640 Washington
11. Gianesella M (2010) Pyrolytic kinetics of ligno-cellulosic biomass. Polithechnic of Turin First School of Engineering
12. Paasikallio V, Lindfors C, Kuoppala E, Solantausta Y, Oasmaa A, Lehto J, Lehtonen J (2014) Product quality and catalyst deactivation in a four day catalytic fast pyrolysis production run. Green Chem 16:3549–3559
13. Alonso DM, Bond JQ, Dumesic JA (2010) Catalytic conversion of biomass to biofuels. Green Chem 12:1493–1513
14. Abatzoglou N, Fernandez JC, Laramee L, Jollez P, Chornet E (1997) Application of gasification to the conversion of wood, urban and industrial wastes. In: Developments in thermochemical biomass conversion, vol 2, pp 960–972
15. Banapurmath TPG (2009) Comparative performance studies of a 4-stroke Cl engine operated on dual fuel mode with producer gas and Honge oil and its methyl ester (HOME) with and without carburetor. Renew Energy 34:1009–1015
16. Bridgwater AV (2003) Renewable fuels and chemicals by thermal processing of biomass. Chem Eng J 91:87–102
17. Ciferno JP, Marano JJ (2002) Benchmarking biomass gasification technologies for fuels, chemicals and hydrogen production. In: U.S. Department of Energy/National Energy Technology Laboratory
18. Sikarwar VS, Zhao M, Clough P, Yao J, Zhong X, Zaki M, Memon A, Shah N, Edward JD, Anthony F, Paul S, Fennell P (2016) An overview of advances in biomass gasification. Energy Environ Sci 9:2939–2977

19. Combustion Gasification and Propulsion Laboratory, IISc, India. http://cgpl.iisc.ernet.in/site/Portals/0/Technologies/Gasification%20Technology.pdf. Accessed Aug 2011
20. Dasappa S, Paul P, Mukunda H, Rajan N, Sridhar G, Sridhar H (2004) Curr Sci 87:908–916
21. Tang L, Huang H, Hao H, Zhao K (2013) Development of plasma pyrolysis/gasification systems for energy efficient and environmentally sound waste disposal. J Electrostat 71: 839–847
22. Willis KP, Osada S, Willeton KL (2010) Plasma gasification: lessons learned at ecovelly WTE facility. NAWTEC18 May 11–13, Orlando, Florida, USA
23. Jin H, Lu Y, Guo L, Zhang X, Pei A (2014) Hydrogen production by supercritical water gasification of biomass with homogeneous and heterogeneous catalyst. Article ID 160565, p 9. https://doi.org/10.1155/2014/160565
24. Sarasuk K, Sajjakulnukit B (2011) Design of a lab-scale two-stage rice husk gasifier. Energy Procedia 9:178–185
25. Milne T (1981) Pyrolysis—the thermal behavior of biomass below 600 °C. In: Reed TB (ed) Biomass gasification: principles and technology, pp 91–111. Noyes Data Corporation
26. Molino A, Chianese S, Musmarra D (2016) Biomass gasification technology: the state of the art overview. J Energy Chem 25:10–25
27. Xu Q (2013) Investigation of co-gasification characteristics of biomass and coal in fluidized bed gasifiers. Ph D thesis, University of Canterbury
28. Kumar A, Jones DD, Hanna MA (2009) Thermochemical biomass gasification: a review of the current status of the technology. Energies 2:556–581. https://doi.org/10.3390/en20300556
29. Rapagna S, di Celso GM (2008) Devolatilization of wood particles in a hot fluidized bed: product yields and conversion rates. In: Biomass & bioenergy, vol 32, pp 1123–1129
30. Basu P, Vichuda M, Leon MA (2009) Gasification of rice husk in supercritical water. In: Proceedings of the 8th world conference on chemical engineering, paper # 971. Montreal, Canada, p 520
31. Mathieu P, Dubuisson R (2002) Performance analysis of a biomass gasifier. In: Energy conversion and management, vol 43, no 9–12, pp 1291–1299
32. Yan Q, Zhang H, Sun B, Guo L (2014) Effect of heating method on hydrogen production by biomass gasification in supercritical water. Adv Condens Matter Phys. Article ID 519389, p 5. http://dx.doi.org/10.1155/2014/519389. Hindawi Publishing Corporation
33. Dhepe PL, Fukuoka A (2008) Cellulose conversion under heterogeneous catalysis. Chem Sus Chem 1(12):969–975
34. Thomas B Reed (ed) (2002) Encyclopaedia of biomass thermal conversion: the principles and technology of pyrolysis, gasification and combustion. The biomass Energy Foundation Press
35. Wender I (1996) Reactions of synthesis. Fuel Process Technol 48:189–297
36. Abdoulmoumine N, Adhikari S, Avanti Kulkarni A, Chattanathan S (2015) A review on biomass gasification syngas cleanup. Appl Energy 155:294–307
37. Woolcock PJ, Brown RC (2013) A review of cleaning technologies for biomass-derived syngas. Biomass Bioenergy 52:54–84
38. Dalai AK, Sasaoka E, Hikita H, Ferdous D (2003) Catalytic gasification of sawdust derived from various biomass. Energy Fuels 17(6):1456–1463
39. Hu G, Xu S, Li S, Xiao CH, Liu S (2006) Steam gasification of apricot stones with olivine and dolomite as downstream catalysts. Fuel Process Technol 87(5):375–382
40. Lopamudra D, Krzysztof J, Ptasinskia F, Janssena JJG, Sander VB, Van Paasenb P, Bergmanb CA, Kielb JHA (2005) Catalytic decomposition of biomass tars: use of dolomite and untreated olivine. Renew Energy 30(4):565–587
41. Aznar MA, Caballero MA, Sancho JA, Frances E (2006) Plastic waste elimination by co-gasification with coal and biomass in fluidized bed with air in pilot plant. Fuel Process Technol 87(5):409–420
42. Xu G, Takahiro M, Toshiyuki S, Shigeru K, Toshiro F (2005) Distinctive effects of CaO additive on atmospheric gasification of biomass at different temperatures. Ind Eng Chem Res 44(15):5864–5868

43. Sutton D, Kelleher B, Ross JR (2001) Review of literature on catalysts for biomass gasification. Fuel Processesing Technol 73(3):155–173
44. Lee W, Nam SS, Kim SB, Lee KW, Choi CS (2000) The effect of Na2CO3 on the catalytic gasification of rice straw over nickel catalysts supported on kieselguhr. Korean J Chem Eng 17(2):174–178
45. Simell PA, Leppalahti JK, Bredenberg JB (1992) Catalytic purification of tarry fuel gas with carbonate rocks and ferrous materials. Fuel 71(2):211–218
46. Juutilainen SJ, Simell PA, Krause AOI (2006) Zirconia: selective oxidation catalyst for removal of tar and ammonia from biomass gasification gas. Appl Catal B-Environ 62(1–2): 86–92
47. Rostrup-Nielsen J, Hansen JB (1993) CO2-reforming of methane over transition metals. J Catal 144(1):38–49
48. Aznar MP, Corella J, Delgado J, Lahoz J (1993) Improved steam gasification of lignocellulosic residues in a fluidized bed with commercial steam reforming catalysts. Ind Eng Chem Res 32(1):1–10
49. Aznar MP, Caballero MA, Gil J, Martin JA, Corella J (1998) Commercial steam reforming catalysts to improve biomass gasification with steam-oxygen mixtures: catalytic tar removal. Ind Eng Chem Res 37(7):2668–2680
50. Baker EG, Mudge LK, Brown MD (1987) Steam gasification of biomass with nickel secondary catalysts. Ind Eng Chem Res 26(7):1335–1339
51. Caballero MA, Aznar MP, Gil J, Martin JA, Frances E, Corella J (1997) Commercial steam reforming catalysts to improve biomass gasification with steam-oxygen mixtures. 1. Hot gas upgrading by the catalytic reactor. industrial and engineering chemistry research 36(12), 5227–5239
52. Elliott DC, Baker EG, Butner RS, Sealock LJ (1993) Bench-scale reactor tests of low temperature, catalytic gasification of wet industrial wastes. J Sol Energy Eng 115(1):52–56
53. Mudge LK, Baker EG, Mitchell DH, Brown MD (1985) Catalytic steam gasification of biomass for methanol and methane production. J Sol Energy Eng 107(1):88–92
54. Olivares A, Aznar MP, Caballero MA, Gil J, Frances E, Corella J (1997) Biomass gasification: produced gas upgrading by in-bed use of dolomite. Ind Eng Chem Res 36 (12):5220–5226
55. Abu El-Rub Z, Bramer EA, Brem G (2004) Review of catalysts for tar elimination in biomass gasification processes. Ind Eng Chem Res 43(22):6911–6919
56. Trimm D (1997) Coke formation and minimisation during stem reforming reactions. Catal Today 37(3):233–238
57. Bangala DN, Abatzoglou N, Chornet E (1998) Steam reforming of naphthalene on Ni-Cr/ Al2O3 catalysts doped with MgO, TiO2 And La2O3. AIChE J 44(4):927–936
58. Bartholomew C, Weatherbee G, Jarvi G (1980) Effects of carbon deposits on the specific acitivity of nickel and nickel bimetallic catalysts. Chem Eng Commun 5(1–4):125–134
59. Bartholomew C, Sorensen W (1983) Sintering kinetics of silica- and alumina-supported nickel in hydrogen atmosphere. J Catal 81(1):131–141
60. Dibbern H, Olesen P, Rostrup-Nielsen J, Tottrup P, Udengaard N (1986) Make low H2/CO syngas using sulfur passivated reforming. Hydrocarbon Processing 65(1):71–74
61. Wang S, Lu GQ (1998) Reforming of methane with carbon dioxide over Ni/Al2O3 catalysts: effect of nickel precursor. Appl Catal A-Gen 169(2):271–280
62. Gadalla AM, Bower B (1988) The role of catalyst support on the activity of nickel for reforming methane with CO2. Chem Eng Sci 43(11):3049–3062

Algal Biomass Generation as Feedstock for Sustainable Bio-oil Production

G Venkata Subhash, Natarajan Mohan, Amar S. Musale, Meghna Rajvanshi, Kshipra Gautam, G. Raja Krishna Kumar, Debanjan Sanyal, and Santanu Dasgupta

1 Introduction

The conventional sources of energy such as coal, natural gas and oil are limited and may last for another 40–50 years [1]. Also, the increasing anthropogenic energy needs because of increasing human population have led the scientific community to identify sustainable alternate energy options [2]. This has paid more thrust on the renewable energy options especially solar energy, hydropower and biomass which generate cleaner and sustainable energy. Biomass such as corn, sugarcane, palm and soybean have been used for the production of first- and second-generation biofuels.

G. Venkata Subhash · A. S. Musale · M. Rajvanshi · K. Gautam · G. R. K. Kumar · S. Dasgupta
Research and Development Centre, Reliance Corporate Park, Ghansoli, Mumbai, India
e-mail: Venkata.Goriparti@ril.com

A. S. Musale
e-mail: Amar.Musale@ril.com

M. Rajvanshi
e-mail: Meghna.Rajvanshi@ril.com

K. Gautam
e-mail: Kshipra.Gautam@ril.com

G. R. K. Kumar
e-mail: Raja.Kumar@ril.com

S. Dasgupta
e-mail: Santanu.Dasgupta@ril.com

N. Mohan · D. Sanyal (✉)
Research and Development, Reliance Industries Ltd, Jamnagar, India
e-mail: debanjan.sanyal@ril.com

N. Mohan
e-mail: Natarajan.Mohan@ril.com

U. K. Bhui (ed.), *Macromolecular Characterization of Hydrocarbons for Sustainable Future*, Green Energy and Technology,
https://doi.org/10.1007/978-981-33-6133-1_17

However, the third-generation biofuels are produced from algae and do not involve any food versus fuel controversy. Microalgae are a very promising option for the production of biofuels owing to their high growth rates and better ability to accumulate lipids [3]. Algae have higher photosynthetic efficiency as compared to land plants. Algae are very efficient in converting light, using water and carbon dioxide, into biomass and are responsible for 40% of global CO_2 fixation [4]. Algae cultivation can be done on any land as it does not require any fertile or agriculture land. There are ample algal species those contain high lipid percentage or carbohydrates those can be easily converted into biofuel such as biodiesel or bioethanol. However, the residual biomass is useful in the production of bio-oil, biohydrogen, bioethanol and biogas. Further, the algal biofuels are carbon neutral and do not emit any harmful gases into the atmosphere. Interestingly, algae-derived biofuel and the by-products serve as an excellent replacement for edible crops.

Different types of biofuels can be produced from algae such as biodiesel, jet fuel, bioethanol, biohydrogen and biomethane. The conversion of algal biomass into end products or algal biofuels requires certain processes, those mainly depend on the properties of the selected algal species. Biodiesel is one of the most anticipated fuel that is obtained by transesterification of microalgal lipid molecules. This process of transesterification includes both chemical and enzymatic (lipases) activities. Biohydrogen can be produced by a process of biophotolysis, which is the action of light on biological systems that results in dissociation of water into molecular hydrogen and oxygen. However, bioethane can be produced via bacterial anaerobic digestion of the protein produced by microalgae or residues produced from oil extraction step. Bioethanol and biobutanol also can be produced by the process of fermentation of the carbohydrates in algae by microbes or yeasts. By applying high pressures and temperatures to algal biomass can be converted into hydrocarbon biofuels through thermochemical conversion processes which is furthermore used as heat, electricity and/or transportation fuels [5].

The present chapter is intended to emphasize more on the processes of biofuel and by-products production from microalgae. The chapter also discusses the different economical processes like algae cultivation, harvesting processes and techniques used to improve the strains for their lipid content. Introduction of algae-based biofuels and their by-products at commercial scale and eventually to the market is only possible by aligning the combination of biological, engineering and downstream processes to the best implementation practices.

2 Algal Cultivation and Processing

2.1 Algae Cultivation

Microalgae are simplest form of photosynthetic organism, which can grow rapidly and survive in extreme environmental conditions due to their simple structure [6].

Microalgae were studied extensively in the last 50 years for their uses in various applications. Many strains show potential for large-scale cultivation, and this cultivation process is the major part of the algae production system. In early 1960s, first large-scale cultivation was started in Japan using *Chlorella*. Microalgae can grow in different cultivation system and in various environmental conditions and have the potential to convert solar energy into chemical energy by photosynthesis. To achieve higher biomass production, it is important to improve cultivation technologies and the biomass areal productivity must be above 30 g m^{-2} d^{-1} [7] for sustainable cultivation. Mainly, microalgae are cultivated (Fig. 1) in two types of cultivation systems, i.e. in open raceway pond and closed photobioreactors (PBRs), in various cultivation conditions of autotrophic, mixotrophic and heterotrophic modes.

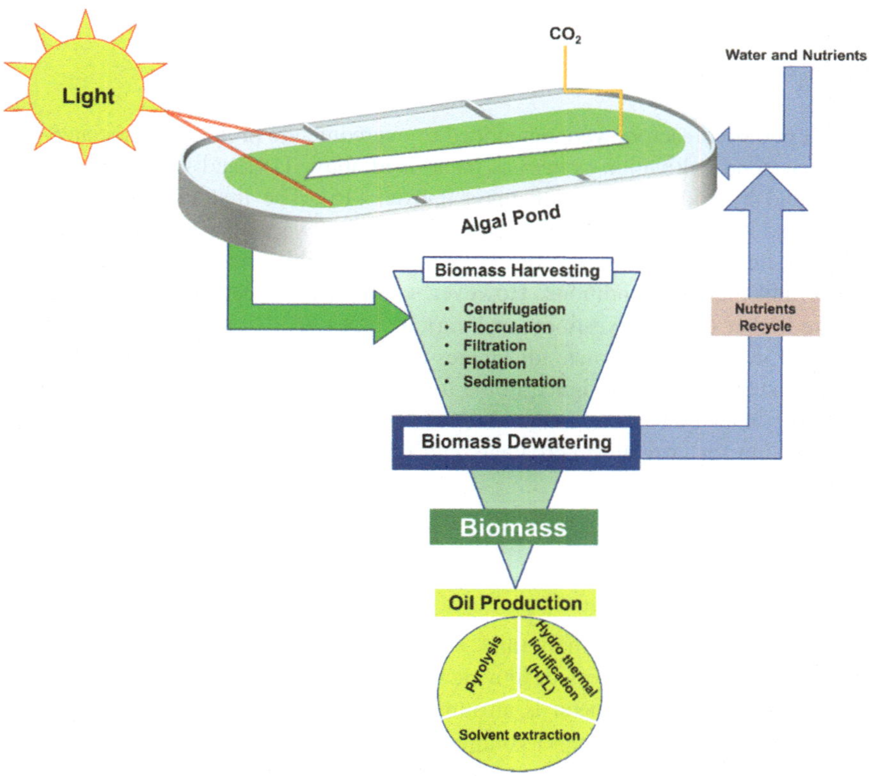

Fig. 1 Steps involved in microalgae cultivation for biomass and oil production

2.1.1 Open Pond System

Open raceway ponds are widely used for algae cultivation due to their lower capital investment and ease of operation for biomass production [8]. Microalgae cultivation in open system is proven since 1950s. Raceway pond is constructed with paddle wheel for the culture agitation [9]. Raceway ponds under open environment will be effected by some kind of contaminants like rotifers consume live algal cells which affects biomass production [10]. Depth of raceway pond is generally designed between 0.2 and 0.5 m to enhance the growth rate of the algae [11, 12]. Most of the raceway ponds are made with concrete and lined with tiles and sheets. The culture is continuously agitated by paddle wheel to avoid sedimentation of microalgae in the pond. Due to continuous agitation, the evaporation rate is higher. Similarly, the open pond has low CO_2 utilization efficiency and difficulties with pH control compared with closed system [13].

2.1.2 Closed System

Closed photobioreactors were developed to overcome the problems associated with the open pond cultivation system as mentioned above [14] and well controlled for gas–liquid hydrodynamics, with mass transfer [15]. The first PBR systems were introduced in the 1940s, making it possible to control variables, including nutrients, pH, light and temperature at a large scale [16]. Photobioreactors are useful in both indoor and outdoor conditions. However, compared to open pond system, PBRs require higher capital investment and various types of PBRs have been developed and each system has its own advantages and disadvantages. Generally, PBRs are closed systems that are designed in the form of plates, tubes or bags made of glass, plastic or other transparent materials [17]. There are several types of photobioreactors viable for growing microalgae, and they are tubular, flat panel and column photobioreactors.

Tubular Photobioreactor

Microalga was cultivated first time in tubular photobioreactor. Tubular photobioreactors are the simplest design developed for industrial purposes [18]. The selection of the tube diameter is an important decision for an ideal design of the PBR, since it affects surface-to-volume ratio, light absorption of the culture, biomass concentration, volumetric productivity, temperature profile of the culture, amount of oxygen in the culture, CO_2 storage capacity of the PBR, head loss for culture recycling and length of the tubes [19]. *Leptolyngbya* is a common contaminant in *Arthrospira* culture, which is highly adhesive that results in fouling formation on the internal tube surface. This affects the light penetration, and hence cleaning is required. Frequent shutdown is a serious issue in aged plastic tubes which reduces smoothness. In PBRs, the major setbacks are the high energy requirement for pumping [19].

Flat Panel Reactor

The flat panel PBR is simple design and significantly reduces the light diffusion depth over the culture surface. Flat panel PBRs comprise a frame enclosed by a transparent plate on both sides, and the culture is circulated through pumping. The important characteristics of flat panel PBRs are high surface-to-volume ratio, the vertical inclination from the horizontal of the channels and absence of mechanical devices for cell suspension. Moreover, the exchange of gases of the culture is achieved through air bubble formation from base of channels. In general, flat panel PBRs are constructed with 16-mm-thick plexiglass alveolar plates, due to the surface-to-volume ratio [20]. Hu et al. have reported that the high photosynthetic efficiencies can be achieved in flat panel reactors, due to the large irradiance surface area. Moreover, the buildup of O_2 concentration is comparatively low in flat panel photobioreactor [21]. The flat panel PBRs are useful for small- scale production, and not suitable for commercial scale, owing to constraints associated with construction parts and materials, lack of control for culture on temperature maintenance and difficulties associated with aeration-induced hydrodynamic stress [22].

Column Photobioreactor

Column photobioreactors provide the most effective mixing, higher volumetric mass transfer rates and high growth [23]. The advantages of column photobioreactor are low-cost, compact and easy to operate. Here, the column is placed in vertical direction and aerated from the bottom, and irradiated over transparent walls [23] or internally. A column photobioreactor aids higher biomass production as compared to horizontal photobioreactor because of efficient hydrodynamics [24]. The gas bubbles rise fast beside the column and distributed well when it has reached to surface of the reactor. Hence, the fluid will flow all through the column upright close to the cylinder axis and downward close to the walls. This actually enhances gas–liquid exchange. The rate of mass transfer can be achieved by controlling the residence time of the gas bubbles. Their performance compares favourably with tubular photobioreactors [25].

2.1.3 Hybrid System

The hybrid system of cultivation is the two-stage cultivation method which includes the cultivation in photobioreactors and in open pond cultivation systems. In the first stage, algae are cultivated in a controlled condition in photobioreactor with higher biomass concentration and less contaminations. The second is the production stage in an open pond system, which enhances production of the desired product, e.g. lipid under nutrients and environmental stresses conditions [26].

2.2 Growth Patterns Under Different Nutritional Sources

2.2.1 Phototrophic Growth

In phototrophic condition, microalgae use solar energy as energy source in the presence of inorganic carbon sources (e.g. carbon dioxide) as the carbon source to form chemical energy through photosynthesis [20]. Generally, microalgae are cultivated in autotrophic conditions [27]. The advantage of autotrophic cultivation is that CO_2 is used as carbon source for the growth and oil production. Though CO_2 is the only carbon source, the cultivation location should be close to industries or power plants to use the CO_2 from flue gas to minimize to CO_2 cost. Furthermore, the contamination issue is minimal in autographic condition. Hence, open ponds and raceway ponds are generally cultivated in phototrophic conditions [28].

2.2.2 Heterotrophic Growth

In heterotrophic mode of cultivation, an organic carbon source is used as both energy and carbon for algal growth in the absence of light. The advantages of heterotrophic cultivation are a high degree of growth control, higher cell density and lower harvesting costs. Some species of microalgae can grow very well in heterotrophic condition. Some microalgae grown in this condition are *Chlorella vulgaris* [29], *Chlorella* protothecoides [30], *Crypthecodinium cohnii* [31] and *Schizochytrium limacinum* [32]. Water and nutrients are potential microbial contamination sources and require systematic control system. The disadvantage of heterotrophic cultivation is that pigmentation and secondary metabolite productions are impaired [28].

2.2.3 Mixotrophic Growth

In mixotrophic growth, microalgae use both organic and inorganic carbon as a carbon source for their growth. They can grow under either phototrophic or heterotrophic conditions or both. They have the ability to assimilate organic compounds and CO_2 as a carbon source, and the CO_2 released during respiration is reused for photosynthesis [28]. However, microalgae cultivated rarely in mixotrophic condition for oil production.

2.3 *Downstream Processing*

2.3.1 Harvesting and Dewatering

Harvesting is a major energy bottleneck in cultivation and processing. Harvesting methods currently involve sedimentation, centrifugation, filtration, ultrafiltration, flocculation and floatation. In total biomass production, cost about 20–30% of cost contributes for recovering biomass from the culture medium [33]. To achieve higher recovery efficiency, selection of a low-cost method is important for large-scale biomass production. Moreover, it is important to combine two or more methods efficient harvesting [34]. The harvesting methods of separating the biomass from the culture depend on the cell size and density. However, it is important to choose suitable harvesting method based on the desired product [35].

Dewatering of microalgae slurry involves various mechanical methods with different types of filtration system and centrifugation. After harvesting, the dewatered algal cake is generally dried to improve the efficiency of the downstream process. There are many methods employed for drying, namely sun drying [36], spray drying [37], drum drying, fluidized bed drying [38], freeze drying and Refractance Window™ technology drying [39]. However, sun drying is the cheapest method, but the disadvantages include long drying times, the requirement for large drying surfaces, and the risk of material loss [36]. There are merits and demerits in each drying technique. Spray drying is used for high value product but it is comparatively expensive, and the major bottle neck is the decomposition of algal pigments. Freeze drying is an expensive method for large-scale operations. Moreover, intracellular components for example, oils are tough to extract from wet biomass through solvents without cell disruption, but easily extractable using freeze dried biomass. Different dewatering techniques might be useful for different microalgae species. However, centrifuge and belt filter are commonly used dewatering methods for microalgae [40].

3 Biofuel Products from Microalgae

After harvest, energy packed wet algal slurry or dry algal biomass was processed by different treatments to get algal-based biofuels like bioethanol, biohydrogen, biogas and bio-oil are described in detail (Fig. 1).

3.1 *Algal Fatty Acids for Bio-oil*

Microalgae-based oil and biomass have several advantages over terrestrial oleaginous crops (fast growth, consume carbon dioxide (CO_2) and release oxygen (O_2)).

Microalgae comprises varying proportions of energy-rich compounds proteins, carbohydrates, nucleic acids and lipids as biochemical components. They contain primarily polar (structural lipids such as glycolipids and phospholipids) and non-polar lipids (neutral or storage lipids), i.e. triglycerides (TAGs), diglycerides (DAGs), monoglycerides (MAGs), free fatty acids (FFAs), hydrocarbons and pigments. Lipids are stored in algae during stationary phases and environmental growth conditions. Very little amounts of TAGs will be accumulated during exponential growth [41]. TAGs can be easily catabolized to supply metabolic energy. They are mainly synthesized in the light, accumulated in cytosolic lipid bodies and then reused for polar lipid synthesis in the absence of light [42, 43]. Generally, the algal oil contains unsaturated fatty acids (16:1, 18:1, 18:2, 18:3), and the saturated fatty acids (16:0, 18:0) are the main components and influence the fuel properties of algal biodiesel [44].

3.1.1 Environmental Effects on Microalgae Lipids

Microalgae highly depend on the culture conditions and unfavourable conditions could change the cellular composition of microalgae like growth rate, lipid accumulation, etc. For example, the culture conditions imposed onto the cells may increase the oil content of the *Chlorella vulgaris* by almost two or three times. Nutrient availability, e.g. nitrogen, phosphorous and silicate (in diatoms), carbon source, salinity, light intensity and temperature in the culture medium are the key for growth and primary production of microalgae. Compared to other nutrients, nitrogen and phosphorous manipulations have been found to be the most effective factor to increase lipid accumulation several folds in microalgae. The sensitivity of microalgae to nutrient manipulation is usually considered through three levels of limitation including starvation, limitation and depletion [42].

3.1.2 Effect of Nitrogen

Proteins have a major pool of nitrogen and are an essential constituent of cell key components (proteins, amino acids, nucleic acids, enzymes, etc.,) [45]. Nitrogen in the growth medium can influence the lipid synthesis by carbon sequestration (organic or inorganic) and growth and the nitrogen depleted conditions result to decreased levels of cellular protein content. Nitrogen depletion in the medium activates acyl hydrolase and increases the intracellular content of fatty acid acyl-CoA followed by activation of diacylglycerol acyltransferase involved in the conversion of acyl-CoA to triglyceride (TAG) [46]. Studies were performed in *Chlorella sp.*, and some other examples of microalgae [43]. Under nitrogen deprived condition, just in hours or days, the accumulation of lipid content of the cell will increase up to 2–3 folds [47, 48]. For example from the studies, 70–72% of lipid accumulated in 9 days of incubation in *Monallantus salina* and the prolonged incubation may suppress the lipid accumulation and reported in algal strains *Scenedesmus* sp. and *Chlorella sorokiniana* (>40% in 14 days) [49–56].

3.1.3 Effect of Phosphorous

Phosphorus (P) is the second most important nutrient after N for the growth of microalgae as it plays a key role in cellular metabolic processes and cell membrane. Inorganic orthophosphate is the main source for the P supply for algal growth and biomass enhancement, and phosphorus depletion shows impact on membrane phospholipids and reported in *Chlorella* sp. [45, 57]. Under P deprived conditions, there will be a reduced cell division and lead to accumulation of TAG [58–63].

3.2 *Algal Bio-oil Production*

There are a variety of processes available to convert microalgae biomass into bio-oil, such as by extracting lipids from wet and dry algal biomass and transesterification, thermochemical and biochemical processes.

3.2.1 Solvent-Based Oil Extraction and Transesterification

Lipids (ranging from 20 to 50%) or fatty acids and the removal of algal oil from the multilayered cell wall envelops by using solvents (chloroform, methanol, etc.) and mechanical (microwave, grinding, bead beating and ultrasound) methods are known as lipid extraction [64]. In present scenario, research on lipid extraction is directed towards free from solvents using supercritical fluid technique (SCF) and accomplishes safe and good quality product.

The conversion of this extracted lipids into C18 range carbon fuels by the process called transesterification is the reaction between one mole of TAG and 3–4 mol of alcohol to produce simple esters (Biodiesel). Overall reaction is catalysed by acid (sulphonic and sulphuric acid) and base catalysts (NaOH, KOH, sodium methoxide, sodium ethoxide and K_2CO_3) [64] (Fig. 2).

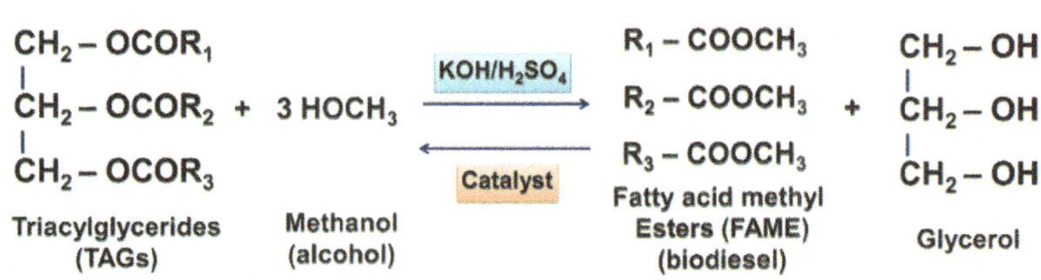

Fig. 2 Steps involved in transesterification process

Table 1 Operating parameters in pyrolysis

Pyrolysis methods	Parameters		
	Temperature (°C)	Residence time	Heating rate (°C/s)
Slow pyrolysis	400	>30 min	0.1–1
Fast pyrolysis	500	10–20 s	1–200
Flash pyrolysis	500	1 s	>1000

3.2.2 Bio-oil by Pyrolysis

In the absence of oxygen, direct thermal decomposition of the biomass at high temperature (400–1000 °C) to produce bio-oil into solid (char and coke), liquid (bio-oil) or gaseous fuels (methane and higher gaseous hydrocarbons) is called as pyrolysis [65]. Pyrolysis is categorized into slow (low process temperatures and longer time) and fast (higher process temperatures and shorter time). The product resulted from the pyrolysis method is called as bio-oil. Along with bio-oil aqueous phase having a varied low mw compounds (primary alcohol, acid and ketone) and non-aqueous phase of oxygen containing compounds (aliphatics, alcohols, carbonyls, acids, phenols and cresols, etc.) and aromatic hydrocarbons [66] (Table 1).

3.2.3 Bio-Oil by Hydro Thermal Liquification (HTL)

Hydro thermal liquification (HTL) of any biomass is a promising way to produce thick and viscous bio-crude. Application of HTL process in algae can convert all nutrients including proteins and carbohydrates and not only lipids to biocrude. Without processing, the algae with high water content and any pretreatment make the method viable than drying of algae. HTL under added catalyst reaction is more viable in conversion rate from one form to other by comparison of non-catalytic reactions. At pressure 25 MPa, below 375 °C whole microalgae can be liquified. HTL of *Nannochloropsis* at a 300 °C with Ni-doped TiO_2 yielded 48% of biocrude at 90% conversion order. The complete combustion of algae released higher % of NOx into atmosphere due to the significant presence of nitrogen in algae (5–8 wt%) [67].

3.3 Algae Aviation Fuel Prospective

Biodiesel seems to be an adequate replacement for diesel fuel used in ground transportation. An important challenge is to find replacement for aviation turbine fuels. The main use of aviation turbine fuels is to power jet and turbo engine aircraft. Biodiesel alone is not suitable as a jet fuel, it must meet the properties of energy density, and the low-temperature fuel for any alternative option is quite important. Algae biomass is one of the bioresources that can produce clean and

Table 2 Comparison of biodiesel versus conventional jet fuel

Fuel property	Biodiesel	Petroleum-based jet fuel
Flash point, °C	100 °C	40 °C–45 °C
Kinematic viscosity at 40 °C	4.7	1.2
Net heat of combustion, MJ/kg	36–39	43.2
Specific gravity, 15 °C	0.87–0.89	0.80
Freezing point, °C	About 0	<−40
Approximate number of carbon atoms	C16 to C22	C8 to C16
Sulphur wt%	<0.05	0.05–0.15

sustainable fuel despite a few challenges. Today, it has technically been proved that there is a real possibility of converting algae or any biomass into bioethanol, bio-jet fuel and biodiesel. High costs in producing algae are the main limitation for the usage of algae oil [68]. The high costs include carbon dioxide costs, the large area to grow and the many steps of the algae to oil process [4, 68, 69] (Table 2).

There are three routes to produce bio-jet fuel from microalgae.

(i) The first route involves using algae oil to produce bio-SPK (Bio-derived Synthetic Paraffinic Kerosene) by cracking and hydro processing. This can be used for kerosene type of fuels include Jet A, Jet A-1, JP-5 and JP-8.
(ii) The second route involves processing solid biomass using pyrolysis to produce pyrolysis oil or gasification to produce a syngas, which is then possessed into FT-SPK (Fisher–Tropsch Synthetic Paraffinic Kerosene).
(iii) The third route to produce bio-jet fuel is blending algae biodiesel with kerosene. This route involves algae growth, harvesting, oil extraction and transesterification (or in situ process) to produce microalgae biodiesel.

Properties of algal kerosene and normal kerosene differ slightly, but aviation infrastructure (e.g. engines) does not need to be adjusted for the use of algal kerosene. It is very complex to determine the economic feasibility for application of algal fuel in the aviation industry due to a number of issues. Currently, the production costs per litre of algal oil are very high, mainly due to small-scale production. Also, the state of the art has shown that there is such a huge variety in available algal species, cultivation systems and processing methods, which makes it very difficult to determine the most suitable algae to production of kerosene.

4 Future Prospective

Microalgae-based lipid and biomass are one of the most promising alternatives to biofuel feedstocks, due to their promising sustainable advantages along with the productivity potential compared to traditional terrestrial feedstocks. This comprehensive review presents the effects of nutrients and environmental stress on microalgae lipid content and productivity. It summarizes the lipid compositions of the most suitable and widely studied strains as well as the effect of environmental factors on their lipid compositions. Although more research has focused on individual effects of the parameters to reach the highest lipid productivity, it has been found that nitrogen deprivation can be one of the most effective strategies. The other useful strategy is to support the microalgae to reach their maximum density (growth) and apply the environmental stress at the stationary phase. Biodiesel production from the algal source plays an important role in environment and transportation. In near future, the major challenges for the researchers would be the improvement of lipid profiles of certain important strains which have high lipid productivity through genetic engineering, and the efforts must be invested in biomass production needed for biofuel production.

5 Conclusion

The review underlines the significance of algal biofuels over the fossil fuel-derived conventional fuels. Microalgae as feedstocks for the production of biofuels such as biodiesel, biomethane, biohydrogen and bioethanol have immense potential especially because of their high growth rates and photosynthetic efficiency which surpasses that of any other photosynthetic organism. This can be coupled with biomitigation of CO_2 emissions, hence biofuels produced from microalgae are clean and extremely promising in countering the problem of climate change and harmful gas emissions. It is the most environmentally sustainable option which can support anthropogenic latent energy demand and a giant step towards carbon neutral world. However, research breakthroughs in algae through genetic modification and engineering techniques are still to be achieved to make the 'biofuels from algae' a reality, which results in a sustainable and cost-effective process. Further, a single integrated process should be designed to utilize all the algal components such as lipids, carbohydrates, proteins, pigments and other value-added products which will make the process highly economical.

Acknowledgements We would like to thank Reliance Industries Limited for all the support. The views expressed in this chapter are based on published literature and not from Reliance Industries Limited (RIL).

References

1. Werner Z, Schindler J (2007) Crude oil the supply outlook
2. Edwards M (2008) Green algae strategy: end biowar I and engineer sustainable food and biofuels. TalentDNA

3. Hu Q, Sommerfeld M, Jarvis E, Ghirardi M, Posewitz M, Seibert M et al (2008) Microalgal triacylglycerols as feedstocks for biofuel production: perspectives and advances. Plant J 54 (4):621–639
4. Hannon M, Gimpel J, Tran M, Rasala B, Mayfield S (2010) Biofuels from algae: challenges and potential. Biofuels 1(5):763–784
5. Darzins AI, Pienkos P, Edye L (2018) Current status and potential for algal biofuels production—a report to IEA bioenergy task 39
6. Li Y, Horsman M, Wang B, Wu N, Lan CQ (2008) Effects of nitrogen sources on cell growth and lipid accumulation of green alga Neochloris oleoabundans. Appl Microbiol Biotechnol 81 (4):629–636
7. Bioenergy technology office: multi-year plan. U.S. Department of Energy (2014)
8. Rawat I, Ranjith Kumar R, Mutanda T, Bux F (2011) Dual role of microalgae: phycoremediation of domestic wastewater and biomass production for sustainable biofuels production. Appl Energy 88(10):3411–3424
9. Schenk PM, Thomas-Hall SR, Stephens E, Marx UC, Mussgnug JH, Posten C et al (2008) Second generation biofuels: high-efficiency microalgae for biodiesel production. BioEnergy Res 1(1):20–43
10. Collet P, Helias A, Lardon L, Ras M, Goy RA, Steyer JP (2011) Life-cycle assessment of microalgae culture coupled to biogas production. Bioresour Technol 102(1):207–214
11. Brennan L, Owende P (2010) Biofuels from microalgae—a review of technologies for production, processing, and extractions of biofuels and co-products. Renew Sustain Energy Rev 14(2):557–577
12. Narala RR, Garg S, Sharma KK, Thomas-Hall SR, Deme M, Li Y et al (2016) Comparison of microalgae cultivation in photobioreactor, open raceway pond, and a two-stage hybrid system. Front Energy Res 4(29)
13. Slade R, Bauen A (2013) Micro-algae cultivation for biofuels: cost, energy balance, environmental impacts and future prospects. Biomass Bioenergy 53:29–38
14. Posten C, Schaub G (2009) Microalgae and terrestrial biomass as source for fuels—a process view. J Biotechnol 142(1):64–69
15. Zemke PE, Sommerfeld MR, Hu Q (2013) Assessment of key biological and engineering design parameters for production of Chlorella zofingiensis (Chlorophyceae) in outdoor photobioreactors. Appl Microbiol Biotechnol 97(12):5645–5655
16. Huntley ME, Redalje DG (2007) CO2 mitigation and renewable oil from photosynthetic microbes: a new appraisal. Mitig Adapt Strat Glob Change 12(4):573–608
17. Lehr F, Posten C (2009) Closed photo-bioreactors as tools for biofuel production. Curr Opin Biotechnol 20(3):280–285
18. Pulz O (2001) Photobioreactors: production systems for phototrophic microorganisms 287–293
19. Richmond A, Boussiba S, Vonshak A, Kopel R (1993) A new tubular reactor for mass production of microalgae outdoors. J Appl Phycol 5:327–332
20. Huang G, Chen F, Wei D, Zhang X, Chen G (2010) Biodiesel production by microalgal biotechnology. Appl Energy 87(1):38–46
21. Hu Q, Kurano N, Kawachi M, Iwasaki I, Miyachi S (1998) Ultrahigh-cell-density culture of a marine green alga Chlorococcum littorale in a flat-plate photobioreactor 655–662
22. Ugwu CU, Aoyagi H, Uchiyama H (2008) Photobioreactors for mass cultivation of algae. Bioresour Technol 99(10):4021–4028
23. Eriksen N (2008) The technology of microalgal culturing. Biotechnol Lett 30
24. Sánchez Mirón A, Contreras Gómez A, García Camacho F, Molina Grima E, Chisti Y (1999) Comparative evaluation of compact photobioreactors for large-scale monoculture of microalgae. J Biotechnol 70(1):249–270
25. Sánchez Mirón A, Cerón García M-C, García Camacho F, Molina Grima E, Chisti Y (2002) Growth and biochemical characterization of microalgal biomass produced in bubble column and airlift photobioreactors: studies in fed-batch culture. Enzyme and Microbial Technology 31(7):1015–1023

26. Rodolfi L, Chini G, Bassi N, Padovani G, Biondi N, Bonini G et al (2008) Microalgae for oil: strain selection, induction of lipid synthesis and outdoor mass cultivation in a low-cost photobioreactor. Biotechnol Bioeng 102
27. Yoo C, Jun S-Y, Lee J-Y, Ahn C-Y, Oh H-M (2010) Selection of microalgae for lipid production under high levels carbon dioxide. Bioresour Technol 101(1, Supplement):S71–S74
28. Mata TM, Martins AA, Caetano NS (2010) Microalgae for biodiesel production and other applications: a review. Renew Sustain Energy Rev 14(1):217–232
29. Liang Y, Sarkany N, Cui Y (2009) Biomass and lipid productivities of Chlorella vulgaris under autotrophic, heterotrophic and mixotrophic growth conditions. Biotech Lett 31 (7):1043–1049
30. Xiong W, Li X, Xiang J, Wu Q (2008) High-density fermentation of microalga Chlorella protothecoides in bioreactor for microbio-diesel production. Appl Microbiol Biotechnol 78 (1):29–36
31. Couto RM, Fernandes J, da Silva MDRG, Simões PC (2009) Supercritical fluid extraction of lipids from spent coffee grounds. J Supercrit Fluids 51(2):159–166
32. Johnson MB, Wen Z (2009) Production of biodiesel fuel from the microalga Schizochytrium limacinum by direct transesterification of algal biomass. Energy Fuels 23(10):5179–5183
33. Kim J, Yoo G, Lee H, Lim J, Kim K, Kim CW et al (2013) Methods of downstream processing for the production of biodiesel from microalgae. Biotechnol Adv 31(6):862–876
34. Shen Y, Yuan W, Pei ZJ, Wu Q, Mao E (2009) Microalgae mass production methods. Trans ASABE 52(4):1275–1287
35. Richmond A (2008) Handbook of microalgal culture: biotechnology and applied phycology. Wiley, New Jersey
36. Prakash J, Pushparaj B, Carlozzi P, Torzillo G, Montaini E, Materassi R (1997) Microalgae drying by a simple solar device. Int J Solar Energy 18(4):303–311
37. Desmorieux H, Decaen N (2005) Convective drying of spirulina in thin layer. J Food Eng 66 (4):497–503
38. Leach G, Oliveira G, Morais R (1998) Spray-drying of Dunaliella salina to produce a β-carotene rich powder. J Ind Microbiol Biotechnol 20(2):82–85
39. Nindo CI, Tang J (2007) Refractance window dehydration technology: a novel contact drying method. Drying Technol 25:37–48
40. Spellman FR (1997) Dewatering biosolids. Technomic Publishing Company
41. Tonon T, Harvey D, Larson TR, Graham IA (2002) Long chain polyunsaturated fatty acid production and partitioning to triacylglycerols in four microalgae. Phytochemistry 61(1):15–24
42. Sajjadi B, Chen W-Y, Abdul Raman AA, Ibrahim S (2018) Microalgae lipid and biomass for biofuel production: a comprehensive review on lipid enhancement strategies and their effects on fatty acid composition 200–232
43. Kaliane SA, Vanessa NB, Tomás GRV, Tiago dSL, Olinto LP, Eduardo SG et al (2018) Diversity of culturable endophytic fungi of *Hevea guianensis*: a latex producer native tree from the Brazilian Amazom. Afr J Microbiol Res 12(42)
44. Xiao-Jun J, Ren LJ, He Huang H (2015) Omega-3 biotechnology: a green and sustainable process for omega-3 fatty acids production. Front Bioeng Biotechnol 3(158)
45. Liang K, Zhang Q, Gu M, Cong W (2013) Effect of phosphorus on lipid accumulation in freshwater microalga Chlorella sp. J Appl Phycol 25(1):311–318
46. Takagi M, Watanabe K, Yamaberi K, Yoshida T (2000) Limited feeding of potassium nitrate for intracellular lipid and triglyceride accumulation of Nannochloris sp. UTEX LB1999, 112–117
47. Goncalves EC, Johnson JV, Rathinasabapathi B (2013) Conversion of membrane lipid acyl groups to triacylglycerol and formation of lipid bodies upon nitrogen starvation in biofuel green algae Chlorella UTEX29. Planta 238(5):895–906
48. Scragg A (2002) Growth of microalgae with increased calorific values in a tubular bioreactor. Biomass Bioenerg 23(1):67–73
49. Adams C, Godfrey V, Wahlen B, Seefeldt L, Bugbee B (2013) Understanding precision nitrogen stress to optimize the growth and lipid content tradeoff in oleaginous green microalgae. Bioresour Technol 131:188–194

50. Fakhry EM, El Maghraby DM (2015) Lipid accumulation in response to nitrogen limitation and variation of temperature in Nannochloropsis salina. Botanical Studies 56(1):6
51. Boussiba S, Vonshak A, Cohen Z, Avissar Y, Richmond A (1987) Lipid and biomass production by the halotolerant microalga Nannochloropsis salina. Biomass 12(1):37–47
52. Aaronson S (1973) Effect of incubation temperature on the macromolecular and lipid content of the phytofalgellate *Ochromonas danica* 1. J Phycol
53. Patterson GW (1970) Effect of culture temperature on fatty acid composition of Chlorella sorokiniana. Lipids 5(7):597–600
54. Singh P, Guldhe A, Kumari S, Rawat I, Bux F (2015) Investigation of combined effect of nitrogen, phosphorus and iron on lipid productivity of microalgae Ankistrodesmus falcatus KJ671624 using response surface methodology. Biochem Eng J 94:22–29
55. Singh Y, Kumar HD (1992) Lipid and hydrocarbon production by *Botryococcus* spp. under nitrogen limitation and anaerobiosis. World J Microbiol Biotechnol 8(2):121–124
56. Breuer G, Lamers PP, Martens DE, Draaisma RB, Wijffels RH (2013) Effect of light intensity, pH, and temperature on triacylglycerol (TAG) accumulation induced by nitrogen starvation in Scenedesmus obliquus. Bioresour Technol 143:1–9
57. Li M, Welti R, Wang X (2006) Quantitative profiling of Arabidopsis polar glycerolipids in response to phosphorus starvation. Roles of phospholipases D zeta1 and D zeta 2 in phosphatidylcholine hydrolysis and digalactosyldiacylglycerol accumulation in phosphorus-starved plants. Plant Physiol 142(2):750–761. Epub 2006/08/08
58. Guschina IA, Harwood JL (2009) Algal lipids and effect of the environment on their biochemistry. In: Arts MT, Brett MT, Kainz M (eds). Springer, New York
59. Michelon W, Da Silva ML, Mezzari MP, Pirolli M, Prandini JM, Soares HM (2016) Effects of nitrogen and phosphorus on biochemical composition of microalgae polyculture harvested from phycoremediation of piggery wastewater digestate. Appl Biochem Biotechnol 178 (7):1407–1419
60. Xin L, Hong-Ying H, Ke G, Ying-Xue S (2010) Effects of different nitrogen and phosphorus concentrations on the growth, nutrient uptake, and lipid accumulation of a freshwater microalga Scenedesmus sp. Bioresour Technol 101(14):5494–5500
61. Khozin-Goldberg I, Cohen Z (2006) The effect of phosphate starvation on the lipid and fatty acid composition of the fresh water eustigmatophyte Monodus subterraneus. Phytochemistry 67(7):696–701
62. Guschina IA, Dobson G, Harwood JL (2003) Lipid metabolism in cultured lichen photobionts with different phosphorus status. Phytochemistry 64(1):209–217 Epub 2003/08/30
63. Karapinar Kapdan I, Aslan S (2008) Application of the Stover-Kincannon kinetic model to nitrogen removal by Chlorella vulgaris in a continuously operated immobilized photobioreactor system. J Chem Technol Biotechnol 83(7):998–1005
64. Venkata Subhash G, Rajvanshi M, Navish Kumar B, Govindachary S, Prasad V, Dasgupta S (2017) Carbon streaming in microalgae: extraction and analysis methods for high value compounds. Bioresour Technol 244:1304–1316
65. Demirbaş A (2003) Biodiesel fuels from vegetable oils via catalytic and non-catalytic supercritical alcohol transesterifications and other methods: a survey. Energy Convers Manag 44(13):2093–2109
66. Araújo AMdM, Lima RdO, Gondim AD, Diniz J, Souza LD, Araujo ASd (2017) Thermal and catalytic pyrolysis of sunflower oil using AlMCM-41. Renewable Energy 101:900–906
67. Wang W, Xu Y, Wang X, Zhang B, Tian W, Zhang J (2018) Hydrothermal liquefaction of microalgae over transition metal supported TiO2 catalyst. Bioresour Technol 250:474–480
68. Huber GW, Iborra S, Corma A (2006) Synthesis of transportation fuels from biomass: chemistry, catalysts, and engineering. Chem Rev 106(9):4044–4098
69. Patil V, Tran KQ, Giselrod HR (2008) Towards sustainable production of biofuels from microalgae. Int J Mol Sci 9(7):1188–1195

CO_2 Bio-sequestration Studies on Microalgae—An Approach Through Sustainable Biofuel Production

M. S. Mahajan, M. A. Rasheed, P. L. S. Rao, P. Bhutiya, S. Z. Hasan, and S. Shah

1 Introduction

The rising temperatures owing to the emission of greenhouse gases are an emerging concern in current world scenario. The emissions of CO_2 into atmosphere have increased due to combustion of fossil fuels, natural gas, coal, etc., where it contributes to 68% of total emissions of greenhouse gases [1–3]. In order to contain or mitigate, this demands for dedicated research towards applicable technologies to capture and store CO_2 emitted from stacks. Though different methods have been developed (geological, physiological, chemical) and applied, but still the issue needs to be addressed efficiently [4–6]. In view of drawbacks of physio-chemical methods and under the current scenario of environmental pollution along with the factors as economic aspects, life cycle analysis and capital investment all such applications have made these technological impacts slower at commercial level [7–10],

M. S. Mahajan · M. A. Rasheed · P. L. S.Rao · P. Bhutiya · S. Z. Hasan (✉)
Petroleum Research Wing, Gujarat Energy Research and Management Institute, First Floor, Energy Building, PDEU Campus, Raisan, Gandhinagar 382007, Gujarat, India
e-mail: zaheer@germi.res.in; szaheerhasan2001@yahoo.com

M. S. Mahajan
e-mail: mayurmahajan528@gmail.com

M. A. Rasheed
e-mail: abdul@germi.res.in

P. L. S.Rao
e-mail: plsrao@germi.res.in

P. Bhutiya
e-mail: bhutiyapriyank@gmail.com

S. Shah
Climate Change Department, Government of Gujarat, Block No. 11, First Floor, Sardar Bhavan, Sachivalaya, Gandhinagar 382010, Gujarat, India
e-mail: spshah@gmail.com

U. K. Bhui (ed.), *Macromolecular Characterization of Hydrocarbons for Sustainable Future*, Green Energy and Technology,
https://doi.org/10.1007/978-981-33-6133-1_18

thereby arising a need to develop sustainable and cost-effective methods for CO_2 mitigation. The biological fixation of CO_2 obtained through photosynthetic microorganisms has attracted much attention as a strategic alternative that is found associated both environmental and economic interests along with the production of biomass based energy [11–13]. CO_2 can be captured naturally using plants, microalgae and cyanobacteria in their photosynthetic process. By comparing with terrestrial plants potential, uptake of CO_2 is 10–50 times greater in microalgae due to its faster growth rate, higher photosynthetic rate, tolerance to alterations in environmental conditions in addition to the profitability of cultivation in areas unsuitable for agriculture [14, 15]. Microalgae are efficient convertors of solar energy into chemical energy and are able to utilize CO_2 as inorganic carbon source for photosynthesis in the presence of sunlight to store energy as lipids, carbohydrates and proteins [16]. The CO_2 incorporated as lipids in microalgae can be extracted and used as biofuel [15]. Potential research has been carried out on different methodologies to grow microalgae [17] depending upon the species, though not dealt in detail for CO_2 sequestration using the biological route. Among the different cultivation systems, PBR appears to be promising as it yields higher biomass productivity with direct conversion of CO_2 to microalgal biomass accumulating lipids that can serve as a potential biofuel feedstock. When CO_2 is passed into culture media, it reduces the pH due to the formation of carbonic acid (free CO_2) and escapes into the atmosphere, thereby resulting in significant CO_2 loss during algae culture [18]. Appropriately designed PBR can reduce the cultivation area increasing the carbon residence time in cultivation medium and CO_2 utilization efficiency. Coupling the cultivation of photosynthetic microorganisms with the biofixation of CO_2 has the potential not only to reduce the costs of culture media for growing such organisms on an industrial scale but also to offset carbon emissions [11]. Earlier, it was assumed that microalgae are susceptible to high CO_2 concentrations. However, some microalgae are now reported to grow rapidly even at very high CO_2 concentrations [19, 20]. A few microalgal species have been studied for trapping CO_2 biologically such as *Chlorella vulgaris* [19], *Dunaliella tertiolecta*, *Botryococcus braunii*, *Spirulina platensis*, *Scenedesmus obliquus* [11, 12]. Also, maximum daily CO_2 biofixation was reported to be 53.29 and 28.08% for *S. obliquus* and *Spirulina sp*., respectively, at input CO_2 concentration of 6% [11]. *Chlorella sp*. and *S. platensis* showed 46 and 39% mean fixation efficiency at input CO_2 concentration of 10% [21]. These studies have mainly been focused on the effect of varying CO_2 concentrations on the biomass production and CO_2 fixation rate and therefore call for establishment of microalgae species that are tolerant to high CO_2 concentrations. *Chlorella* is a freshwater, unicellular, green algae belonging to the family *Chlorellaceae*, whose name is derived from two words, ‘Chloros’ meaning green and Latin suffix ‘ella’ meaning small. Each cell of chlorella is 2–8 μ thick and does not have flagella. *Chlorella* consists of green photosynthetic pigments chlorophyll a and chlorophyll b courtesy to which it is the richest source of chlorophyll available [22] while *S. platensis* is a blue-green algae (Cyanobacterium) belonging to the family of *Oscillatoriaceae*. It forms unbranched, multicellular helicoidal filaments of 200–300 μm length and 5–10 μm widths

[23]. The study aims to investigate the efficiency of *Chlorella pyrenoidosa* NCIM 2738 and *S. platensis* NCIM 5143 for pure CO_2 fixation in PBR and to determine its effect on different microalgal growth parameters for increased lipid yield and in turn biofuel production.

2 Materials and Methods

2.1 Microalgae and Growth Conditions

Chlorella pyrenoidosa NCIM 2738 and *S. platensis* NCIM 5143 were used for experimental analysis, obtained from National Collection of Industrial Microorganisms (NCIM), NCL Pune, India. The cultures were grown in 1000 ml Erlenmeyer flasks fed with 500 ml media at 24 ± 2 °C with 12:12 h light-dark period as inoculum. The media used for cultivation was BG 11 prepared in tap water (fresh water) [24]. The flasks were hand shaken thrice daily to avoid cell adherence.

2.2 PBR Setup

20 L Grade 1 (PET) polyethylene terephthalate was used as PBR having diameter of 0.25 m, height 0.37 m and working effective area of 0.1649 m^2, respectively. The CO_2 gas flow connections were made out by 8 mm diameter polyurethane tubes and individual valve assembly for each PBR to control the flow. A schematic diagram (Fig. 1) demonstrates the PBR setup that was inoculated with algal species grown in BG 11 media to a total volume of 10 L. Pure 99.999% CO_2 (source—gas cylinder) was bubbled into the PBR, and the pH was lowered to 5.00. The experiment was performed under direct sunlight with daily measurement of pH. PBR got agitated thrice a day to avoid adherence and the biomass was harvested after 13th day.

2.3 Analytical Procedures

2.3.1 Determination of Growth and Biomass Productivity

Algal growth was monitored everyday by measuring optical density (O.D.) at 750 nm using UV–Visible Spectrophotometer (Model—Specord 2000). The biomass productivity (mg/L/day) has been calculated according to the equation

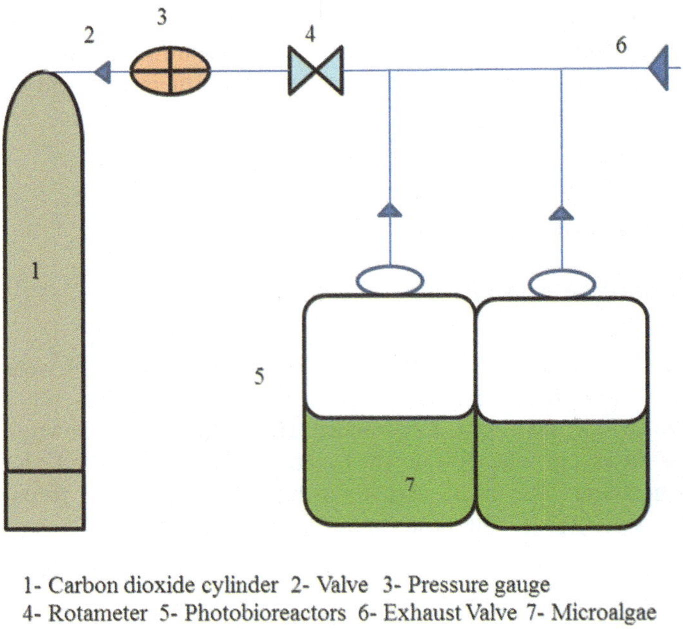

Fig. 1 Schematic diagram of CO_2 sequestration assembly

$P = (X_2 - X_1)/(t_2 - t_1)$, where X_2 and X_1 are the dry cell weight (DCW) concentration (mg/L) at time t_2 and t_1, respectively [25].

2.3.2 Measurement of Dissolved Carbon Dioxide

Measurement of dissolved CO_2 was done using 0.0227 N sodium hydroxide as titrant. 100 ml sample was titrated against 0.0227 N sodium hydroxide with phenolphthalein as an indicator. Carbon dioxide concentration was calculated according to the following equation. Concentration of CO_2 (mg/L) = mL of titrant*10.

2.3.3 Determination of Pigments

For determination of total pigment content, 99% methanol was used, and the concentration was determined spectrophotometrically. 2 ml of culture was taken in an Eppendorf tube and centrifuged at 5000 rpm for the duration of 10 min. Further, the supernatant was discarded and 2 ml of 99% methanol was added to the pellet, mixed well and incubated at 45 °C for 24 h in dark. The total pigment has been calculated according to the following equations [26].

$$\text{Chlorophyll a : Chl-a } (\mu g/ml) = 16.72\, A_{665.2} - 9.16\, A_{652.4}$$

$$\text{Chlorophyll b; Chl-b } (\mu g/ml) = 34.09\, A_{652.4} - 15.28\, A_{665.2}$$

$$\text{Carotenoids} = (1000\, A_{470} - 1.63 \text{ Chl-a} - 104.9 \text{ Chl-b})/221$$

2.3.4 Determination of Lipids

Lipid was extracted from the dry biomass by using Bligh and Dyer method [27]. The cells were harvested by centrifugation at 5000 rpm for 20 min, washed twice with distilled water and oven dried at 60 °C for 2 h. The chloroform and methanol mixture were added to 50 mg of dried biomass at 50 °C for 1 h. 50 mg dry biomass was first microwave digested with 6 ml distilled water and later mixed thoroughly with chloroform: methanol (1:2 v/v), centrifuged for 5 min at 3000 rpm, oven dried and measured gravimetrically.

3 Results and Discussion

3.1 Effect on Dry Cell Weight, Biomass Productivity and Growth

The OD of algae was measured to study the tendency of growth of microalgal biomass. OD was measured at 750 nm with UV–Visible spectrophotometer. The microalgal cultures of *C. pyrenoidosa* NCIM 2738 and *S. platensis* NCIM 5143 were cultivated at 33–40 °C temperature under the presence of direct sunlight.

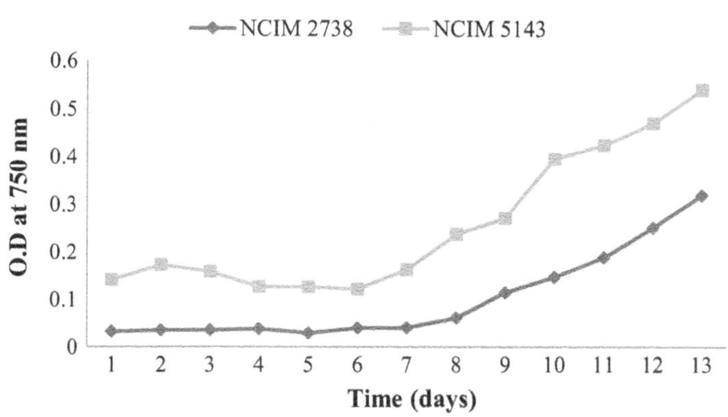

Fig. 2 Growth curve of *C. pyrenoidosa* NCIM 2738 and *S. platensis* NCIM 5143 (BG-11 medium under CO_2 supplementation)

Figure 2 gives the daily growth profile of the microalgae grown in BG 11 medium under CO_2 supplementation.

During the first 6 days of cultivation, no or little growth shows the lag phase of the microalgae with average growth rate of zero. Further, growth of microalgae was observed with increase in linear pattern. It is observed that as the incubation time increased the maximum growth was obtained on 13th day of cultivation for these microalgae with 415 and 440 mg/L DCW along with 31.90 and 33.80 mg/L/day biomass productivity of *C. pyrenoidosa* NCIM 2738 and *S. platensis* NCIM 5143, respectively. Thus, inferring that under optimum sunlight and CO_2 the photosynthetic rate increases which is observed from Fig. 2.

The biomass yield significantly improved once the microalgae adapted to the environmental conditions and achieved its log phase of growth increasing the overall biomass productivity. Further, from the observations, it can be seen that *C. pyrenoidosa* NCIM 2738 and *S. platensis* NCIM 5143 can tolerate high amount of CO_2, but its growth rate decreases due to the pH shift towards acidic condition.

3.2 Effect on pH Value

The microalgal growth is affected by the pH variations in the culture medium. They usually have an optimum pH around 7.00 but are also tolerant to low pH or high pH conditions. Figure 3 shows the relationship between the pH values of culture medium and the cultivation time for the present study. The study demonstrates the growth of *C. pyrenoidosa* NCIM 2738 and *S. platensis* NCIM 5143 for a period of 13 days with a pH range between 5.00 and 8.00. The microalgae grew well in pH range of 6.00–8.00. It has been noticed that the growth enhances when the pH value is slightly moving towards alkaline region. It is critical to maintain the pH of culture medium as the cell functioning processes may get disrupted or degraded due to extremes of pH leading to the loss of culture [28]. Usually, pH value depends on the dissolved CO_2 concentration which is present in the form of bicarbonates (HCO_3^-) in algal solutions due to the bubbling of CO_2 with a relation inferring the carbon formation in the medium. CO_2 firstly combines with water to form carbonic acid which dissociates into HCO_3^- and H^+. The HCO_3^- further dissociates into CO_2 or CO_3^{2-} depending on the pH of medium [20, 29]. CO_2 can enter into the microalgal cells through the active transport mechanism and later gets bio-fixed. Also, the feeding of CO_2 increases the pH value of the culture medium due to the formation of carbonates and bicarbonates [30]. During the process of photosynthesis, the microalgae uptake the CO_2 from water and the pH value is found gradually increasing. *C. vulgaris* optimal growth occurs in pH range of 7.50–8.00 and temperature range of 30–35 °C [31]. It can now be inferred that the algal biomass can be produced having good tolerance with low pH but not below 5.00.

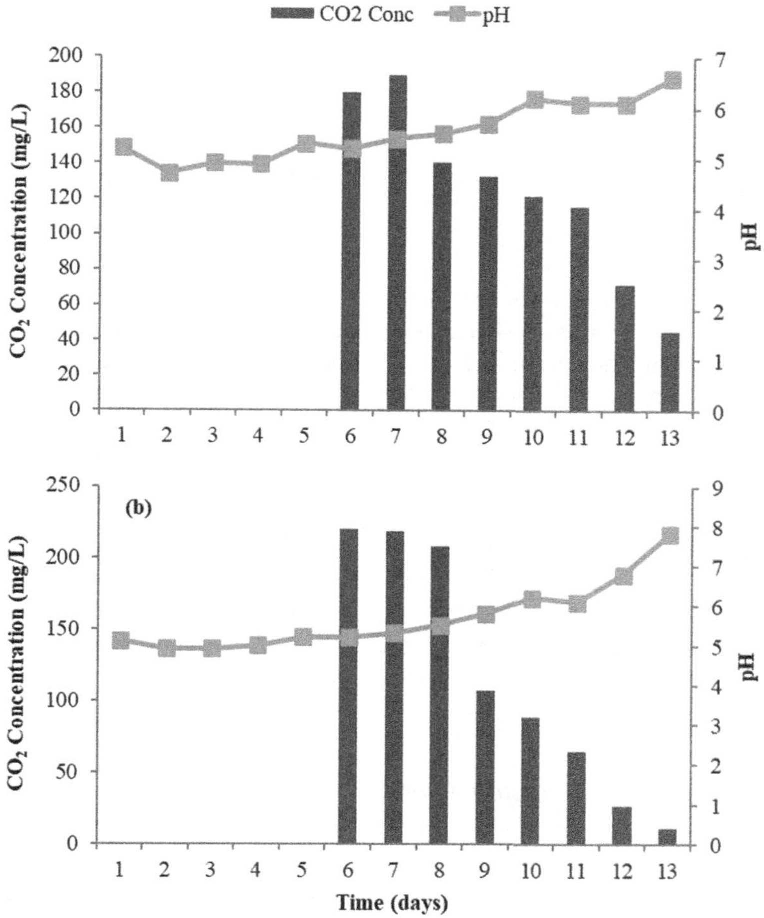

Fig. 3 Time course relationship between pH value and dissolved CO_2 concentration for **a** *C. pyrenoidosa* NCIM 2738 and **b** *S. platensis* NCIM 5143

3.3 CO_2 Bio-sequestration Profile

The concentration of dissolved CO_2 depends on the concentration of CO_2 at the inlet and the amount of time for the contact. During the course of study, CO_2 was supplemented to microalgae by bubbling it in the photobioreactor. After, a sufficient rise in pH value, the amount of dissolved CO_2 present in the photobioreactor was determined. Figure 3 shows the dissolved CO_2 profile for the microalgae *C. pyrenoidosa* NCIM 2738 and *S. platensis* NCIM 5143. The values of dissolved CO_2 were observed to be decreasing with increase in the biomass growth rate over the period of cultivation time. When the CO_2 is transferred from the gas phase into the culture medium, some of the CO_2 dissolved in water is converted in soluble

Table 1 Measurement of dissolved CO_2 concentration in the culture medium

Species	Days							
	6	7	8	9	10	11	12	13
Chlorella pyrenoidosa NCIM 2738 (mg/L)	180	190	141	133	122	116	72	45
Spirulina platensis NCIM 5143 (mg/L)	220	218	208	107	89	65	27	11

HCO_3^- phase which gives the pH value of the culture as shown in Fig. 3. This presence of organic carbon source in the culture medium shifts the growth of microalgae from photoautotropic to mixotropic as the microalgae cells prefer HCO_3^- uptake over CO_2 although the former is poor source carbon than later [32]. The possible explanation for this decrease in the dissolved CO_2 concentration is the photosynthetic mechanism of the microalgae cell. During the photosynthesis, microalgae utilizes the light and CO_2 for growth and organic photosynthetic production [33]. Therefore, the CO_2 in dissolved is utilized and consumed as a carbon source inferring the bio-sequestration [34]. Table 1 shows for the decrease in dissolved CO_2 concentrations from 6th day of cultivation. Further, it can be observed from Table 1 that *S. platensis* NCIM 5143 consumed the dissolved CO_2 at a much higher rate as compared to *C. pyrenoidosa* NCIM 2738 having a minimal average rate of CO_2 consumption of about 18.00 mg/L.

3.4 Pigment Composition Profile

The changes in the photosynthetic pigments were measured to study the effects of carbon dioxide on photosynthesis rate. Chlorophyll is a nitrogen-rich compound that supports the cell growth. CO_2 supplementation leads to increase in photosynthetic pigments, viz. Chl-a (Chlorophyll a), Chl-b (Chlorophyll b) and Caro (total carotenoid) content which further increases the biomass productivity. Further, the ratio of Chl a/b and Caro/Chl (a + b) was found decreased indicating the increase in photosystem II (PS II) activity and light harvesting complexes. Higher ratios of Chl a/b have been considered to decrease light collection in relation to the rate of PS II photochemistry [35], and carotenoids could have an antioxidative role in protecting unsaturated lipids in oil bodies from peroxidation serving as a protective function against oxidative stress-induced lipid production in microalgae, and could participate in the screening and trapping of excessive light otherwise absorbed by the chloroplast [36]. A similar pattern of pigment profile and photosynthetic activity was observed in experiments wherein the cultures were supplemented with CO_2 increasing the photosynthetic activity. Thus, the results indicate that chlorophyll is a nitrogenous compound and its composition was greatly influenced by CO_2 concentration in the medium as shown in Table 2.

Table 2 Effect of CO_2 supplementation of pigment composition of *C. pyrenoidosa* NCIM 2738 and *S. platensis* NCIM 5143

Microalgae	NCIM 2738	NCIM 5143
Chl-a (μg/ml)[a]	2.6015240	6.9244480
Chl-b (μg/ml)[b]	1.3080490	1.3361480
Caro (μg/ml)[c]	0.5499782	1.7866571
Chl (a + b) (μg/ml)	3.9095730	8.2605960
Chl (a/b)	1.9888582	5.1823960
Caro/Chl (a + b)	0.1406747	0.2162867

[a]Chl-a: Chlorophyll-a
[b]Chl-b: Chlorophyll-b
[c]Caro: Carotenids

3.5 *Effect of CO_2 on Biochemical Composition of Microalgae*

For economical production of biofuel from microalgae, biomass as well as lipid content plays an important role. As CO_2 is one of the most important components for photosynthesis, increasing its concentration in growth medium and making its availability has enhanced the biomass and lipid production. Lipid was extracted by Bligh dyer method with few modifications. The algal cells were disrupted prior to lipid extraction by Microwave Digester (Titan MPS, Perkin Elmer) for maximum lipid extraction. In the present study, under CO_2 supplementation *C. pyrenoidosa* NCIM 2738 and *S. platensis* NCIM 5143 had 21.80 and 45.20 % of lipid content, respectively. Thus, the amount of lipid content for these microalgae can be enhanced by further studies regarding the biomass growth rate and appropriate amount of CO_2 supplementation which can serve to be a potential feedstock for sustainable biofuel production.

4 Conclusion

The study concludes that the microalgae *C. pyrenoidosa* NCIM 2738 and *S. platensis* NCIM 5143 are robust microorganisms that can tolerate high concentration of CO2 with a range of pH 5.00 to ~8.00 at temperatures between 33.00 and 40.00 °C while giving rise to higher biomass productivity thorough photosynthesis. The biomass productivity was calculated as 31.90 mg/L/day for *C. pyrenoidosa* NCIM 2738 and 33.8 mg/L/day for *S. platensis* NCIM 5143. The CO_2 sequestration process can bring the rapid and normal growth in the microalgae that can be achieved at large-scale production subject to the harvesting technology. Moreover, 21.80 and 45.20% lipid was found in these microalgae, respectively, with average CO_2 consumption rate of 18.00 mg/L. Finally, studies can be conducted to optimize the process by implementing suitable techniques to have the enhanced cost-effective approach for sustainable biofuel production while reducing the emissions made from the burning of fossil fuels.

Acknowledgements The authors gratefully acknowledge the support of the authorities of Climate Change Department, Government of Gujarat, for providing the financial support as part of this study.

References

1. Brennan L, Owende P (2010) Biofuels from microalgae—a review of technologies for production, processing, and extractions of biofuels and co-products. Renew Sustain Energy Rev 14:557–577. https://doi.org/10.1016/j.rser.2009.10.009
2. Ho S-H, Chen C-Y, Lee D-J, Chang J-S (2011) Perspectives on microalgal CO_2-emission mitigation systems—a review. Biotechnol Adv 29:189–198. https://doi.org/10.1016/j.biotechadv.2010.11.001
3. Kumar A, Ergas S, Yuan X, Sahu A, Zhang Q, Dewulf J, Malcata FX, van Langenhove H (2010) Enhanced CO_2 fixation and biofuel production via microalgae: recent developments and future directions. Trends Biotechnol 28:371–380. https://doi.org/10.1016/j.tibtech.2010.04.004
4. Hussain F, Shah SZ, Zhou W, Iqbal M (2017) Microalgae screening under CO_2 stress: growth and micro-nutrients removal efficiency. J Photochem Photobiol B 170:91–98. https://doi.org/10.1016/j.jphotobiol.2017.03.021
5. Tsai DD-W, Ramaraj R, Chen PH (2016) Carbon dioxide bio-fixation by algae of high rate pond on natural water medium. Ecol Eng 92:106–110. https://doi.org/10.1016/j.ecoleng.2016.03.021
6. Tsai DD-W, Chen PH, Chou CM-J, Hsu C-F, Ramaraj R (2015) Carbon sequestration by alga ecosystems. Ecol Eng 84:386–389. https://doi.org/10.1016/j.ecoleng.2015.09.024
7. Abinandan S, Shanthakumar S (2016) Evaluation of photosynthetic efficacy and CO_2 removal of microalgae grown in an enriched bicarbonate medium. 3 Biotech 6:9. https://doi.org/10.1007/s13205-015-0314-5
8. Majolagbe A, Oketola A, Osibanjo O (2016) Vulnerability assessment of groundwater pollution in the vicinity of an active dumpsite (Olusosun), Lagos, Nigeria. Chem Int 2:232–241
9. Majolagbe A, Oketola A, Osibanjo O, Adams A, Ojuri O (2017) Pollution vulnerability and health risk assessment of groundwater around an engineering landfill in Lagos, Nigeria. Chem Int 3:58–68
10. Pires JCM, Martins FG, Alvim-Ferraz MCM, Simões M (2011) Recent developments on carbon capture and storage: an overview. Chem Eng Res Des 89:1446–1460. https://doi.org/10.1016/j.cherd.2011.01.028
11. de Morais MG, Costa JAV (2007) Biofixation of carbon dioxide by Spirulina sp. and Scenedesmus obliquus cultivated in a three-stage serial tubular photobioreactor. J Biotechnol 129:439–445. https://doi.org/10.1016/j.jbiotec.2007.01.009
12. Sydney EB, Sturm W, de Carvalho JC, Thomaz-Soccol V, Larroche C, Pandey A, Soccol CR (2010) Potential carbon dioxide fixation by industrially important microalgae. Biores Technol 101:5892–5896. https://doi.org/10.1016/j.biortech.2010.02.088
13. Wang B, Li Y, Wu N, Lan CQ (2008) CO_2 bio-mitigation using microalgae. Appl Microbiol Biotechnol 79:707–718. https://doi.org/10.1007/s00253-008-1518-y
14. Kürsten E, Burschel P (1993) CO_2-mitigation by agroforestry. In: Wisniewski J, Sampson RN (eds) Terrestrial biospheric carbon fluxes quantification of sinks and sources of CO_2. Springer, Netherlands, Dordrecht, pp 533–544

15. Mondal M, Goswami S, Ghosh A, Oinam G, Tiwari ON, Das P, Gayen K, Mandal MK, Halder GN (2017) Production of biodiesel from microalgae through biological carbon capture: a review. 3 Biotech 7:99. https://doi.org/10.1007/s13205-017-0727-4
16. Jalalizadeh M (2012) Development of an integrated process model for algae growth in a photobioreactor
17. Tredici MR (2007) Mass production of microalgae: photobioreactorshandbook of microalgal culture. Blackwell Publishing Ltd, pp 178–214
18. Chi Z, O'Fallon JV, Chen S (2011) Bicarbonate produced from carbon capture for algae culture. Trends Biotechnol 29:537–541. https://doi.org/10.1016/j.tibtech.2011.06.006
19. Chinnasamy S, Ramakrishnan B, Bhatnagar A, Das KC (2009) Biomass production potential of a wastewater alga Chlorella vulgaris ARC 1 under elevated levels of CO_2 and temperature. Int J Mol Sci 10:518–532. https://doi.org/10.3390/ijms10020518
20. de Morais MG, Costa JAV (2007) Isolation and selection of microalgae from coal fired thermoelectric power plant for biofixation of carbon dioxide. Energy Convers Manag 48:2169–2173. https://doi.org/10.1016/j.enconman.2006.12.011
21. Ramanan R, Kannan K, Deshkar A, Yadav R, Chakrabarti T (2010) Enhanced algal CO_2 sequestration through calcite deposition by Chlorella sp. and Spirulina platensis in a mini-raceway pond. Biores Technol 101:2616–2622. https://doi.org/10.1016/j.biortech.2009.10.061
22. Rani K, Sandal DN, Sahoo PSP (2018) A comprehensive review on chlorella-its composition, health benefits, market and regulatory scenario
23. Benelhadj S, Gharsallaoui A, Degraeve P, Attia H, Ghorbel D (2016) Effect of pH on the functional properties of Arthrospira (Spirulina) platensis protein isolate. Food Chem 194:1056–1063. https://doi.org/10.1016/j.foodchem.2015.08.133
24. Rippka R, Deruelles J, Waterbury JB, Herdman M, Stanier RY (1979) Generic assignments, strain histories and properties of pure cultures of cyanobacteria. Microbiology 111:1–61. https://doi.org/10.1099/00221287-111-1-1
25. Pancha I, Chokshi K, George B, Ghosh T, Paliwal C, Maurya R, Mishra S (2014) Nitrogen stress triggered biochemical and morphological changes in the microalgae Scenedesmus sp. CCNM 1077. Biores Technol 156:146–154. https://doi.org/10.1016/j.biortech.2014.01.025
26. Lichtenthaler HK (1987) Chlorophylls and carotenoids: pigments of photosynthetic biomembranes. In: Methods in enzymology. Academic Press, pp 350–382
27. Bligh EG, Dyer WJ (1959) A rapid method of total lipid extraction and purification. Can J Biochem Physiol 37. https://doi.org/10.1139/o59-099
28. Razzak SA, Ilyas M, Ali SAM, Hossain MM (2015) Effects of CO2 concentration and pH on mixotrophic growth of Nannochloropsis oculata. Appl Biochem Biotechnol 176:1290–1302. https://doi.org/10.1007/s12010-015-1646-7
29. Zhao B, Zhang Y, Xiong K, Zhang Z, Hao X, Liu T (2011) Effect of cultivation mode on microalgal growth and CO_2 fixation. Chem Eng Res Des 89:1758–1762. https://doi.org/10.1016/j.cherd.2011.02.018
30. Miller AG, Espie GS, Canvin DT (1990) Physiological aspects of CO2 and HCO3-transport by cyanobacteria: a review. Can J Bot 68:1291–1302. https://doi.org/10.1139/b90-165
31. Rachlin JW, Grosso A (1991) The effects of pH on the growth of Chlorella vulgaris and its interactions with cadmium toxicity. Arch Environ Contam Toxicol 20:505–508. https://doi.org/10.1007/BF01065839
32. Carvalho AP, Meireles LA, Malcata FX (2006) Microalgal reactors: a review of enclosed system designs and performances. Biotechnol Prog 22:1490–1506. https://doi.org/10.1021/bp060065r
33. Chojnacka K, Noworyta A (2004) Evaluation of Spirulina sp. growth in photoautotrophic, heterotrophic and mixotrophic cultures. Enzyme Microb Technol 34:461–465. https://doi.org/10.1016/j.enzmictec.2003.12.002

34. Moraes L, Rosa GMD, Cardias BB, Santos LOD, Costa JAV (2016) Microalgal biotechnology for greenhouse gas control: carbon dioxide fixation by Spirulina sp. at different diffusers. Ecol Eng 91:426–431. https://doi.org/10.1016/j.ecoleng.2016.02.035
35. Demmig-Adams B, Adams WI (1996) Chlorophyll and carotenoid composition in leaves of *Euonymus kiautschovicus* acclimated to different degrees of light stress in the field. Funct Plant Biol 23:649–659. https://doi.org/10.1071/PP9960649
36. Edge R, McGarvey DJ, Truscott TG (1997) The carotenoids as anti-oxidants—a review. J Photochem Photobiol B 41:189–200. https://doi.org/10.1016/S1011-1344(97)00092-4

Role of Advance Carbon Materials in the New Paradigm of Energy and Environment

Harender Bisht, Nibedita Sanyal, Sukumar Mandal, and Asit K. Das

1 Introduction

Global warming and deterioration of air quality are some of the consequences of indiscriminate exploitation of natural resources. To undo the environmental degradation and improve quality of life, a paradigm shift has already initiated in the way we use energy and resources today. Fossil fuels, being one of the major sources of energy today, may not enjoy the same status in near future with emission norms being tightened and tax being imposed on CO_2 emission in different parts of the world. Thus, greater use of renewable and greener energy such as solar, wind, hydrothermal, geothermal, etc., is becoming competitive and acceptable in terms of cost, efficiency and environmental impact. In particular, a major shift is already happening in the area of mobility and transportation where conventional fuels are being challenged to be replaced by different forms of renewable energies. Internal combustion engine (ICE) technology has found a new competitor in terms of electric vehicle (EV) and is trying to improve itself to survive the new paradigm shift. Significant improvement in solar photovoltaic (PV) efficiency and reduction in cost of production of solar cells have made solar power more and more competitive to conventional coal-based power. Advanced carbon materials can play a

H. Bisht · N. Sanyal · S. Mandal · A. K. Das (✉)
Refining R&D, Reliance Industries Limited, Jamnagar, Gujarat, India
e-mail: asit.das@ril.com

H. Bisht
e-mail: Harender.Bisht@ril.com

N. Sanyal
e-mail: Nibedita.Sanyal@ril.com

S. Mandal
e-mail: sukumar.k.mandal@ril.com

U. K. Bhui (ed.), *Macromolecular Characterization of Hydrocarbons for Sustainable Future*, Green Energy and Technology,
https://doi.org/10.1007/978-981-33-6133-1_19

decisive role in this new paradigm of energy and environment in both improving existing technologies and facilitating emergence new technologies.

1.1 India's Energy Sector

By 2040, India's energy demand is expected to rise 2.7–3.2 times over 2012 levels. By 2022, clean and renewable electricity would account for more than 15% of total electricity produced in India.

However, in spite of increase in renewable energy, the share of coal, oil and natural gas may remain the major source of energy (~78%) in near future [1].

By 2040, improvement in existing technologies and development of new technologies in transport, industry and construction sectors can cut energy demand by ~17%. Although a distant dream, however, if most Indian vehicles were electric by 2030, pollution levels in cities could drop by >80%, and India could save $100 billion, a sum over two times larger than the current defence budget on account of reduction in import of crude oil. Indian National Energy Policy [1] focuses on four major objectives:

(i) Affordable energy access for all.
(ii) Reducing dependence on fossil fuel imports.
(iii) Low carbon economy through growth in renewable energy.
(iv) Sustaining economic growth.

1.2 Hydrocarbon Sector in India

At present, India has 23 refineries (Fig. 1) having combined refining capacity of around 248 MMTPA and imports >85% of crude oil worth ~100 billion$/year [2]. India is the 2nd largest importer of LPG amounting to 12 MMTPA. India's natural gas production is only ~26 MMTPA, whereas consumption is 78 MMTPA. India is highly dependent on import for both crude oil and LNG. Indian refining industry is growing at a CAGR of 3.5% [1]. This trend is expected to continue in future although electric vehicle will compete with internal combustion engine for the preferred mode of transport due to cleaner energy footprint. It is expected that demand of refining capacity will reach 360 MMTPA by 2030. Even in 2040, when the total number of passenger cars in India is expected to grow from existing ~30 to ~180 million, about 70% vehicles would be running on conventional fossil fuels [3]. Considering the significant increase in number of vehicles based on conventional technology, new refineries will come into existence. There is a plan to build mega-refinery at western coast by joint venture of public sector oil companies.

As per experts [3], global demand of transport fuel may peak in 2035. Gasoline and diesel will continue to grow in developing countries including India up to 2040,

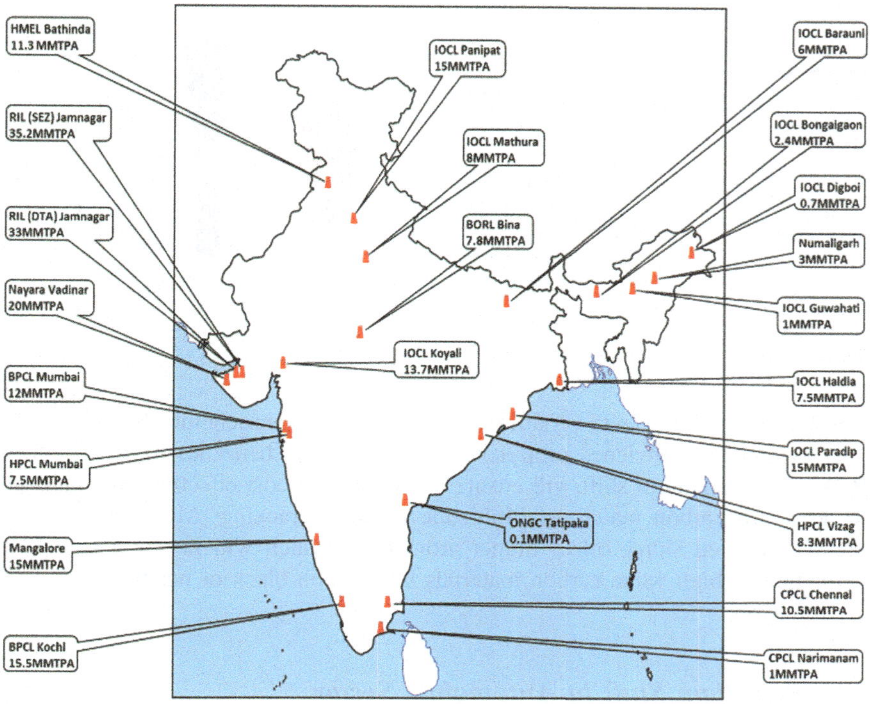

Fig. 1 Oil refineries in India

while the same will decline in developed countries after 2030. Quality of fossil fuels will improve in developing countries to meet high standards of emission and efficiency. LPG demand is expected to grow at CAGR of 3.5% globally, and it has potential to address some of the concerns of internal combustion engine (ICE). LPG may provide economical and environmentally attractive option if promoted as automobile fuels.

Petrochemical demand is expected to grow very rapidly in Asia, which will take care of lower gasoline and naphtha demand. Jet fuel demand is expected to grow at CAGR of 5% in developing countries which may provide some respite to falling demand for diesel fuel. Low sulphur specification (<0.5 wt%) of marine fuel will promote hydrodesulphurization (HDS) of fuel oil and blending of heavier diesel to fuel oil (FO). High sulphur fuel oil (HSFO) and hydrocarbon residues are less likely to find buyers and will be used for power production in the refinery with flue gas desulphurization (FGD), to cater high demand of power with controlled pollution.

Future refineries will transform to produce higher yields of petrochemicals with better water, energy and environment management. Refineries will adapt to digital technologies and artificial intelligence. Processes which directly convert crude oil into chemicals will be favoured, e.g. multizone catalytic cracking (MCC) of Reliance Industries Limited [4] over conventional technologies (Fig. 2).

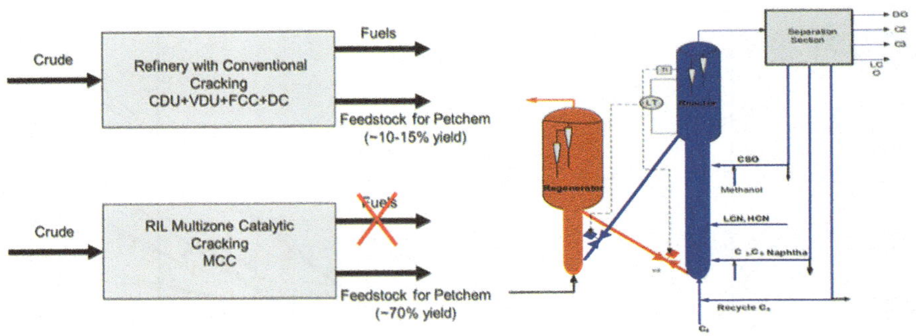

Fig. 2 Crude to chemicals: MCC process of Reliance Industries Limited

Refineries of future will produce higher amount of petrochemicals such as olefinic monomers (ethylene, propylene, butylene) and benzene, toluene, xylene (BTX). This paradigm shift will ensure availability of cost-effective raw materials for advanced carbon materials. Multizone catalytic cracking (MCC) process can produce clarified slurry oil of higher aromaticity which will be more suited for production of high-value carbon materials like carbon black or needle coke.

1.3 Paradigm Shift in Automobile Sector

A question comes to mind that why mobility technology will change so drastically and go beyond internal combustion engine (ICE) in future? This is mainly due to very low engine efficiency (<25% on road) of gasoline ICEs which results in high CO and CO_2 emission per km than diesel engines ($\sim$35%). Diesel ICEs on the other hand produce more NOx and particulate matter (PM) [5]. Serious research is on to improve diesel engines and meet stringent emission norms set by environmental agencies. Some of the major technologies in this area are diesel oxidation catalyst (DOC), selective catalytic reduction (SCR), diesel particulate filter (DPF), etc. They reduce NOx and particulate emission; however, installation of such devices has penalty on engine performance. Hence, the benefits of lower CO and CO_2 emission are compromised on the cost of reducing NOx and PM [5].

Diesel ICEs although more efficient than gasoline engines are still far less efficient than electric motors which operate at $\sim$90% efficiency. Therefore, future of mobility technology is expecting a paradigm shift which is summarized in Fig. 3 [6].

After 2030, the global vehicle fleet is expected to witness major phase change with substantial plug-in hybrid electric vehicles (PHEVs), battery electric vehicles (BEVs) and a few fuel cell vehicles (FCVs) coming into existence.

In the near foreseeable future, number of vehicles based on both existing internal combustion engines and emerging electric motor technology will grow. Advanced carbon martials with improved qualities and lower weight will be in high demand for both these technologies to make them more energy efficient and safer.

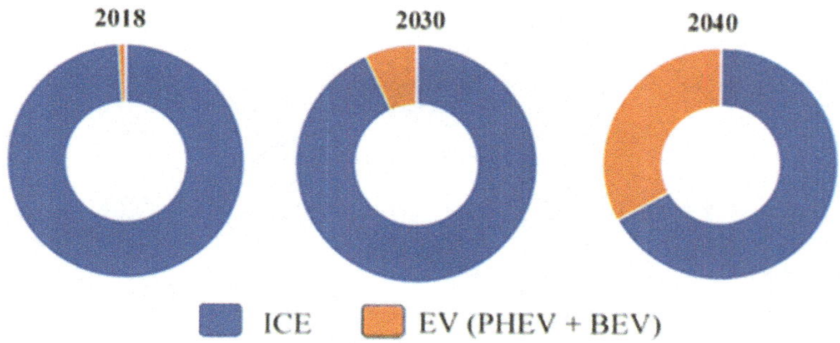

Fig. 3 Breakdown of global light duty vehicle fleet

2 Advance Carbon Materials

Higher energy efficiency, lower CO_2 emission and safety are few of the critical parameters that will govern the design of future technologies either using fossil fuels or electric energy. Carbon-based advanced materials will play a major role in attaining these prerequisites. Advance carbon materials, e.g. specialty petroleum coke, carbon black, carbon fibre, carbon foam, carbon nanotubes, graphene, etc., offer excellent thermal, mechanical, electrical, electronic and chemical properties with much lower density than metal-based conventional materials. Thus, there is great opportunity for replacement of conventional materials with high-quality advanced carbon materials for various new applications in the new paradigm of energy and environment.

For chemical industry, graphene and carbon nanotubes offer advanced material with very high surface area and chemical inertness which are desirable for catalyst supports in many chemical reactions. Also, these materials have great potential to improve the efficiency of solar PV cells. Advanced carbon composites are already being used as the material of choice in wind turbine blades. In Fig. 4, we have tried to classify advanced carbon materials into wide range of applications based on their physico-chemical properties.

Carbon has extraordinary capability to exhibit highly tunable thermal, mechanical, electrical and electronic properties. Carbon material such as graphite exhibits very high thermal and electrical conductivity with very low coefficient of thermal expansion. Similarly, carbon nanotubes show tunable electronic properties ranging from highly conducting to semiconducting to insulating.

In the following section, we have tried to summarize some of the advanced carbon materials with chronology of their development. Calcined petroleum coke (CPC) was developed in the 1920s with development of delayed coking process [7]. The early 1960s witnessed development of Rayon-based carbon fibre [8] and polyacrylonitrile (PAN)-based carbon fibre [9]. Mesophase pitch developed in the late 1960s [10, 11] provided alternate raw material for carbon fibre and other high-value carbon materials. Although multiwall CNT was observed in the 1950s

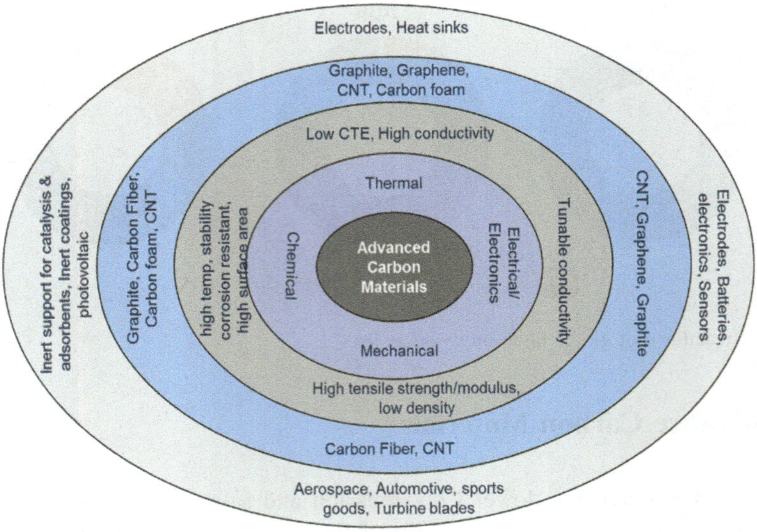

Fig. 4 Atlas of advanced carbon materials

by Roger Bacon of Union Carbide, it got major interest after publication of Iijima [12]. Single-wall carbon nanotubes were also reported by Iijima and Ichihashi [13]. Graphene was physically produced and identified in 2004 by Novoselov et al. [14]. All these discoveries helped to realize the potential of advanced carbon materials for highly demanding applications.

2.1 *Petroleum Coke*

2.1.1 Raw/Green Petroleum Coke

Petroleum coke produced in refinery is called raw pet coke (RPC) or green pet coke (GPC). Three types of petroleum coke are produced in the delayed coking unit of refineries, namely

(i) Fuel grade coke (shot coke).
(ii) Anode grade coke (sponge coke).
(iii) Needle grade coke.

Fuel grade petroleum coke is produced from vacuum residue having very high amount of asphaltenes, metals and sulphur. It has very small crystalline regions called '*mosaics*' ranging in size from 1 to 10 μm. More than 80% of total petroleum coke produced in refineries is shot coke having high amount of sulphur and metal impurities and is generally used without any further processing as fuel in power and cement industry. Green anode grade coke is produced from petroleum residues having moderate amount of asphaltenes, metals and sulphur. Anode coke has crystalline regions of 10–50 μm called '*domains*' with lot of sponge-like pores, whereas needle coke is produced from petroleum residues like clarified slurry oils or coal tar distillates having very low amount of asphaltenes, metals and sulphur. Needle coke has long crystalline regions called '*flow domains*' which can range from few tens of microns to hundreds of microns in length and 10–50 μm in width.

Table 1 shows comparison of typical properties of raw/green fuel grade, anode grade and needle grade petroleum coke as produced in delayed coker before calcination.

2.1.2 Calcined Petroleum Coke (CPC)

When the sulphur and metal impurities are below 3.5 wt% and 500 ppm, respectively, raw petroleum coke (RPC) can be calcined at 1350 °C to produce calcined petroleum coke (CPC) [15]. Figure 5 shows the naked eye and optical microscope views of anode coke. CPC is used for making carbon anode for aluminium industry; therefore, this type of coke is also called anode coke.

Carbon anode is used to reduce Al_2O_3 to metallic aluminium, and in this process, carbon is oxidized to CO_2. About 0.4 kg CPC is consumed for production of 1 kg aluminium.

$$2AI_2O_3 + 3C \rightarrow 4AI + 3CO_2$$

Global production of raw anode coke is ~35MMTPA; during calcination, most of the volatile components present in coke get removed and global production of calcined pet coke (CPC) remains around 28MMTPA. China is the major producer of calcined anode coke and covers about 50% global market. USA has about 17% market share. Indian demand for calcined coke is about 1.5MMTPA, and most of this demand is met by import. Approximately 0.9MMTPA CPC was imported in 2016.

Table 1 Typical properties of a different raw/green petroleum coke

Petroleum Coke → Properties↓	Green fuel grade pet coke	Green anode grade pet coke	Green needle grade pet coke
C, wt%	85–90	90–95	95–98
H, wt%	3–4	2–3	1–2
S, wt%	4–8	0.5–3.5	0.1–0.5
Real density (RD), g/cc	1.3–1.5	1.3–1.5	1.4–1.5
Hardgrove grindability index (HGI)	35–65	65–95	55–85
ASH, wt%	0.3–1.0	0.1–0.5	0.05–0.3

Fig. 5 **a** Anode coke by naked eye and **b** cross-polarized optical image

Major aluminium producers in India are Hindalco, Sterlite Industries (Balco and Malco) and Nalco. Indian CAGR of aluminium is 9% which is higher than world CARG of 5%. Rain calcining is the largest calciner of anode coke in India. Since CPC is consumable in aluminium production, its demand is linked with growth of aluminium industry. Calcined anode coke is also used in TiO_2, steel, metallurgical, chemicals and other carbon consuming industries. Table 2 shows the typical specifications of anode grade petroleum coke.

Commercial process for production of raw or green anode coke is similar to conventional delayed coking process used to make fuel grade coke where the heavy hydrocarbon residue is heated in a furnace to 480–500 °C and routed to a coke drum where thermal cracking takes place. Depending upon the Conradson carbon residue (CCR) of feed, about 25–35% petroleum coke is produced along with coker distillates comprising gaseous and liquid products. The yield and quality of coke depend upon the amount of impurities, e.g. sulphur, metals and asphaltenes present

Table 2 Typical properties of calcined anode grade coke (CPC)

Specification	Calcined anode coke
Sulphur, wt%	0.5–2.5
Ash, wt%	<0.4
Volatile matter, wt%	<0.25
Nickel, ppm	<200
Vanadium, ppm	<300
Silicon, ppm	<200
Iron, ppm	<250
Calcium, ppm	<200
Resistivity, μΩ.m	950
Real density, g/cc	2.06
Bulk density, g/cc	0.8
Coefficient of thermal expansion (CTE)/°C	2×10^{-6}

in coker feed. For anode grade coke production, the feed should have significantly lower amount of these impurities.

The change in marine fuel oil sulphur specification from existing 3.5 to 0.5 wt% after 2020 as directed by International Maritime Organization (IMO) is going to create demand and price hike for low-sulphur fuel oil. Those refineries which are using such feed for making anode grade coke will prefer to sell it as low-sulphur fuel oil.

2.1.3 Calcined Needle Coke (CNC)

Needle coke is the most valuable petroleum coke and is primarily used for making graphite electrode for electric arc furnace (EAF) in steel industry. Steel manufacturers are switching to EAF from conventional blast furnace to process higher amount of scrap iron. Global needle coke demand is about 1.5 MMTPA. Indian demand of needle coke is approximately 200 KTPA which is entirely met by import.

The production of needle grade petroleum coke is also similar to that of anode coke production. The feed for making needle coke is either coal tar distillate or clarified slurry oil (CSO) produced in fluidized catalytic cracking (FCC) unit of refinery. The temperature and pressure for production of needle coke are optimized based on the quality of feedstock. Generally, much higher coke drum pressure is maintained for needle coke production in comparison with fuel grade or anode grade coke [16]. Figure 6 shows the naked eye and optical microscope views of needle coke.

World CAGR of needle coke is ~3.9%, whereas China has shown CAGR of 17%. China is one of the major producers of needle coke. Presently, most of the needle coke is consumed for making graphite electrodes for electric arc furnaces.

In the new paradigm of energy and environment, electric vehicle and solar energy industry are expected to grow very fast. The growth of batteries for energy storage in both transportation and stationery power sector will also increase

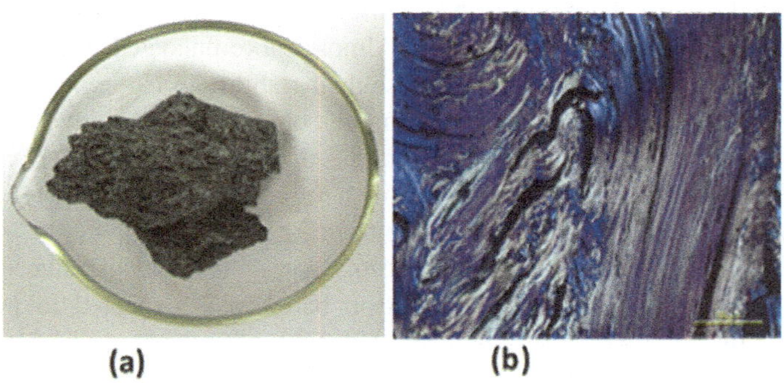

Fig. 6 **a** Needle coke by naked eye and **b** cross-polarized optical image

Table 3 Typical specifications of super-premium and premium needle coke

Target spec.	Typical values	
Grade→	Super-premium	Premium
Sulphur, wt%	<0.50	<0.55
Coefficient of thermal expansion (CTE), 10^{-6}/°C	<0.8	<1.2
Real density (RD), g/cc	>2.13	>2.12
Dry ash, wt%	<0.3	<0.3
Moisture, wt%	<0.1	<0.1
Volatile combustible matter (VCM), wt%	<0.3	<0.3
R6 (>6.7 mm), wt%	>50	>50
Vibrated bulk density (VBD), g/cc	>0.75	>0.75

accordingly. Graphite is one of the major raw materials for production of lithium-ion batteries where it is used as porous anode. Therefore, demand of synthetic graphite is also expected to increase in future. In the new scenario, needle coke will also find its application in production of high-purity synthetic graphite for making anode of lithium-ion batteries. Table 3 shows the typical specifications of needle grade petroleum coke.

2.2 *Spherical Graphite for Lithium-Ion Battery*

Lithium-ion batteries use very high-purity graphite having total graphitic content (TGC) of >99.9% for making anodes. Due to lower cost, currently, natural graphite is used for this application. Graphite flakes are classified into various sizes. Generally, flakes having 100–150 μm size are used to make uncoated spherical graphite (uSPG) for battery application. The spherical particles are made by proprietary techniques of micronizing and rounding [17, 18]. Uncoated spherical graphite particles should have graphitic content of >94% and size of approximately 40 μm. HF/H_2SO_4 is used as purification agents to increase the graphitic content to >99.95% and size in the range of 5–20 μm. Yield of uncoated spherical graphite particles from graphite flakes is ~40%. Almost entire uncoated spherical graphite (~90%) is produced in China. To increase their stability in battery, these uncoated spherical graphite particles are coated with pitch or asphalt to provide a hard surface coating. After coating, these spherical particles are called coated spherical graphite (cSPG). Hard carbon shell created by pitch or asphalt protects exfoliation and inhibits reaction of electrolyte with graphite in Li-ion batteries.

Typically, 1kWh battery requires 2 kg of coated spherical graphite or 5 kg of graphite flakes. The demand of lithium-ion batteries is expected to increase to >100 GWh by 2021 which will require 250KTPA of coated spherical graphite or ~600KTPA of graphite. In future scenario of high graphite demand, needle coke is expected to also become competitive to produce high-purity synthetic graphite for

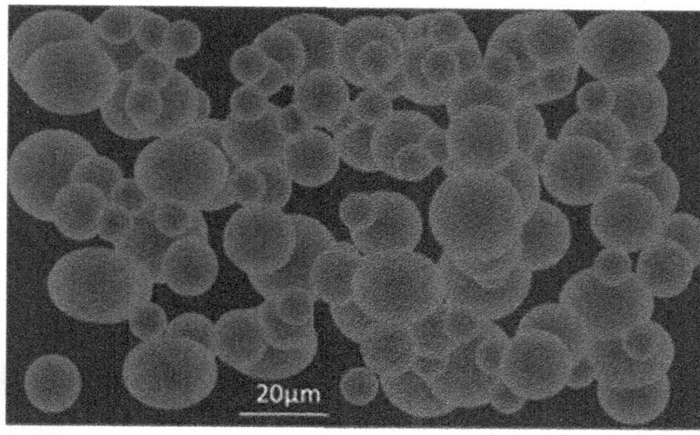

Fig. 7 Coated spherical graphite for anode of Li-ion battery (artistic depiction)

battery applications. High-purity graphite is also used in nuclear power plants and aerospace industry, but the volume is very small and quality standards are extremely stringent. Figure 7 depicts the image of spherical graphite. To further improve the lithiation capacity of lithium-ion battery anode, composite of spherical graphite with Si, graphene oxide and carbon are also being explored [19].

2.3 *Carbon Black*

Carbon black is carbonaceous material having short- and medium-range crystalline order. Its surface area is higher than carbon shoot but lower than activated carbon. Carbon black is mainly used as reinforcing filler in tyre and rubber industry. It is also used as pigment in plastics, paints and inks. Worldwide demand of carbon black in 2016 was about 12MMT and is expected to grow to 15MMT by 2021. Indian carbon black demand was 0.8MMT in 2016 which is expected to grow to 1MMT by 2021 [20]. Table 4 shows the typical properties of carbon black.

Figure 8 shows majority of carbon black is used in tyre and rubber industry [20]. Other industries such as pigment, paint and ink although consume much small amount of carbon black, they fetch higher value due to specialized application.

Table 4 Typical physico-chemical properties of carbon black

Parameter	Typical value
Carbon, wt%	96–99.5
Hydrogen, wt%	0.2–1.3
Oxygen, wt%	0.2–0.5
Nitrogen, wt%	0–0.7
Sulphur, wt%	0.1–1.0
Residual ash, wt%	<1
Density, g/cc	1.7–1.9
Particle size, nm	10–80

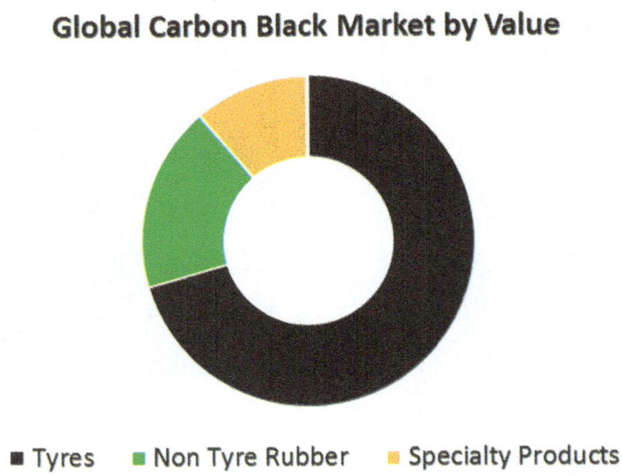

Fig. 8 Distribution of carbon black in various industries

Carbon black is produced either by thermal oxidative decomposition (TOD) or by thermal decomposition (TD) processes. Acetylene black process is a thermal decomposition process which uses acetylene gas as feed to produce carbon black with higher crystallinity. Most of the carbon black is produced by furnace black process (Fig. 9) which is a thermal oxidative decomposition process [21] where carbon black feedstock (CBFS) having high concentration of polyaromatic compounds is partially oxidized to make carbon black particles. Other thermal oxidative decomposition processes are channel process and lampblack process. High value (>135) of Bureau of mines correlation index (BMCI), a measure of aromaticity of carbon black feedstock, is critical for making good-quality carbon black in high yield. Reliance Industries Limited has developed a process to improve BMCI of clarified slurry oil by selective extraction of less polar components by paraffinic solvents [22].

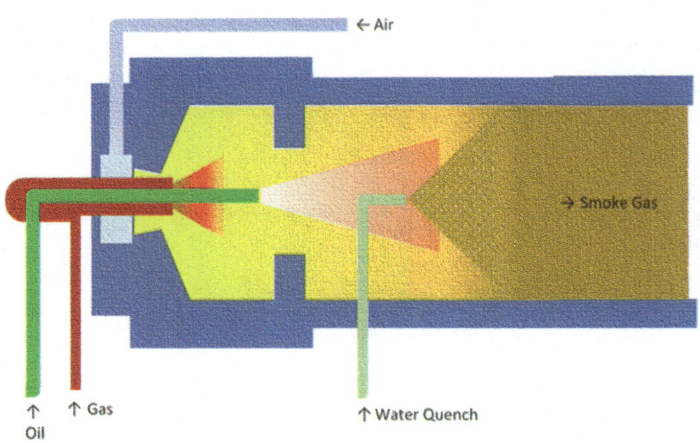

Fig. 9 Furnace black process

Due to cheap availability in China, coal tar distillate was major raw material for carbon black during 2004–2016. However, environmental concerns have forced to shut down several coal tar production facilities in China. Therefore, clarified slurry oil (CSO) has again become the choice of raw material for carbon black production.

No matter which technology (ICE or EV) emerges as winner in future, tyre demand is bound to increase; therefore, carbon black demand is also expected to grow in the new paradigm of energy and environment. However, carbon black manufacturing process has to improve to minimize the emission of pollutants, e.g. CO_2, SO_2, NOx and carbon shoot. Plasma process along with solar energy has been reported to produce carbon black with much lower CO_2 emission [23]. Another threat to carbon black is development of superior quality silica-based filler materials.

2.4 Carbon Fibre

Graphitic carbon arranged in the form of fibres of 5–10-μm diameter is called carbon fibre. Carbon fibres show very high stiffness, tensile strength, chemical resistance, temperature tolerance; in addition, they also exhibit very low density and thermal expansion. Combination of these properties makes carbon fibre a superior replacement of metals in many demanding applications. Being a lightweight material, the environmental impact of carbon fibre was evaluated using life cycle (from digging of material to scrapping of carbon fibre product) assessment method by Toray Industries [24]. They have estimated that a car made of 30% lighter using carbon fibre will emit 50 tons lower CO_2 for every 1 ton of carbon fibre over a life cycle of 10 years. Similarly, when the fuselage structure of aircraft is made of 20% lighter using carbon fibre, 1400 tons of CO_2 will be reduced under the same condition.

Table 5 shows the demand and supply estimates of carbon fibre. As shown in this table, most of the time the actual supply of carbon fibre is much lower than nameplate capacity due to lower demand. This is mainly due to higher cost of carbon fibre, which is uneconomical for low-value bulk applications.

Some of the important applications of carbon fibre in the area of alternative energy are (i) wind turbine blades, (ii) compressed natural gas storage,

Table 5 Demand and supply estimates of carbon fibre [25]

Year	Carbon fibre demand, metric ton	Carbon fibre supply (nameplate), metric ton	Carbon fibre supply (actual), metric ton
2010	48370	79650	47790
2015	82400	143595	93171
2020 (Est.)	150200	180600	129965

(iii) automobiles and (iv) fuel cells. In addition, repair and maintenance in construction industry is also a fast growing market for carbon fibres.

General electric introduced its 100-m-long next-generation wind blades made with composite of carbon fibre for 1.6 MW wind turbines which are 20–30% lighter than corresponding glass fibre blades. Carbon fibre-based ultra-lightweight, high-capacity compressed natural gas storage systems are being used in commercial vehicles [26]. In fuel cells, carbon fibre-based paper or cloth is used to make gas diffusion layer (GDL) to achieve even distribution of reactants throughout the catalyst layer with efficient removal of water and CO_2. It also acts as a substrate for the catalyst and provides connection between catalyst layer and charge collector plate. In automobile industry, carbon fibre is currently being used only in high-performance automobiles. In future, with lower cost, it has great potential for use in normal vehicles also. In construction, earthquake protection and infrastructure sector carbon fibre are finding applications as lightweight precast concrete. Recycling of manufacturing waste in carbon fibre industry is expected to lower costs and improve circular economy.

Most of the carbon fibre currently produced is made from polyacrylonitrile (PAN) polymer and to some extent from Rayon and phenolic resins. Coal tar and petroleum pitch are also used to make carbon fibre although at much smaller scale. Table below compares the properties of PAN-based and pitch-based carbon fibres. As shown in Table 6, currently pitch-based carbon fibre is mainly confined to defence and aerospace applications.

In the new paradigm of energy and environment, pitch-based carbon fibres are also expected to compete with PAN-based fibres due to development of better processes and availability of good-quality coal tar pitches and petroleum mesophase pitches. Pitch-based carbon fibre is made from hydrocarbon streams rich in tri- and tetra-aromatic hydrocarbons. One of the most suitable feedstocks for pitch-based carbon fibre is mesophase pitch derived from fluidized catalytic cracking (FCC) clarified slurry oil (CSO). In this process, CSO is heated to polymerize into

Table 6 Properties of PAN- and pitch-based carbon fibres

	PAN-based CF (T-700)	Pitch-based CF
Density	Lower (1.8 g/cc)	Higher (2.1 g/cc)
Oxidation resistance	Lower	Higher
Strength	Higher (4.9 GPA)	Lower (brittle) (2–3 GPA)
Modulus	Lower (240 GPA)	Higher (500–800 GPA)
Stain	Higher (2%)	Lower (0.5–1.5%)
Thermal conductivity	Lower (10–40 W/m K	Higher ($\sim$1000 W/m K) along a-axis
Diameter	7 μm	5–10 μm
Applications	Areas where high strength is required, e.g. automobiles, sports, etc.	Areas where high modulus and high thermal conductivity are required, e.g. rocket tips, defence equipment, etc.

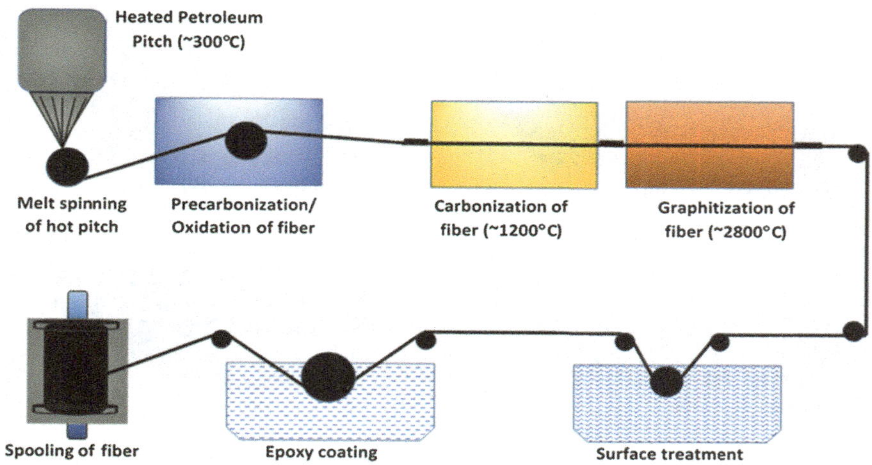

Fig. 10 Process of pitch-based carbon fibre

higher polyaromatic hydrocarbons (PAH). The lighter hydrocarbons produced in this process are distilled out to obtain mesophase pitch as residue. Some of the important properties of pitch are softening point (SP), quinoline insoluble (QI), toluene insoluble (TI). After production of good-quality mesophase pitch, melt spinning is performed at slightly higher temperature than SP of pitch to make fibre of desired diameter. The spun fibres are stabilized at 200–300 °C for 0.5–3 h under controlled air environment. Stabilization is a process where cross-linking takes place to increase the stability of fibre and convert it into thermoset material. After stabilization of fibre, carbonization and graphitization are done at 1000–3000 °C for few minutes under inert atmosphere. Surface treatment and sizing of graphitized fibre are done by coating with epoxy/polyester/nylon/urethane [27]. Figure 10 shows typical steps involved in pitch-based carbon fibre production [28].

2.5 Carbon Foam

Carbon foam is highly porous cellular material having random distribution of spherical pores ranging from 50 to 500 microns. Coal tar pitch, petroleum pitch, mesophase pitch, phenolic resins, graphite, sucrose and polyurethanes are some of the common precursor materials of carbon foams. Carbon foam is made by sudden increase in volume of carbonaceous material during carbonization process. Various methods such as blowing and carbonization, template carbonization and exfoliation are used to make carbon foam. Specific thermal conductivity of graphitic foam (>150 W/m K) is ~3 times higher than copper (45 W/m K) and aluminium (54 W/m K). Density of carbon foam ranges from 0.1 to 0.6 g/cm^3, and specific surface ranges from 5000 to 50000 m^2/m^3. Non-graphitic carbon foam on the other

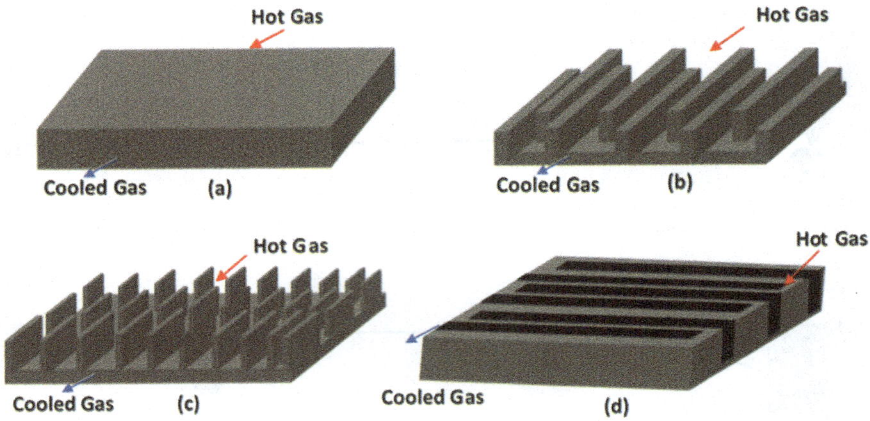

Fig. 11 Graphite foam heat sinks: **a** block, **b** staggered, **c** baffle, **d** corrugated configurations (artistic depiction)

hand is highly insulating in nature and exhibits very low specific thermal conductivities (~1 W/m K). Graphitic carbon foams have very high thermal conductivity (faster movement of heat), high thermal diffusivity (quick response to sudden change in temperature) and low density; all these properties make them better material for heat sink and heat exchanger applications. The major limitation of carbon foam is their lower mechanical strength and brittle nature. Figure 11 shows different types of heat sinks made from graphitic foam [29, 30].

In new paradigm of energy and environment, making the components lighter will further improve the energy efficiency and carbon foams can provide a better option in both heat dissipation (graphitic foam) and insulation (non-graphitic foam)-related applications.

2.6 Carbon Nanotubes

Carbon nanotubes (CNTs) are formed when sheet of graphene is rolled. If only one sheet is rolled, then single-walled CNT (SWCNT) is formed, and if multiple sheets are rolled together, then multiwall CNT (MWCNT) is formed (Fig. 12). Typical diameters of SWNT and MWNT are 0.8–2 nm and 5–50 nm, respectively. CNT lengths can range from 100 nm to 0.5 m [31, 32]. In 2006, industrial production of MWCNT and SWCNT was 300TPA and 7TPA, respectively [33].

Most common production processes are (i) arc discharge (AD), (ii) laser ablation (LA) and (iii) chemical vapour deposition (CVD). AD and LA processes require high energy and vacuum, whereas CVD operates at mild conditions with better control of CNT structure. Most common raw materials of CNT are ethylene and acetylene gases. Last 20 years have seen huge development in control of CNT

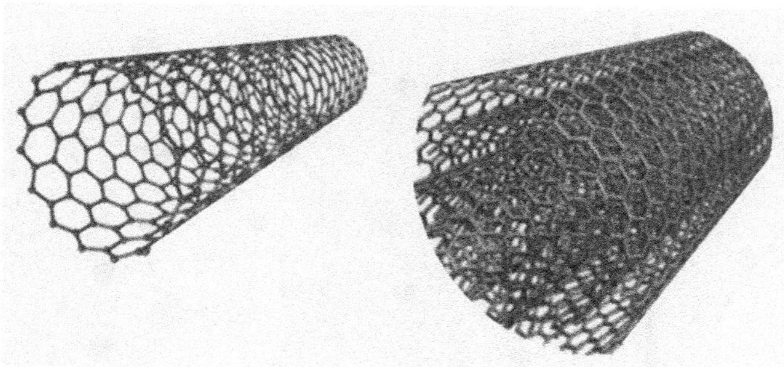

Fig. 12 Single-walled (SWCNT) and multiwalled carbon nanotubes (MWCNT)

structure and yield. Various transition metal catalysts have been studied for efficient CNT growth. Fe is one of the most studied catalysts for CTN production. Researchers have reported up to 55 g of SWCNT with 1 g of Fe/Mo/MgO catalyst and 170 g of MWCNT with 1 g of $Ni/Fe/Al_2O_3$ catalyst by CVD process [34].

For industrial production, fluidized bed is one of the most suitable reactors for carbon nanotube with moderate length and alignments. For production of carbon nanotubes with higher length and better alignment, transport bed-II type reactors are more suitable. In a fluidized bed reactor, the CNT agglomerates are fluidized, and the flow pattern is similar to that of a continuous stirred tank reactor [34].

A US-based company has shown polymeric membranes with carbon nanotube having extremely high number of highly precise opening of <1.27 nm [35]. As depicted in Fig. 13, these membranes provide new avenues for advanced separation of various gases such as N_2, CO_2, CH_4, C_2H_4 and C_2H_6 based on difference in their molecular sizes. Functionalized CNT membranes are capable of separating therapeutic antibodies and similar medically useful biological materials. CNT-based membranes have potential to provide less energy-intensive solution for desalination of seawater and purification of drinking water. Composites made by mixing small amount of carbon nanotubes with plastic significantly improve the thermal/ electrical conductivity and mechanical strength.

In the new paradigm of energy and environment, carbon nanotubes especially MWCNT have great potential for industrial application in the area of composites, lithium-ion batteries, supercapacitors, solar cells, catalysis, electronics, membranes, etc [31]. However, consistency in CNT quality at industrial-scale production and very high price are still big deterrents for widespread application of CNT.

2.7 *Graphene*

In 2004, Andre Geim and team from University of Manchester, UK, and Institute for Microelectronics Technology, Chernogolovka, Russia [14], first isolated

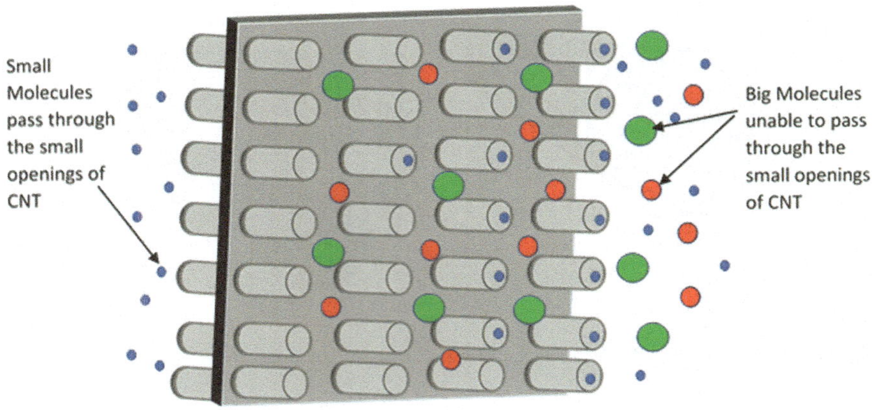

Fig. 13 Highly aligned CNT fixed within flexible polymer sheet to provide precise openings (artistic depiction)

single-layer graphene from graphite. As shown in Fig. 14, graphene is a flat monolayer of sp^2 bonded carbon atoms into a 2D honeycomb lattice. Graphene is the basic building block of other important allotropes, e.g. if stacked it forms graphite (3D), if rolled it forms nanotubes (1D) and if wrapped it forms fullerenes (0D). Graphene has extremely high thermal conductivity (~ 5000 W/m K), charge carrier mobility (>2000 cm^2/V s), intrinsic strength (~ 130 GPa), Young's modulus (~ 1.0 TPa) and surface area (~ 2600 m^2/g).

Graphene can be produced from graphite in top-down approach by mechanical, electrochemical, chemical, photo-exfoliation techniques. It can also be produced from a carbon precursor in bottom-up approach by chemical vapour deposition and chemical synthesis [36].

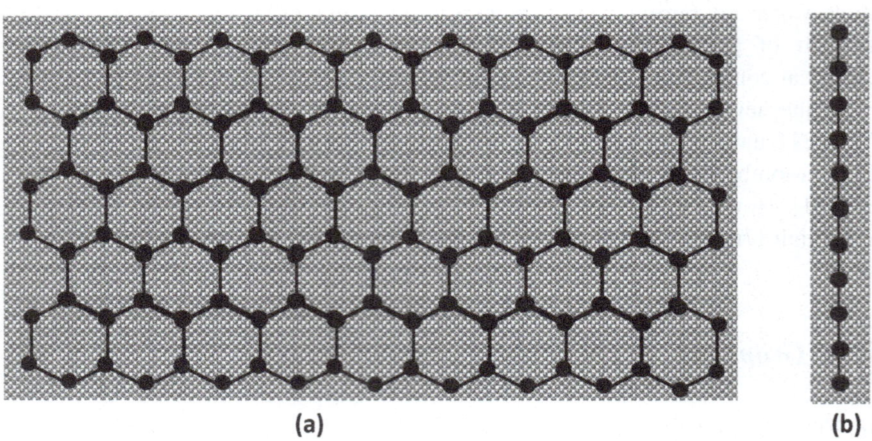

Fig. 14 Single-layer graphene with electron cloud: **a** front view and **b** side view

Majority of graphene is produced by top-down approach where it is isolated from graphite by layer-by-layer exfoliation methods such as mechanical exfoliation, photo-exfoliation, chemical exfoliation or electrochemical exfoliation. Andre Geim and co-workers [14] used mechanical exfoliation (repeated peeling) of highly oriented pyrolytic graphite (HOPG) to produce first physical sample of graphene. Various aqueous and non-aqueous solvents are also used for exfoliation of graphene layers from graphite dispersion followed by purification. Bottom-up approach of graphene production is mostly based on chemical vapour deposition techniques.

Very low coefficient of light absorption (2.3%) makes graphene almost transparent to light [37]. Therefore, solar cells made of graphene can significantly expand the wavelengths of absorbed spectrum of light. Incorporation of graphene also improves the mechanical strength of silicon and organic substrates. Graphene electrode improves the conversion efficiency by 3.5% compared to indium tin oxide (ITO) in organic solar cell and up to 15.6% in graphene/TiO_2/perovskite solar cell. Greater flexibility, transparency and conducting behaviour of graphene compared to ITO result in better conversion efficiency [37].

3 Challenges

Although significant efforts have gone in development of advanced carbon materials, still several challenges and roadblocks are yet to be overcome to make their widespread use in the new paradigm of energy and environment. Some of the common challenges are summarized below:

1. Most of the technologies for production of advanced carbon materials are confined to laboratory or pilot scale; hence, availability of proven and credible technology is not readily available for commercialization. Even for relatively mature technologies like needle coke and carbon fibre, very few licensors are available for commercialization.
2. Crude oil or coal is some of the cheapest sources of raw materials for advanced carbon materials which vary significantly in quality with change of source/supplier. Furthermore, most of the raw materials are composed of multicomponents; hence, quality monitoring is also challenging.
3. Multiple sophisticated characterization tools such as SEM, TEM, XRD, dilatometer, thermal conductivity analyser and pycnometer are required for monitoring and quality control of advance carbon materials which are time consuming and need expert analysts. In commercial-scale production, consistent monitoring of products becomes a challenge as it requires huge number of analysis and man-hours.

4. In some of the emerging technologies, the final applications and requirements are also constantly changing; hence, commercial process requires good flexibility to produce carbon material with different specifications. This requires process optimization and hardware changes in commercial plant.
5. High price and inconsistent quality at industrial scale is one of the deterrents for mass applications of advanced carbon nanomaterials, e.g. graphene and CNT.

4 Conclusions

High energy density batteries are the future of energy storage. Advanced carbon material like synthetic graphite is already being used as anode in Li-ion batteries. Constant developments are on to improve the porosity, conductivity, durability, density and cost of advanced carbon materials such as spherical graphite for development of better anode material for Li-ion batteries. Supercapacitors are another area of energy storage where high-quality activated carbon electrodes are being developed for electrostatic double layers.

Major petrochemicals, e.g. olefins, aromatics and poly aromatic hydrocarbons, are being produced by hydrocarbon sector to make conventional polymers such as PE, PP and PET. Excellent opportunity exists to further improve the quality of these conventional polymers by making composites with advance carbon materials such as carbon nanotubes and carbon fibres. Hydrocarbon industry has huge potential to supply the raw material for synthetic graphite and pitch-based carbon fibre. Carbon fibre has found a wide range of applications with its excellent thermal, electrical and mechanical properties. Today, about 150 KTPA is the worldwide demand of carbon fibre, most of which, however, is based on high-cost raw material, e.g. PAN. Refining industry is already supplying large quantity of anode coke (34 MMTPA). Refinery pitch with adequate purification can be potentially converted to high-quality needle coke, graphite, carbon foam and carbon fibre.

Cost of all new technologies is inversely related to their volumes. Therefore, the new materials which look exorbitantly expensive today may become affordable by intensifying the manufacturing steps and using cheaper raw materials.

References

1. Draft national energy policy, NITI Aayog, Government of India, Version as on 27.06.2017
2. https://www.ppac.gov.in/WriteReadData/userfiles/file/RefineriesMap.pdf
3. World oil outlook 2040, organization of the petroleum exporting countries (2017) In: Griffin J, Fantini, AM(eds)
4. US patent 9550708 (2017) A process for catalytic conversion of low value hydrocarbon streams to light olefins

5. Johnson T (2008) Diesel engine emissions and their control—an overview. Platinum Metals Rev 52(1):23–37
6. https://about.bnef.com/blog/electric-vehicles-accelerate-54-new-car-sales-2040/
7. Edwards L (2015) The history and future challenges of calcined petroleum coke production and use in aluminum smelting. J Miner Met Mater Soc 67(2), 308–321
8. Bacon Roger (1960) Growth, structure and properties of graphite whiskers. J Appl Phys 31:283
9. Akio Shindo (1962) Japan Patent 374405
10. Ōtani S (1965) On the carbon fiber from the molten pyrolysis products. Carbon 3(1), 31–34
11. Barr J, Chwastiak S, Didchenko R, Lewis I, Lewis S, Singer L (1976) Appl Polym Symp 29:161
12. Iijima S (1991) Helical microtubules of graphitic carbon. Nature 354:56–58
13. Ijima S, Ichihashi T (1993) Single-shell carbon nanotubes of 1-nm diameter. Nature 363:603–605
14. Novoselov KS, Geim AK, Morozov SV, Jiang D, Zhang Y, Dubonos SV, Grigorieva IV, Firsov AA (2004) Electric field effect in atomically thin carbon films. Science 306 (5696):666–669
15. Jones SS (1986) Anode-Carbon usage in the aluminum industry. In: ACS symposium series, vol 303, Petroleum-Derived Carbons, pp 234–250
16. Charles A (1976) Stokes, manufacture of industrial carbons from petroleum raw materials. In: ACS symposium series, vol 21, Petroleum-Derived Carbons, pp 1–17
17. US Patent 6939526 (2005) Graphite particles and process for production thereof
18. US Patent 9184437 (2015) Potato-shaped graphite particles with low impurity rate at the surface, method for preparing the same
19. Huang Y, Peng J, Luo J, Li W, Wu Z, Shi M, Li X, Li N, Chang B, Wang X (2020) Composite as high-performance anode material for lithium-ion batteries. Energy Fuels 34 (6):7639–7647
20. Jani J. Report of edelweiss investment consulting, https://www.edelweiss.in/ewwebimages/WebFiles/Rearch/7d1c4952-c207-4ae7-82bc-c6d8e835de3a.pdf
21. https://pentacarbon.de/en/wiki/
22. Indian Patent 278224 (2011) A process for improving aromaticity of heavy aromatic hydrocarbons
23. Solhycarb's day, CNRS-PROMES—1 MW Solar Furnace-Odeillo Font Romeu, Franc September 28, 2009
24. http://www.torayca.com/en/aboutus/abo_003.html
25. http://www.compositesworld.com/
26. https://www.compositesworld.com/news/quantum-wins-follow-on-contract-for-carbon-fiber-natural-gas-tanks
27. Buckley JD, Edie DD (1993) Carbon-carbon material and composites. Noyes Publications
28. Diefendorf RJ (1987) Carbon/graphite fibers, engineered materials handbook. vol 1, Composites, ASM International, pp 49–53
29. Almajali MR. Engineered carbon foam for temperature control applications, Ph.D. thesis, University of Dayton
30. Lin W, Yuan J, Sundén B (2011) Review on graphite foam as thermal material for heat exchangers. In: World Renewable Energy Congress
31. Michael FL, De Volder, Sameh HT, Ray H, Baughman A, John Hart J (2013) Carbon nanotubes: present and future commercial applications. Science 339, 535–539
32. Choudhary V, Gupta A (2011) Poly/Carbon Nanotube Nanocomposites. https://doi.org/10.5772/18423 IntechOpen
33. Eklund P et al. In: WTEC Panel Report on International assessment of research and development of carbon nanotube manufacturing and applications. World Technology Evaluation Center, Inc. 4800 Roland Avenue Baltimore, Maryland 21210
34. Zhang Q, Huang J-Q, Zhao M-Q, Qian W-Z, Wei F (2011) Carbon nanotube mass production: principles and processes. Chem Sus Chem 4:864–889

35. McGinnis RL, Reimund K, Ren J, Xia L, Chowdhury MR, Sun X, Abril M, Moon JD, Merrick MM, Park J, Stevens KA, McCutcheon JR, Benny D, Freeman BD (2018) Large-scale polymeric carbon nanotube membranes with sub–1.27-nm pores, Science Advances, 4: e1700938
36. Bonaccorso F, Lombardo A, Hasan T, Sun Z, Colombo L, Andrea C, Ferrari AC (2012) Production and processing of graphene and 2D crystals. Mater Today 15(12):564–589
37. Czerniak-Reczulska M, Niedzielska A, Jędrzejczak A (2015) Graphene as a material for solar cells applications. Adv Mater Scie 15(4) (46), 67–81

Synthetic Production, In-house Utilization and Subsurface Disposal of Sodium Carbonate: A Novel Carbon Dioxide Capture and Storage Strategy

Dayanand Saini

1 Introduction

India, the third-largest emitter of the greenhouse gas (GHG) emissions and one of the first ratifying countries of the Paris Agreement, is committed to following the low carbon path to progress. Similarly, the state of California, which is the second-largest producer of GHG emissions in the USA, is on the path to a low carbon future. It has taken a leading role in achieving the Paris Agreement goals; even the USA has begun the formal withdrawal from the Paris Agreement. Anthropogenic carbon dioxide (CO_2) is the primary GHG in countries like India and the US states like California.

In California, 83% of GHG emissions in 2017 [1] were CO_2 emissions. CO_2 emissions contributed to almost 70% of India's total GHG emissions in 2016 [2]. Electrical production (in-state and import) and industrial sources accounted for 41% of California's total CO_2 emissions in 2017 [1]. In contrast, power and other industrial combustion sectors accounted for around 70% of India's total CO_2 emissions in 2016 [2].

India's continuing reliance on fossil fuel-based energy is one of the most significant contributors to India's rising CO_2 emissions. However, it also presents an enormous opportunity for deploying the geologic carbon storage (GCS)-based carbon capture, utilization and storage (CCUS) strategies for sequestering CO_2 from stationary sources in India. Similarly, in the US states like California, CCUS can lead to a measurable reduction in its GHG emissions from its major industrial subsectors [3].

There are two leading GCS-based CCUS strategies currently used to lower anthropogenic CO_2 emissions at large and commercial scales. One relies on the

D. Saini (✉)
Department of Physics and Engineering, California State University,
Bakersfield, CA 93311, USA
e-mail: dsaini@csub.edu

U. K. Bhui (ed.), *Macromolecular Characterization of Hydrocarbons for Sustainable Future*, Green Energy and Technology,
https://doi.org/10.1007/978-981-33-6133-1_20

injection of captured CO_2 in depleted hydrocarbon (i.e. oil or oil and gas fields) reservoirs for both EOR and storage purposes, which are commonly referred to as simultaneous CO_2-EOR and CO_2-storage strategy. The other is the injection of captured CO_2 into deep saline aquifers for storage purposes, which is considered a dedicated GCS strategy. The use of captured CO_2 for recovering stranded oil while leaving it behind (i.e. storing) in depleted oil or oil and gas fields is a proven and commercial CCUS technology. In contrast, the dedicated GCS strategy has also been tested and deployed around the world [4].

However, the implementation of GCS-based CCUS projects in places like India and California is still in scoping, planning or design phases (e.g. [5–9]). In countries like India or the US states like California, there are some inherent techno-economic challenges and barriers as well, which need to be addressed before its potential and role in a future low carbon economy could be fully realized.

In the case of storing CO_2 in deep saline aquifers (DSAs), there is no value addition from CO_2 storage in saline aquifers [7] except as a means of storing CO_2. The extraction of water from deep saline aquifers or an underlying oil-free water zone (aquifer) that may be present in hydrocarbon reservoirs to increase the CO_2 storage capacity of targeted aquifers has been reported in the published literature (e.g. [10, 11]). Such a scheme (i.e. extraction of water from saline aquifers during CO_2 injection and storage operations) is often dubbed as enhanced water recovery (EWR) to draw a parallel with the term "EOR." As noted by the Global CCS Institute [12], costs associated with DSA storage operations may significantly increase with the application of EWR. On the other hand, being first of its kind projects, CCUS-focused CO_2-EOR plans (which are still either in planning (India, [5]) or design (California, [8]) phases) are faced with many inherent techno-economic challenges and barriers.

As outlined by Mishra et al. [5], capturing CO_2 from the flue gas stream and its compression and transportation is a cost and energy-intensive process, whereas, at the sink end, CO_2 being acidic and corrosive gas will need retrofit modifications in terms of special corrosion-resistant metallurgy for existing processing facilities. Further, the target of retaining 99% of the injected CO_2 in the designated trap (i.e. no spill out of the reservoir [i.e. injection zone(s)]) over 1000 years (Intergovernmental Panel on Climate Change [13]) and continued maintenance and monitoring of the integrity of cap and reservoir rocks and wellbores (i.e. active injection and production wells and pre-existing idle or abandoned wells) prone to unintended migration of stored CO_2 (surface, near-surface and subsurface) are another prime barriers for a CCUS-focused CO_2-EOR project.

California is the most geologically diverse and the second most seismically active state in the USA [14]. Hence, deployment of a GCS-based CCUS project in California is inherently a challenging job. It is evident from the fact that California has yet to launch a commercial CO_2-EOR (either pure [i.e. no CO_2 storage component] or CCUS-focused) project; even the majority of the world's commercial CO_2-EOR, as well as CCUS-focused CO_2-EOR projects, are in the USA [4]. Only CO_2 injection pilots have been conducted in the past [15, 16]. On the other hand, countries like India have limited experience with commercial CO_2-EOR technology. In India, there are only a few ongoing CO_2 injection-based commercial EOR projects [17].

The inherent techno-economic challenges and prime barriers, as mentioned above, need to be overcome before the GCS-based CCUS strategies can start to play a crucial role in low carbon economies of countries like India and the US states like California. The present study proposes a novel CCUS strategy. The present study investigates the potential and feasibility of the proposed CCSU strategy in tackling some of the inherent techno-economic challenges and barriers associated with leading GCS-based CCUS technologies. Both of the leading GCS-based CCUS technologies (i.e. simultaneous CO_2-EOR and storage in depleted hydrocarbon reservoirs and storing CO_2 in saline aquifers) have been employed on large and commercial scales [4]. They are of particular importance to the petroleum industry as they inherently rely on the expertise and experience of exploration and production (E&P) professionals. However, they can assist in a rapid transition of the petroleum industry to a greener energy industry with a sustainable future.

The proposed CCUS strategy relies on the synthetic production of sodium carbonate (Na_2CO_3) while capturing CO_2 either from typical flue gas stream emitted by stationary sources present in power and other industrial combustion sectors or from ambient air, utilization of produced Na_2CO_3 to further capture CO_2 from the existing flue gas stream or ambient air, and disposal of effluent (i.e. aqueous Na_2CO_3 solution) in suitable geological formation(s) such as saline aquifers.

2 The Proposed CCUS Strategy

The novel CCUS strategy proposed and presented here relies on the following three main steps:

1. Use of saline water stream (either water associated with the production of oil and gas operations (i.e. produced water) or water extracted from saline aquifers) and captured CO_2 (either from flue gas streams emitted by a stationary source or from ambient air) to synthetically produce Na_2CO_3 powders (either dry or hydrated) while converting feed stream (i.e. saline water) into low-salinity water for beneficial reuses.
2. In-house use of synthetically produced Na_2CO_3 powders from step 1 to further capture CO_2 from the flue gas stream (new and vented from step 1).
3. Subsurface disposal of aqueous Na_2CO_3 solution (the combined effluent stream of step 1 and step 2).

The critical aspects of the three main steps, as mentioned above, are briefly described and discussed next.

2.1 Step 1: Synthetic Production of Na_2CO_3

Typically, Na_2CO_3 is either mined from naturally occurring minerals (e.g. trona and nahcolite) and Na_2CO_3-rich brines occurring or is manufactured from sodium chloride and calcium carbonate via the ammonia soda (Solvay) process [18]. Recently, Saini [19] described an apparatus and process in which saline water stream (i.e. produced water) or water extracted from saline aquifers and anthropogenic CO_2 (either from typical flue gas stream emitted by stationary sources present in power and other industrial combustion sectors or from ambient air) can also be used to produce solid Na_2CO_3. The apparatus and process described by Saini [19] explicitly use the produced water as a feed stream for generating NaOH solution, which in turn works as CO_2 absorbent and produces solid Na_2CO_3 at low temperature (<50 °C [122 °F]). In the case of Saini [19], the generation of NaOH solution, which is one of the raw materials needed for producing Na_2CO_3, can be achieved by a customized version of electrochemical membrane separation cell offered by vendors such as an Australia-based company—namely Australian Biorefining Private Limited (ABR) operating from California and Montana in the USA [20].

In a recent study, Salmón et al. [21] reported that the use of 1 mol per litre (mol L^{-1}) NaOH solution is to remove up to 97% CO_2 from a flue gas stream containing 10% (v/v) CO_2 using an ambient condition novel membrane contactor. A fine spray column-based process, as described by Yincheng et al. [22], can remove up to 90% CO_2 from a gas stream containing 15% CO_2 (v/v) at 54 °C (129.2 °F) when a 5% (w/w) aqueous NaOH solution is used as an absorbent. In the laboratory setting, use of a spray dryer-based process using three wt% NaOH solution has been found sufficient to remove up to 65% CO_2 at 100 °C (212 °F) from a gas stream (air + CO_2) containing 4% (volume by volume [v/v]) CO_2 [23].

The cost of capturing CO_2 from ambient air using a spray-based contactor and NaOH solution, as reported by Stolaroff et al. [24], ranges from $53 to $127/ton-$CO_2$. It is noted that the process is evaluated by Stolaroff et al. [24], as pointed out by Yincheng et al. [22], an energy-intensive process. It involves the capture of CO_2 from the atmosphere, where CO_2 concentration is significantly lower (approximately 0.04% [v/v]) compared to a typical flue gas stream (8–10% [v/v] [25]). Further, the complex nature of the process (i.e. using produced Na_2CO_3 to regenerate NaOH by addition of CaO (calcium oxide) in the causticizer and regeneration of CaO by heating in the calciner from resulting $CaCO_3$ solids while capturing the CO_2 from the calciner flue gas and compressing by conventional means) appears to be highest cost contributing factors, whereas a fine spray column-based process, as reported by Yincheng et al. [22], potentially requires about half of the energy requirement compared to the processes such as Stolaroff et al. [24], which involves conventional causticization process using CaO.

The experimental studies, like Tavan and Hosseini [23] and Yincheng et al. [22], being conducted in laboratory conditions, did not discuss the potential source(s) of NaOH solutions. However, Salmón et al. [21] noted that NaOH could be taken from mercaptan oxidation (Merox) towers or subproduct flows of the chlor-alkali industry. Shim et al. [26] have also reported capturing CO_2 from flue gas using an aqueous solution of NaOH, which could be converted to sodium bicarbonate ($NaHCO_3$). According to them, $NaHCO_3$ has an extensive range of uses, including baking soda or a pH buffer for the chemical industry. The economic evaluation of a commercial-scale (100 tons CO_2/day) plant concluded that the process is a cost-effective CO_2 capturing technology [26]. As noted by the authors, NaOH required for CO_2 capture can be produced by a chlor-alkali process.

Nevertheless, these studies demonstrate the efficacy of NaOH solutions as CO_2 absorbent (i.e. in capturing CO_2 either from ambient air or from the flue gas stream). Further, processes described by Yincheng et al. [22] and Salmón et al. [21] rely on recovering absorbent (i.e. NaOH solutions), whereas processes described by Tavan and Hosseini [23] and Saini [19] produce solid Na_2CO_3 without any emphasis on recovering the absorbent (i.e. NaOH).

In the case of Saini [19], one of the effluent streams is low-salinity water. It is noted here that both the salinity of this effluent water stream and concentration of aqueous NaOH solution would be dictated by the salinity of water used as a feed stream. There are other potential sources (e.g. water extracted from shallow or deep saline aquifers or from oil-free water zone(s) (aquifer(s)) that may be present below the oil–water contact (OWC) in hydrocarbon reservoirs). They could also serve as the feed stream for generating Na_2CO_3 using an apparatus and process like Saini [19]. It is worth mentioning that detailed technical parameters on the apparatus and process described by Saini [19] are not available in the public domain yet.

Depending on the feed water stream's source, an effluent water stream can be used for various beneficial reuses. Such reuses may include the preparation of injection water stream of a specific salinity to use in low-salinity water flooding projects, agricultural irrigation, surface discharge for recreational and water banking purposes, groundwater recharging and meeting industrial water demands. Such potential beneficial reuses of effluent water stream bring additional value to the proposed CCUS strategy. On the other hand, reliance on the chlor-alkali process for NaOH solutions [21, 26] may not be viable from the CCUS point of view. The chlor-alkali industry is among the highest energy-consuming processes with pollutant emissions that have a severe effect on the environment and human life [27].

If CO_2 capturing strategy for producing Na_2CO_3, as outlined above, to be implemented on large and commercial scales, effective use of the produced solid Na_2CO_3 is a must. Hence, in-house use of produced Na_2CO_3 (dry as well as hydrated) in capturing CO_2 further either from a new flue gas stream or vented from step 1 is proposed. This forms the basis of the second step of the CCUS strategy presented in this study. The same is briefly discussed next.

2.2 *Step 2: Utilization of Produced Na_2CO_3*

Na_2CO_3 (also called soda ash) is one of the largest volume mineral products in the inorganic chemical industry [28] and has many industrial applications (e.g. glass manufacturing, soap and detergents). On the other hand, Na_2CO_3 can also be used to capture CO_2 from the flue gas stream using the process [29]. The process described by Cai et al. [29] relies on the use of support-free (i.e. free-flowing) hydrated Na_2CO_3 powders (HSCPs) for an enhanced CO_2 sorption capacity and kinetics.

According to Cai et al. [29], the elimination of supports reduces the overall cost of raw materials but also increases the CO_2 sorption capacity, both of which make HSCPs-based CO_2 capturing technology more competitive for large-scale applications. They further noted that 70% (w/w) Na_2CO_3 and 30% (w/w) water (i.e. HSCP-70) were able to absorb as much as 254 mg/g CO_2 for a 16-min absorption time at 30 °C (86 °F), which is significantly higher than the other Na_2CO_3-based CO_2 sorbents reported in the literature. The process described by Cai et al. [29] is comparatively superior to 30% (w/w) monoethanolamine (MEA) aqueous solution, which is a proven and commercial CO_2 capturing technology. Cai et al. [29] found that, compared to 30% MEA aqueous solution, HSCP-70 was able to absorb almost three times more CO_2 at 30 °C (86 °F) during 16-min absorption time.

It is worth noting here that the process described by Cai et al. [29] needs to rely on an external source for its primary raw material (i.e. Na_2CO_3 powder). Hence, to reduce the consumption of Na_2CO_3 powder in the process, it employs the regeneration and recycling (multiple cycles) of spent HSCP-70 at 250 °C (482 °F). As per Cai et al. [29], HSCPs should also have high potential in capturing other acid gases, including SOx, NOx, H_2S and Cl_2. Having such potential and being a near ambient condition process make HSCP-based CO_2 capturing processes an attractive process from CCUS point of view.

In step 2 of the proposed CCUS strategy, Na_2CO_3 generated in step 1 can generate an optimal HSCP (i.e. Na_2CO_3 powder mixed with the desired amount of water) for further capturing CO_2 either from new flue gas stream or vented from step 1. Principally, all the Na_2CO_3 powder and a fraction of low-salinity water can generate an HSCP. Such in-house use eliminates the potential problems associated with finding an alternative and cost-effective way of utilizing the Na_2CO_3 produced in step 1. Also, instead of regenerating and recycling the HSCP, as it may be the case in the processes like Cai et al. [29], all of the spent HSCP (i.e. $Na_2CO_3 \cdot 3NaHCO_3$) can be mixed with any of the unused NaOH solutions from step 1 for producing an aqueous Na_2CO_3 solution. Further, the vented gas stream can be mixed with the feed gas stream and recycled until almost all the CO_2 is captured.

As evident from the discussion, as provided above and provided earlier, the proposed CCUS strategy can remove almost all the CO_2 from a feed flue gas or ambient air stream while generating aqueous Na_2CO_3 solution an effluent stream. Industrial or commercial use of this effluent stream may end up releasing either all or a fraction of CO_2 back into the atmosphere, which is captured during the generation of aqueous Na_2CO_3 solution. However, a subsurface disposal scheme which involves the injection of aqueous Na_2CO_3 solution produced in step 2 into suitable geological formation(s) can ensure the safe and long-term storage of CO_2 captured during steps 1 and 2.

This scheme is another unique aspect of the proposed CCUS strategy. The same is discussed next.

2.3 Step 3: Subsurface Disposal of Aqueous Na_2CO_3 Solution

The subsurface disposal of aqueous Na_2CO_3 solution generated in step 2 into suitable geological formation(s) can provide several advantages that are very attractive from the CCUS point of view.

First and foremost, the petroleum industry around the world has enough experience and expertise in injecting water of varied salinities for either water disposal purposes or pressure maintenance (water flooding) or for EOR (e.g. water alternating gas [WAG]). Hence, the design, operation and management of the aqueous Na_2CO_3 solution injection project as a CCUS strategy should neither be too technically tricky nor economically cumbersome.

Secondly, the natural occurrence of Na_2CO_3-rich brines in geological formations and reported extraction of them by chemical companies in the USA [30] could be used as a proxy to evaluate the suitability of proposed disposal scheme. It means that the aqueous Na_2CO_3 solution can be injected into suitable geological formation (s). However, it can also be stored in them without special metallurgy for equipment associated with the injection programme (e.g. pipeline, wellbore, pumps).

The saline aquifers, which are in the vicinity of large stationary CO_2 emitters (e.g. power plants) and are unused, are ideal candidates for disposal as transportation costs be minimal in this scenario of source-sink proximity. The use of saline aquifers is especially beneficial for countries like India, where the maximum saline aquifers and major thermal power plants are in the same areas [7].

The subsurface disposal of aqueous Na_2CO_3 solution in saline aquifers (either shallow or deep) is advantageous in many ways. The injection of aqueous Na_2CO_3 solution in saline aquifers can also be used to extract native saline water, which can serve as a feed stream (i.e. saline water) required in step 1. It is worth to mention here that the volume of feed stream used in step 1 is the deciding factor while determining the disposal volume of aqueous Na_2CO_3 solution. Nevertheless, a large-scale project that uses the proposed CCUS strategy is expected to produce a large volume of aqueous Na_2CO_3 solution. However, it will still be significantly lesser than the volume of CO_2 injected in leading GCS-based CCUS technologies.

For disposing of the anticipated large volume of aqueous Na_2CO_3 solution, a conventional water flooding pattern (e.g. line drive) can be employed. Further, the injection of an aqueous Na_2CO_3 solution can also be used as a drive mechanism itself. Because viscosity and density of aqueous Na_2CO_3 solution are not going to be very different from that of native saline water, an efficient displacement of native saline water by aqueous Na_2CO_3 solution can be engineered.

Because both aqueous Na_2CO_3 solution and native saline water are expected to have little to no compressibility, a little compression will be more than enough in maintaining an injection pressure higher than formation pressure. Also, the majority of saline aquifers can be expected to be normally pressured. Hence, the extraction of native saline water can ensure that the pressure in the injection zone never exceeds beyond initial hydrostatic pressure, ensuring the integrity of caprock and reservoir rock while eliminating any remote possibility for an injection-induced seismic activity. It is a desirable feature of the proposed CCUS strategy for a geologically diverse and seismically active US state like California.

The effluent stream low-salinity water can also be proven a boon to the regions around the world. The US states like California regularly face severe drought. Hence, the additional availability of low-salinity water is always desired in California. The oil-rich regions such as several middle-eastern countries and certain parts of India (e.g. the states of Gujarat and Rajasthan) can benefit from the availability of additional low-salinity water as the availability of potable water is often an issue.

As outlined above, the subsurface disposal scheme's attributes make the proposed CCUS strategy more competitive compared to leading GCS-focused CCUS technologies. Both the simultaneous CO_2-EOR and CO_2 storage and dedicated GCS strategies require capture, compression, transportation, injection and utilization (either for EOR in depleted hydrocarbon reservoirs with or without storage component or as dedicated storage with or without EWR component) of captured CO_2. They also require detailed site characterization and multiple monitoring, verification and accounting (MVA) tools and techniques to ensure the safe and long-term storage of captured CO_2 in the injection zone(s). However, in the case of CCUS strategy proposed here, a significant reduction in compression cost can be expected because the effluent stream is not a highly CO_2, which needs significant compression before it is ready for injection. Compared to leading GCS-focused CCUS technologies, the cost associated with the MVA programme for the proposed CCUS strategy should also be relatively less. The tools and techniques used in conventional water flooding projects can be used for active surveillance and monitoring of injected aqueous Na_2CO_3 solution and reservoir response of the targeted saline aquifers.

3 Summary

The proposed novel CCUS strategy appears to have the potential of playing a significant role in advancing the deployment of GCS-based CCUS technologies in countries like India and the US states like California. In the absence of limited to no presence of commercial CO_2-EOR projects, the petroleum industry in places like India and California can use its water flooding- and steam flooding-related expertise and experience for design, operation, management, monitoring, verification and accounting of the CCUS strategy presented here.

Based on the discussion provided in previous sections, it can be said that the proposed CCUS strategy is expected to be a competitive strategy that can be deployed on large as well as commercial scales. Also, in the US states like California, which have a highly regulated petroleum industry, the proposed CCUS strategy, a GCS-based CCUS strategy without a CO_2-EOR component, should provide a viable option for achieving a measurable reduction in CO_2 emissions.

Similarly, in countries like India, high CO_2 flue gas concentration resulting from significant reliance on coal for electrical generation can further make the proposed CCUS strategy a lucrative option.

Conceptually, the proposed CCUS strategy has all the elements for being competitive and deployable on large and commercial scales. The future research and development efforts need to be focused on determining various operational parameters and scalability of the CCUS strategy conceptualized, described and discussed in the present study.

References

1. California Air Resources Board (CA ARB) (2020) GHGs descriptions & sources in California. https://ww2.arb.ca.gov/ghg-descriptions-sources. Accessed 26 Jun 2020
2. Timperley J (2020) The carbon brief profile: India. https://www.carbonbrief.org/the-carbon-brief-profile-india. Accessed 26 Jun 2020
3. Plant Engineering (2020) Cutting carbon in California by gas technology. https://www.plantengineering.com/articles/cutting-carbon-in-california/. Accessed 26 Jun 2020
4. Saini D (2017) Engineering aspects of geologic CO_2 storage: synergy between enhanced oil recovery and storage. Springer International Publishing AG, Cham, Switzerland
5. Mishra GK, Meena RK, Mitra S et al (2019) Planning India first CO_2-EOR project as carbon capture, utilization & Storage: A step towards sustainable growth. Soc Pet Eng. https://doi.org/10.2118/194629-MS
6. Datta A, Krishnamoorti R (2019) Opportunities for a low carbon transition-deploying carbon capture, utilization, and storage in northeast India. Front Energy Res 7:12. https://doi.org/10.3389/fenrg.2019.00012
7. Bhandari AK (2019) Geological sequestration of CO_2 in saline aquifers—an Indian perspective. In: Goel M, Sudhakar M, Shahi RV (eds.) Carbon capture, storage and utilization: a possible climate change solution for energy industry. CRC Press
8. Powering California (2020) CRC designing California's first carbon capture and sequestration project. https://poweringcalifornia.com/newsletter/crc-designing-californias-first-carbon-capture-and-sequestration-project/. Accessed 26 Jun 2020
9. Trautz R, Swisher J, Chiaramonte L et al (2018) California CO_2 storage assurance facility enterprise (C2SAFE): final technical report. United States. https://www.osti.gov/servlets/purl/1452864
10. IEAGHG (2012) Extraction of formation water from CO_2 storage. International Energy Agency (IEA) and Greenhouse Gas (R&D) Programme. https://ieaghg.org/docs/General_Docs/Reports/2012-12.pdf. Accessed 28 Jun 2020
11. Saini D et al (2013) A simulation study of simultaneous acid gas EOR and CO_2 storage at Apache's Zama F Pool. Energy Procedia 37:3891–3900
12. Global CCS Institute (2016) Enhanced water recovery: brine extraction with CO_2 geological storage as part of CCS. https://www.globalccsinstitute.com/news-media/insights/enhanced-water-recovery-brine-extraction-with-co2-geological-storage-as-part-of-ccs/. Accessed 28 Jun 2020
13. IPCC (2007) Climate change 2007: Impacts, adaptation, and vulnerability. In: Parry ML et al (eds) Contribution of working Group II to the fourth assessment report of the Intergovernmental Panel on Climate Change. Cambridge University Press, Cambridge, UK, p 976
14. California Department of Conservation (2020) California Geological Survey. https://www.conservation.ca.gov/cgs. Accessed 26 Jun 2020

15. Winslow D (2012) Industry experience with CO_2 for enhanced oil recovery. Workshop on California opportunities for CCUS/EOR. United States Energy Association. Accessed 26 Jan 2014
16. Jeschke PA, Schoeling L, Hemmings J (2000) CO_2 flood potential of California oil reservoirs and possible CO_2 sources. Soc Pet Eng. https://doi.org/10.2118/63305-MS
17. Oil and Natural Gas Corporation Limited (ONGC) (2017) EOR in Indian context. http://www.dghindia.gov.in/assets/downloads/594b77be75491EOR_in_Indian_Context_by_ONGC.pdf. Accessed 26 Jun 2020
18. The Essential Chemical Industry—Online (2020) Sodium carbonate. https://www.essentialchemicalindustry.org/chemicals/sodium-carbonate.html. Accessed 26 Jun 2020
19. Saini D (2019) Apparatus and process for removal of carbon dioxide from a gas flow and treatment of brine/waste water from oil fields. US Pat App # US20190160395A1. https://patents.google.com/patent/US20190160395A1/en. Accessed 26 Jun 2020
20. ABR (2020) Technologies. http://www.abrprocess.com/technologies.html. Accessed 26 Jun 2020
21. Salmón IR, Cambier N, Luis P (2018) CO_2 capture by alkaline solution for carbonate production: a comparison between a packed column and a membrane contactor. Appl Sci 8 (996)
22. Yincheng G, Zhenqi N, Wenyi L (2011) Comparison of removal efficiencies of carbon dioxide between aqueous ammonia and NaOH solution in a fine spray column. Energy Procedia 4:512–518
23. Tavan Y, Hosseini SH (2017) A novel rate of the reaction between NaOH with CO_2 at low temperature in spray dryer. Petroleum 3(1):51–55
24. Stolaroff JK, Keith DW, Gregory V, Lowry GV (2008) Carbon dioxide capture from atmospheric air using sodium hydroxide spray. Env Sci & Tech 42(8):2728–2735. https://doi.org/10.1021/es702607w
25. Song C, Pan W, Srimat ST et al (2004) Tri-reforming of methane over Ni catalysts for CO_2 conversion to syngas with desired H2/CO ratios using flue gas of power plants without CO_2 separation. In: Park S-E, Chang J-S, Lee K-W (eds) Studies in surface science and catalysis, Elsevier 153, pp 315–322
26. Shim J-G, Lee DW, Lee HJ et al (2016) Experimental study on capture of carbon dioxide and production of sodium bicarbonate from sodium hydroxide. Env Eng Res. https://doi.org/10.4491/eer.2016.042
27. Hou M, Chen L, Guo Z et al (2018) A clean and membrane-free chlor-alkali process with decoupled Cl2 and H2/NaOH production. Nat Commun 9:438
28. Speight JG (2017) Industrial inorganic chemistry. In: Speight JG (ed) Environmental inorganic chemistry for engineers. Butterworth-Heinemann, pp 111–169
29. Cai Y, Wang W, Li L et al (2018) Effective capture of carbon dioxide using hydrated sodium carbonate powders. Materials 11(2):183. https://doi.org/10.3390/ma11020183
30. Dyni JR (1996) Sodium carbonate resources of the Green River formation. US Department of the Interior. US Geological Survey. https://pubs.usgs.gov/of/1996/ofr-96-0729/ofr-96-0729.pdf. Accessed 23 Jun 2020

The manufacturer's authorised representative in the EU is Springer Nature Customer Service Centre GmbH, Europaplatz 3, 69115 Heidelberg, Germany. If you have any concerns regarding our products, please contact ProductSafety@springernature.com

Printed and bound by CPI Group (UK) Ltd, Croydon, CR0 4YY
30/09/2026
02225294-0012